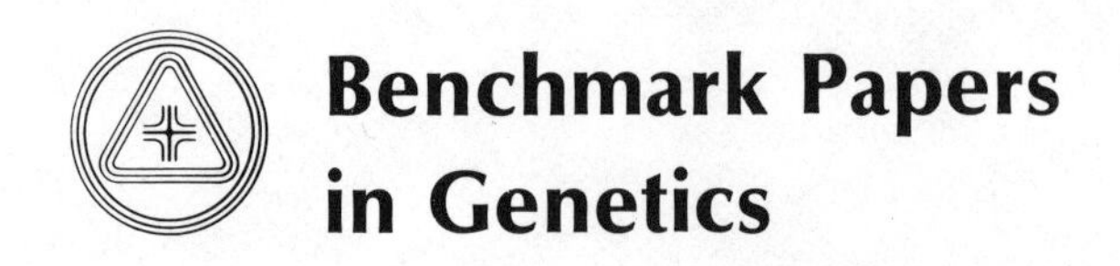

Series Editor: David L. Jameson
University of Houston

PUBLISHED VOLUMES

GENETICS AND SOCIAL STRUCTURE / *Paul A. Ballonoff*
GENES AND PROTEINS / *Robert P. Wagner*
DEMOGRAPHIC GENETICS / *Kenneth M. Weiss and Paul A. Ballonoff*
MUTAGENESIS / *John W. Drake and Robert E. Koch*
EUGENICS: Then and Now / *Carl Jay Bajema*
CYTOGENETICS / *Ronald L. Phillips and Charles H. Burnham*
STOCHASTIC MODELS IN POPULATION GENETICS / *Wen-Hsiung Li*
EVOLUTIONARY GENETICS / *D. L. Jameson*
GENETICS OF SPECIATION / *D. L. Jameson*

RELATED TITLES IN OTHER BENCHMARK SERIES

MICROBIAL GENETICS / *Morad Abou-Sabé*
CONCEPTS OF SPECIES / *C. N. Slobodchikoff*
MULTIVARIATE STATISTICAL METHODS: Among-Groups Covariation / *William R. Atchley and Edwin H. Bryant*
MULTIVARIATE STATISTICAL METHODS: Within-Groups Covariation / *Edwin H. Bryant and William R. Atchley*

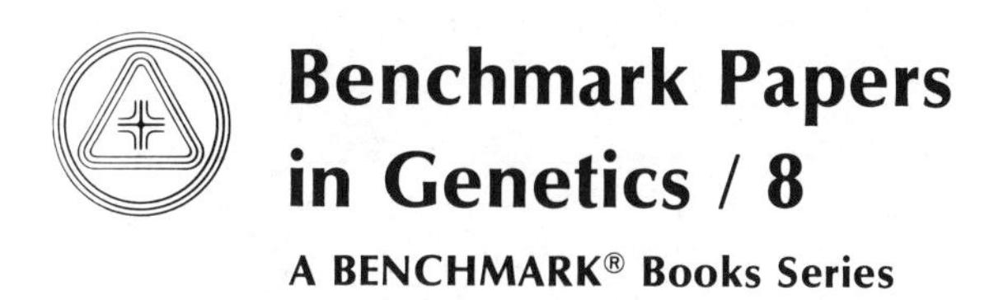

EVOLUTIONARY GENETICS

Edited by
D. L. JAMESON
University of Houston

Benchmark Papers in Genetics, Volume 8
Library of Congress Catalog Card Number: 77-7601
ISBN: 0-87933-295-6

79 78 77 1 2 3 4 5
Manufactured in the United States of America.

LIBRARY OF CONGRESS CATALOGING IN PUBLICATION DATA

Main entry under title:
Evolutionary genetics.
(Benchmark papers in genetics; v. 8)
Includes index.
1. Genetics—Addresses, essays, lectures. 2. Evolution—Addresses, essays, lectures. I. Jameson, David L. [DNLM: 1. Selection (Genetics)—Collected works. 2. Genetics—Collected works. W1 BE516 v. 8 / QH366.2 E93]
QH430.E95 575.1 77-7601
ISBN 0-87933-295-6

Exclusive Distributor: **Halsted Press**
A Division of John Wiley & Sons, Inc.
ISBN: 0-470-99232-8

SERIES EDITOR'S FOREWORD

The study of any discipline assumes the mastery of the literature of the subject. In many branches of science, even one as new as genetics, the expansion of knowledge has been so rapid that there is little hope of learning of the development of all phases of the subject. The student has difficulty mastering the textbook, the young scholar must tend to the literature near his own research, the young instructor barely finds time to expand his horizons to meet his class-preparation requirements, the monographer copes with a wider literature but usually from a specialized viewpoint, and the textbook author is forced to cover much the same material as previous and competing texts to respond to the user's needs and abilities.

Few publishers have the dedication to scholarship to serve primarily the limited market of advanced studies. The opportunity to assist professionals at all stages of their careers has been recognized by the publishers of the Benchmark series and by a distinguished group of Benchmark Volume Editors knowledgeable in specific aspects of the literature of genetics. These editors have selected papers and portions of papers that demonstrate both the development of knowledge and the atmosphere in which that knowledge was developed. There is no substitute for reading great papers. Here you can learn how questions are asked, how they are approached, and how difficult and essential it is to obtain definitive answers and clear writing.

This volume represents a selection of literature from the time when the genetic basis of the evolutionary process was being identified, developed, and established. The volume starts with Darwin and Wallace and traces the development to the early 1930s when Fisher, Haldane, and Wright established integrated and complementary theories which provided the basis for the work of the years since.

D. L. JAMESON

ACKNOWLEDGMENTS

As each new student learns the exciting history of the discovery and the conflicts of a field of knowledge, he is able to relive some of the moments of failure, disillusionment, and disaster which accompany the occasional successes. Great pleasure comes from reading the original literature and from attempting to understand the thought process of these giants of evolutionary genetics. My own introduction to the early history of evolution and of genetics was provided by Samuel Wood Geiser, scientist, historian, and teacher, and participator in a full expanse of scholarship for more than eighty years and, at this writing, still at work. Many missing components have been filled in by other teachers and by colleagues, and I am particularly indebted to J. T. Patterson, M. J. D. White, T. Dobzhansky, and recently to S. Wright, but there were many others. I have had the pleasure of passing this information on to numbers of students far larger than was available to my teachers because of the tremendous expansion of the numbers of students attending universities. To the giants, to Dr. Geiser, to my teachers and colleagues, and particularly to my students, this volume is dedicated.

A faculty development leave from the University of Houston allowed me time to pursue the literature. The Department of Medical Genetics at the University of Wisconsin provided me with space to work and the faculty and students contributed considerable intellectual stimulus. This is contribution number 2,118 from the Laboratory of Genetics of the University of Wisconsin. Pleasant discussion and valuable comments were received from S. Wright, James Crow, W. R. Atchley, and Carter Denniston, each of whom read the manuscript in detail.

CONTENTS

PART III: HAMMER AND TONGS

PART IV: HOW NATURE WORKS

CONTENTS BY AUTHOR

EVOLUTIONARY GENETICS

INTRODUCTION

The early history of the concepts of evolution and that of the concepts of inheritance go hand in hand. Indeed, an understanding of evolution requires knowledge of the mechanisms of heredity. Zirkle (1935, 1946, 1959) has traced the history of these concepts since the Bronze Age. Clearly the ancients understood the inheritance of striking characteristics such as hair or eye color and of diseases such as gout, and even of general form and physical appearance—it takes no fancy measurements to note that offspring do resemble parents. Zirkle notes that Theophrastus described how plants changed when they were transplanted to different soils and climates and that Aristotle described changes in animals and believed that hybridization produced new species. The ancients' belief of ephemeral and mutable species provided no basis for an understanding of evolution or of genetics. Species must be understood as essentially stable units before the processes by which species change can be investigated. A considerable portion of this stability was provided to the species concept by Linnaeus and his system of classification and nomenclature (Ramsbottom, 1938).

In this introduction we will trace the history of genetics as it led to the development of the theories of Haldane, Fisher, and Wright, starting at the time of Linnaeus and ending in the early 1920s. Of particular assistance in the development of this volume have been the studies of Darwin (1868), Roberts (1929), Zirkle (1935, 1946, 1959), Dunn (1965), Sturtevant (1965), Olby (1966), and Provine (1971). Nevertheless, I have relied as much as possible, within the limits of time and availability, on the original literature itself and what it says rather than on what historians say about the people and the papers. The discussion is centered on the papers selected and the references in other papers about these studies.

Not so long ago evolution was a concept outside the established doctrine of Western civilization and its development has been accompanied by considerable conflict. The men involved in evolutionary studies and in genetic studies were strong personalities with intense imaginations, vivid insights into nature, and great ambitions, a considerable stimulus to additional conflicts. Thus the selection of these papers has been influenced by the individual scholars and their interactions with each other as well as by the perceived specific contribution to our understanding of the nature of the gene and its function at various levels.

The threads of the story proceed from species immutability to species transmutability, from continuous slow change of continuous characters to discontinuous change by steps, to an integration of the concepts of particulate Mendelism with the blending of quantitative inheritance and to the development of the initial theories of factor interaction and balance between forces. The story is connected in the final chapter to later developments in plant and animal breeding, quantitative genetics, stochastic models, population genetics, ecological genetics, and speciation.

The early biological experimentalist Joseph Koelreuter (1733–1806), who hybridized many related forms, was particularly impressed with the intermediate nature of the offspring and with the tendency of the F_2 generation to include individuals like both parents and also intermediates. From these results Koelreuter concluded that nature prevented hybrids from breeding true and caused them to revert to the parental forms rather than produce new species. (A Biography of Koelreuter is found in *Genetics,* volume 5, with a frontispiece.)

Carl Friedrich von Gaertner (1772–1850; see *Genetics,* volume 9) was a patient and energetic bibliophile who carefully recorded and summarized the work of the time (von Gaertner, 1849). Additionally he had the resources to perform extensive hybridization studies of many forms and to note that the various characters of the hybrids were frequently mixtures of characters of the different parent forms. Even with these results, the fixity of the species was sufficiently established that he maintained the essential stability of the species characteristics as a whole.

The concept of the inheritance of acquired characters accompanied by a belief that these characters can accumulate indefinitely provides one explanation for evolution. Along with many earlier workers Erasmus Darwin (1796), grandfather of Charles Darwin, supported this position. The great French naturalist Lamarck (1744–1829; see Conklin, 1944; *Genetics,* volume 29, for frontispiece and biography), accepted the fixity of species but proposed the gradual transformation of species—that habits may become instincts and that the effects of use and disuse may be inherited.

Naudin (1865; *Genetics,* volume 6) identified the principle of segregation to explain the reversion of the hybrid to the two original forms. Darwin (1868) recognized the difference between variation which results from changes within individuals (now called mutations) and that which results from recombinations between different individuals. Darwin's theory of pangenesis proposed that the units of inheritance came from all cells in the body and passed through the germ cells from generation to generation. He thought these units were subject to environmental influence.

Francis Galton (1876) felt that the units of heredity received very little environmental influence but did accept the blending of the characters. He made several attempts to develop a quantitative theory of inheritance (1876; 1877a, b; 1889; and Paper 2) and suggested the law of ancestral heredity: "the mean character of the offspring can be calculated with more exactness, the more extensive our knowledge of the corresponding characters of the ancestry." Galton proposed that the individual was related to each generation of its ancestry by some constant proportion of that in the previous generation. The series can be made to explain all the variation. Considerable modification is required to explain other observed phenomena such as dominance. This mathematical approach to blending inheritance was picked up and developed by Karl Pearson (1904, Papers 9 and 10) and W. F. R. Weldon (1902). The procedure is essentially one of explanation and description and fails to provide an understanding of mechanisms.

De Vries (1889; see *Genetics,* volume 4, for biography and frontispiece) postulated the existence of "pangenes," a term adopted from Darwin and later used by Johannsen (1909; see *Genetics,* volume 8, for biography and frontispiece) to form the term gene. De Vries' pangenes differed from those of Darwin in that they were separate and independent factors which passed through the germ cells. He also proposed that all variations were the result of mutation, and that species arose from a single mutation. His theory of mutation suffered because it was all inclusive and his mutants were not species and many of his species were varieties. However, his contribution of the mutability of the hereditary unit was a major and necessary step to the understanding of variation and evolution.

Weismann (1893; see *Genetics,* volume 7, for biography and frontispiece) developed a cytological basis for the continuity of germ plasm which led him to reject the inheritance of acquired characters. His theory separated the somatic action of the hereditary unit from that of the germ cells.

During the last third of the nineteenth century many workers supported the view that evolution proceeded by jumps or steps rather than continuously. These included Huxley, Galton, de Vries, and Bateson. Other scientists of the time supported the position that evolution occurred as a slow continuous process as had been described by Darwin. Pearson, Weldon, and most naturalists maintained this position. The dichotomy between the philosopher-naturalist and the biological experimenter was strong and the effects of this dichotomy still persist.

Thus the scene was set for the rediscovery of Mendel's work. Three workers, de Vries (1900a,b), Correns (1900), and von Tschermak (1900) were each performing experiments which led to Mendel's conclusions, and eventually to Mendel's papers. Their papers are translated in the 1950 supplement of *Genetics* and have been discussed by the historians listed previously. Their respective biographies are found in volumes 4, 13, and 37 of *Genetics.* Their studies, and Mendel's, emphasized the particulate nature of inheritance and thus provided support to those who felt evolution occurred in steps, rather than as a slow and continuous process.

The primary adversaries in the argument on the nature of evolution were W. F. R. Weldon (1894, 1895, 1902, 1903) and William Bateson (1894, 1902). Weldon was born in 1860, studied at London and Cambridge, and in 1884 was appointed lecturer in St. John's College, Cambridge. One of his students was William Bateson. Following Galton's (1889) book, Weldon initiated a series of studies which eventually interested Pearson and Galton and led to the formation of the Evolution Committee of the Royal Society. Weldon died in 1906.

William Bateson was born in 1861, studied at St. John's, became a fellow there in 1885, and was appointed to a newly created chair of biology at Cambridge in 1908. He became director of the John Innes Horticultural Institute in 1910. In 1894 he brought the scattered information on discontinuity in variation together in a book called *Materials for the Study of Variation*. He was the prime proponent of Mendelism during the first decade of the twentieth century. He established the *Journal of Genetics* and named the field of study "Genetics." He died in 1926. A biography and frontispeice is found in *Genetics,* volume 12. The contrasts in the positions of the two adversaries is reviewed by Yule (1902) at the height of the arguments. Provine (1971) provides a detailed analysis of the step-by-step history of the period.

Papers selected from this early period for inclusion in this volume are those of Darwin and Wallace (Paper 1), a later paper of Galton (Paper 2), extracts from a paper by Baldwin (Paper 3), and the paper of Mendel (Paper 4). Darwin and Wallace proposed the theory of natural selection and emphasized that the inherited variation was that portion which contributed to the evolutionary process. Galton's paper is an attempt to demonstrate his theory of inheritance. Baldwin's paper introduces the concept of the inheritance of the ability to respond to changing conditions thus linking developmental processes and inheritance to evolutionary studies, and Mendel's paper provides rules for the particulate basis of the hereditary process.

The exciting rediscovery of the genetic studies of Mendel and the conflict which grew from the diverse opinions about the nature of inheritance and of the evolutionary process was accompanied and followed by a process of resolution. Castle (Paper 5) compared the two theories and described selection of a Mendelian character. Yule (1902 and Paper 6) attempted to explain the inheritance of quantitative characters with Mendelism. Hardy (Paper 7) and Weinberg (Paper 8) identified equilibrium principles and Pearson (1904 and Papers 9 and 10) resolved the two approaches except for the universality of the principle of dominance. East (Paper 11) provided an experimental demonstration of multiple Mendelian factors in the inheritance of a quantitative character.

Following this initial background development a large base of data, experiment, and theory was required to produce an understanding of the mechanisms of evolution. The development of such a data base is diverse and sporadic and the paths of progress are not always clearly visible at the time or later. The sum total is the necessary component for the next step. The development of the theoretical components appears identifiable, and fortunately these papers refer extensively to experimental and observational data which parallel, provide a basis for, or stimulate the development of the theory. These theoretical papers include Norton's (Paper 12) production of a table of the rate of change of gene frequency by selection, Fisher's paper (Paper 13) on the dominance ratio, and one of Wright's early papers (Paper 14) providing a way to measure inbreeding. Papers 15, 16, and 17 of Haldane flesh out the initial theoretical skeleton.

Tracing the applications of the evolutionary genetic theory to evolutionary studies in nature during the first third of this century is more difficult because of the scattered nature of the literature. Two papers by S. S. Chetverikov (Papers 18 and 19) are selected as representative of the literature. A small portion of the literature is discussed.

In the early 1930s Haldane, Fisher, and Wright presented their views on the evolutionary process and made them available in a form which was more accessible than in their separate numerous contributions to the literature. Fisher's *The Genetical Theory of Natural Selection* (1930, see Paper 20), Haldane's *The Causes of Evolution* (1932), and Wright's three papers, "Evolution in Mendelian Populations" (1931, Paper 24), "The Roles of Mutation, Inbreeding, Crossbreeding and Selection in Evolution" (1932, Paper 25), and "The distribution of gene frequencies in populations" (1937) constitute the benchmarks for much of modern studies of how populations change in nature, particularly in the fields of plant and animal breeding, quantitative genetics, behavioral genetics, population genetics, ecological genetics, and speciation. Without these the synthesis works by Dobzhansky, Mayr, Huxley, Stebbins, and Simpson in the late 1930s and 1940s would have lacked the necessary integrated foundation. This volume concludes with a summary of some of the links between these papers and later developments in specific fields of genetics.

Part I

THE BEGINNINGS

Editor's Comments on Papers 1 Through 4

1 DARWIN and WALLACE
On the Tendency of Species to Form Varieties; and on the Perpetuation of Varieties and Species by Natural Means of Selection

2 GALTON
The Average Contribution of Each Several Ancestor to the Total Heritage of the Offspring

3 BALDWIN
A New Factor in Evolution

4 MENDEL
Experiments in Plant Hybridisation

The most famous papers in evolutionary biology, and among the most famous and world-changing articles of all time are those which appeared in the *Journal of the Proceedings of the Linnean Society* in the year 1859. More than a hundred years later we have science writers who convert papers of general public interest in a few journals (e.g., *Science, Nature, PNAS*) into instant news. Mostly these attract public attention for a few days even though they may greatly affect the lives of thousands, or, as in the case of the Salk vaccine, millions of people. In the middle of the last century the activities of science were not as isolated from those of the rest of the educated public as they are today. A smaller portion of the public was educated and knowledge itself was much more general and less specialized. The papers of Darwin and Wallace were communicated to the Linnaean Society by Charles Lyell and J. D. Hooker and included an introduction by the communicators which provided a historical footnote to the mid-nineteenth century workings of science. Darwin, over a period of several decades, contemplatively perused and pursued supportative facts to a theory developed as a result of a voyage around the world. This is hard to understand in these days of rapid publication of preliminary and intermediate results and seldom read final results. Darwin wished to defer in favor of Wallace (Paper 1), but others interceded to assure the historical accuracy of priority and to emphasize the importance of a foundation based "on a wide deduction from facts and matured by years of reflection" (p. 46). This gentle and genteel approach hardly portends the violent nature of later adversary events in evolutionary genetics at approximately twenty-five-year intervals.

In the pages in the Linnean Society Proceedings are presented and interrelated the following significant themes:

1. The amount of resources available to each species is, on the average constant and limiting.
2. The number of individuals in each species tends to increase geometrically.
3. Any minute variation adapting the individual to existing or changing conditions provides that individual with a better chance of survival.
4. ". . . and those of its offspring which inherited the variation, be it ever so slight, would also have a better chance" (p. 49).

These words, written in 1839 but not published until 1859 (p. 47), provide the link between genetics and evolution. Huxley, quickly grasping the central idea of the origin, said "How extremely stupid not to have thought of that" (1887:551).

Darwin (1809–1882) was an accomplished scholar, the author of many papers and a dozen previous volumes, and well known to the general public for his "Voyage of a Naturalist." The original paper was followed (1859) by the first edition of *On the Origin of Species by Means of Natural Selection, or the Preservation of Favoured Races in the Struggle for Life*, which sold the entire printing of 1,250 on the first day, a fate claimed for the next six printings. A number of other books and papers by Darwin followed, including *Variation of Animals and Plants under Domestication* (1868) and *The Descent of Man, and Selection in Relation to Sex* (1871). Most scientists, particularly those attached to the cleric, were unable to accept the transmutation of species, and this certainly included great men of the day, for example Louis Agassiz and W. H. Harvey. Huxley reviewed the reception of the central idea of the origin in 1887 and indicated that the supporters of Darwin in 1860 were numerically extremely insignificant. They did include Lyell, Lubbock, Gray, and, of course, Huxley. The Bishop Wilberforce and Mr. Huxley debated publicly at the meeting of the British Association in Oxford in June of 1860. An unsigned review in the *Quarterly Review,* later acknowledged by Wilberforce, was severely critical of Darwin. In his diaries Darwin wrote "These very clever men think they can write a review with a very slight knowledge of the book reviewed or subject in question" (p. 116, vol. 2, of F. Darwin, 1887). In public he said little, leaving the adversary role to Huxley and Hooker; they in turn obviously enjoyed the experience and were successful in winning converts. Darwin was born in 1809, studied medicine at Edinburgh, religion and botany at Cambridge, and joined the vessel HMS *Beagle* in 1831 for a five-year voyage around the world as a naturalist. He died in 1882.

Alfred Russel Wallace was born in 1823 and served as a land surveyor, architect, and school master until he joined an expedition to the Amazon from 1848 to 1852 and to the Malay Archipelago from 1854 to 1862. He was the author of many books on travels, natural selection (1889), spiritualism, and zoography. He was an active scholar until very near the end of his life in 1913. Wallace and Darwin became friends because of their common experiences as world travelers. He arrived at the theory of natural selection independently but several decades later than Darwin and submitted his thoughts first to his friend.

Francis Galton (1822–1911) was a grandson of Erasmus Darwin and a cousin to Charles Darwin. He contributed to geography, meteorology, anthropology, psychology, and statistics. His great contribution was to effuse quantitative procedures throughout all the fields he studied. In the 1860s he became interested in biology and made initial studies on the problems of nature and nurture, and the inheritance of intelligence, size, and fingerprints in man. Galton made several attempts to develop a theory of heredity. Additionally he formed the Biometric Laboratory and the Eugenics Laboratory of the University of London. He endowed the Galton Chair of Eugenics, first occupied by Karl Pearson. Volume 2 of *Genetics* has a short biography and frontispiece.

In *Variation of Plants and Animals under Domestication*, Darwin proposed a provisional hypothesis of pangenesis which provided for blending of gemmules able to circulate throughout the individual and later collect into sex cells for transmission. Galton (1876) suggested that sex cells were independent of somatic cells, an idea taken up and expanded by Weissman (1893). Galton made significant contributions to statistics and developed a theory of inheritance (1889) which provided for both continuous (stature) and alternative (eye color) characteristics and which had a particulate basis. Thus his work suggests the essential elements in the integration of Mendelian genetics with statistical genetics. At the same time his views formed a basis of great disagreement because those supporting statistical inheritance felt Galton was the "father" of their studies while those supporting discontinuous theories noted that, in fact, Galton supported their opinion. Paper 2 is Galton's (1897) attempt to verify experimentally the statistical law of heredity which he had developed and presented earlier, (Galton, 1889). Galton's concluding remarks on the importance of his work might still be used these many years later. He was a great stimulator of those round him and even though his views on inheritance were not directly adopted, they certainly contributed to the understanding of those that were.

The contribution by Baldwin (Paper 3) is often quoted. The study takes the position that, on the average, those individuals capable of responding to change leave more offspring in future generations than do those not capable of responding to change. Thus the paper forms a link between physiological adjustment, the genetically endowed ability to make some physiological adjustment, and the developmental processes involved in the maturation of that ability. The paper contributes to the understanding of processes later called genetic homeostasis and genetic assimilation. In this volume the paper serves the additional function of describing the state of the art as looked at by a scientist and experimenter of the time just before the rediscovery of Mendel. The paper is long and the terminology well laced with terms from psychology, the implications of the nervous system, and the development of intelligence. For our purposes the initial statement is sufficient. We may note that Baldwin felt that his factor "completely disposes of the Lamarckian factor" (Paper 3, p. 448). Thus, Baldwin did not feel his work was Lamarckian even though some later workers have taken that position about his study.

James Mark Baldwin (1861-1934) trained at Princeton in psychology, receiving the Ph.D. in 1889. He taught at Lake Forest University and the University of Toronto, Princeton, Johns Hopkins, and the National University of Mexico. He received the first honorary degree ever given by Oxford in 1900. He was the author of many books on genetics, evolution, psychology, and philosophy.

Mendel's paper (Paper 4) represents the central point of the origin of modern genetics and thus is the basis not only of this volume but of the "Benchmark Papers in Genetics" series. One can take C. Darwin's position that truth will prevail and that if one investigator does not make a specific contribution, a later one will (F. Darwin, 1887). Mendel's work demonstrates an additional principle; a good idea must have its audience, it must come at the right time and place. Previous workers had advanced the idea of particulate inheritance, found elementary ratios among offspring, and even grasped at the concept of dominance and recessiveness (e.g., see Glass, 1947, Olby, 1966). Mendel also set out to discover rules or patterns in the data obtained over several generations of crosses. Fisher (1936) has suggested that Mendel selected his data to confirm his hypothesis while Wright (1966) and Olby (1966) have responded in Mendel's defense. In any case, hypothesis testing is an integral component of the scientific method and Mendel's work has led to the development of more concrete testable hypotheses than any other biological contribution.

Mendel, as had others before him, noted that when pure breeding lines are crossed the results are variable and often difficult to interpret. Mendel proposed to cross a number of such lines and see if he could find a pattern in the results. More important, he proposed to cross the hybrids both with the pure bred lines and with themselves and determine patterns in the offspring obtained. From such crosses he was able to discern the laws of segregation and of independent assortment, and the world has changed.

Mendel was born in 1822 in Moravia and in 1843 entered the monastery of St. Thomas, a center of intellectual activity near Brunn in present day Czechoslovakia. After becoming a priest in 1847, he encountered difficulties with his duties as a parish leader, later as a high school teacher, and still later in his attempts to pass examinations for teaching certificates. He studied at the University of Vienna in the early 1850s. There followed more than fifteen years of studies on plant hybridization, during which time he was acquainted with the various scholarly activities of the world, read Darwin's *Origin of the Species*, and traveled in France, Italy, and England. The report of the experiments "On Plant Hybridization" were published in 1866. The journal was distributed to major libraries and some reprints were distributed by Mendel, so that the material was available and possibly even read by investigators who were unable to discern its importance. Mendel was later elected abbot of the monastery and spent much of his time struggling with the administrative duties of his position. He died in 1884. Iltis (1932) has provided an interesting biography.

At the turn of the century two interrelated problems confronted biologists: the nature of the evolutionary process, and the nature of inheritance. Darwin (Paper 1, 1859) stated the interrelation:

> Now, can it be doubted, from the struggle each individual has to obtain subsistence, that any minute variation in structure, habits, or instincts, adapting that individual better to the new conditions, would tell upon its vigour and health? In the struggle it would have a better *chance* of surviving; and those of its offspring which inherited the variation, be it ever so slight, would also have a better *chance.* (p. 49)

Darwin recognized that the theory of natural selection was weak in the absence of an adequate theory of inheritance. He supported the inheritance of acquired characters and proposed the blending of characters which were passed from all parts of the body to the germ cells only to be redistributed again in the next generation. The loss of variation which results from blending made Darwin's pangenesis theory incompatible with his theory of evolution.

Galton (1876) proposed a theory of inheritance based on quantitative characters, then later (1889) felt that the theory of natural selection should not be based only on "severally minute" processes. He noted "that the steps *may* be small and that they *must* be small are very different views" (p. 32). Later (Paper 2) he noted that a correct law of heredity would "throw light" on the theory of evolution.

Baldwin (Paper 3) felt that his contribution provided a basis for the development of positive and beneficial inherited characteristics. He felt this was necessary because natural selection only served to eliminate the unfit and not to provide the required expansion of variation.

Many workers at the turn of the century felt the factors of inheritance were particulate and produced particulate results which could be measured and identified. Others felt that inheritance was a statistical process resulting from the accumulation of the effects of a large number of small determinants. The two perceived modes of inheritance were reflected in diverse approaches to the evolutionary process. Experimentalists tended to support the first view while naturalists retained the tradition of the Darwinian views. The positions appeared to produce greater polarity in England than elsewhere (Wright, 1977). The species must be recognized as an essentially stable unit before it can be perceived as evolving. Similarly, the gene must be identified and understood as a stable unit which is transmitted from generation to generation before its instability can be understood. The stage was set for the rediscovery of Mendel's work and the major benchmarks of the next decades.

1

Reprinted from the *J. Linn. Soc. Lond. (Zool.)* 3:45–62 (1859)

ON THE TENDENCY OF SPECIES TO FORM VARIETIES; AND ON THE PERPETUATION OF VARIETIES AND SPECIES BY NATURAL MEANS OF SELECTION

Charles Darwin and Alfred Wallace

My Dear Sir,—The accompanying papers, which we have the honour of communicating to the Linnean Society, and which all relate to the same subject, viz. the Laws which affect the Production of Varieties, Races, and Species, contain the results of the investigations of two indefatigable naturalists, Mr. Charles Darwin and Mr. Alfred Wallace.

These gentlemen having, independently and unknown to one another, conceived the same very ingenious theory to account for the appearance and perpetuation of varieties and of specific forms on our planet, may both fairly claim the merit of being original thinkers in this important line of inquiry; but neither of them having published his views, though Mr. Darwin has for many years past been repeatedly urged by us to do so, and both authors having now unreservedly placed their papers in our hands, we think it would best promote the interests of science that a selection from them should be laid before the Linnean Society.

Taken in the order of their dates, they consist of:—

1. Extracts from a MS. work on Species*, by Mr. Darwin, which was sketched in 1839, and copied in 1844, when the copy was read by Dr. Hooker, and its contents afterwards communicated to Sir Charles Lyell. The first Part is devoted to "The Variation of Organic Beings under Domestication and in their Natural State;" and the second chapter of that Part, from which we propose to read to the Society the extracts referred to, is headed, "On the Variation of Organic Beings in a state of Nature; on the Natural Means of Selection; on the Comparison of Domestic Races and true Species."

2. An abstract of a private letter addressed to Professor Asa Gray, of Boston, U.S., in October 1857, by Mr. Darwin, in which

* This MS. work was never intended for publication, and therefore was not written with care.—C. D. 1858.

he repeats his views, and which shows that these remained unaltered from 1839 to 1857.

3. An Essay by Mr. Wallace, entitled "On the Tendency of Varieties to depart indefinitely from the Original Type." This was written at Ternate in February 1858, for the perusal of his friend and correspondent Mr. Darwin, and sent to him with the expressed wish that it should be forwarded to Sir Charles Lyell, if Mr. Darwin thought it sufficiently novel and interesting. So highly did Mr. Darwin appreciate the value of the views therein set forth, that he proposed, in a letter to Sir Charles Lyell, to obtain Mr. Wallace's consent to allow the Essay to be published as soon as possible. Of this step we highly approved, provided Mr. Darwin did not withhold from the public, as he was strongly inclined to do (in favour of Mr. Wallace), the memoir which he had himself written on the same subject, and which, as before stated, one of us had perused in 1844, and the contents of which we had both of us been privy to for many years. On representing this to Mr. Darwin, he gave us permission to make what use we thought proper of his memoir, &c.; and in adopting our present course, of presenting it to the Linnean Society, we have explained to him that we are not solely considering the relative claims to priority of himself and his friend, but the interests of science generally; for we feel it to be desirable that views founded on a wide deduction from facts, and matured by years of reflection, should constitute at once a goal from which others may start, and that, while the scientific world is waiting for the appearance of Mr. Darwin's complete work, some of the leading results of his labours, as well as those of his able correspondent, should together be laid before the public.

We have the honour to be yours very obediently,

CHARLES LYELL.
JOS. D. HOOKER.

J. J. Bennett, Esq.,
Secretary of the Linnean Society.

I. *Extract from an unpublished Work on Species, by* C. DARWIN, Esq., *consisting of a portion of a Chapter entitled, "On the Variation of Organic Beings in a state of Nature; on the Natural Means of Selection; on the Comparison of Domestic Races and true Species."*

De Candolle, in an eloquent passage, has declared that all nature is at war, one organism with another, or with external nature.

Seeing the contented face of nature, this may at first well be doubted; but reflection will inevitably prove it to be true. The war, however, is not constant, but recurrent in a slight degree at short periods, and more severely at occasional more distant periods; and hence its effects are easily overlooked. It is the doctrine of Malthus applied in most cases with tenfold force. As in every climate there are seasons, for each of its inhabitants, of greater and less abundance, so all annually breed; and the moral restraint which in some small degree checks the increase of mankind is entirely lost. Even slow-breeding mankind has doubled in twenty-five years; and if he could increase his food with greater ease, he would double in less time. But for animals without artificial means, the amount of food for each species must, *on an average*, be constant, whereas the increase of all organisms tends to be geometrical, and in a vast majority of cases at an enormous ratio. Suppose in a certain spot there are eight pairs of birds, and that *only* four pairs of them annually (including double hatches) rear only four young, and that these go on rearing their young at the same rate, then at the end of seven years (a short life, excluding violent deaths, for any bird) there will be 2048 birds, instead of the original sixteen. As this increase is quite impossible, we must conclude either that birds do not rear nearly half their young, or that the average life of a bird is, from accident, not nearly seven years. Both checks probably concur. The same kind of calculation applied to all plants and animals affords results more or less striking, but in very few instances more striking than in man.

Many practical illustrations of this rapid tendency to increase are on record, among which, during peculiar seasons, are the extraordinary numbers of certain animals; for instance, during the years 1826 to 1828, in La Plata, when from drought some millions of cattle perished, the whole country actually *swarmed* with mice. Now I think it cannot be doubted that during the breeding-season all the mice (with the exception of a few males or females in excess) ordinarily pair, and therefore that this astounding increase during three years must be attributed to a greater number than usual surviving the first year, and then breeding, and so on till the third year, when their numbers were brought down to their usual limits on the return of wet weather. Where man has introduced plants and animals into a new and favourable country, there are many accounts in how surprisingly few years the whole country has become stocked with them. This increase would necessarily stop as soon as the country was fully stocked; and yet we have every reason to believe, from what is known of wild animals, that *all* would pair in the spring. In the majority of cases it is most difficult to imagine where the checks fall—though generally, no doubt, on the seeds, eggs, and young; but when we remember how impossible, even in mankind (so much better known than any other animal), it is to infer from repeated casual observations what the average duration of life is, or to discover the different percentage of deaths to births in different countries, we ought to feel no surprise at our being unable to discover where the check falls in any animal or plant. It should always be remembered, that in most cases the checks are recurrent yearly in a small, regular degree, and in an extreme degree during unusually cold, hot, dry, or wet years, according to the constitution of the being in question. Lighten any check in the least degree, and the geometrical powers of increase in every organism will almost instantly increase the average number of the favoured species. Nature may be compared to a surface on which rest ten thousand sharp wedges touching each other and driven inwards by incessant blows. Fully to realize these views much reflection is requisite. Malthus on man should be studied; and all such cases as those of the mice in La Plata, of the cattle and horses when first turned out in South America, of the birds by our calculation, &c., should be well considered. Reflect on the enormous multiplying power *inherent and annually in action* in all animals; reflect on the countless seeds scattered by a hundred ingenious contrivances, year after year, over the whole face of the land; and yet we have every reason to suppose that the average percentage of each of the inhabitants of a country usually remains constant. Finally, let it be borne in mind that this average number of individuals (the external conditions remaining the same) in each country is kept up by recurrent struggles against other species or against external nature (as on the borders of the Arctic regions, where the cold checks life), and that ordinarily each individual of every species holds its place, either by its own struggle and capacity of acquiring nourishment in some period of its life, from the egg upwards; or by the struggle of its parents (in short-lived organisms, when the main check occurs at longer intervals) with other individuals of the *same* or *different* species.

But let the external conditions of a country alter. If in a small degree, the relative proportions of the inhabitants will in most cases simply be slightly changed; but let the number of inhabitants be small, as on an island, and free access to it from other countries be circumscribed, and let the change of conditions continue progressing (forming new stations), in such a case the original inhabitants must cease to be as perfectly adapted to the changed conditions as they were originally. It has been shown in a former part of this work, that such changes of external conditions would, from their acting on the reproductive system, probably cause the organization of those beings which were most affected to become, as under domestication, plastic. Now, can it be doubted, from the struggle each individual has to obtain subsistence, that any minute variation in structure, habits, or instincts, adapting that individual better to the new conditions, would tell upon its vigour and health? In the struggle it would have a better *chance* of surviving; and those of its offspring which inherited the variation, be it ever so slight, would also have a better *chance*. Yearly more are bred than can survive; the smallest grain in the balance, in the long run, must tell on which death shall fall, and which shall survive. Let this work of selection on the one hand, and death on the other, go on for a thousand generations, who will pretend to affirm that it would produce no effect, when we remember what, in a few years, Bakewell effected in cattle, and Western in sheep, by this identical principle of selection?

To give an imaginary example from changes in progress on an island:—let the organization of a canine animal which preyed chiefly on rabbits, but sometimes on hares, become slightly plastic; let these same changes cause the number of rabbits very slowly to decrease, and the number of hares to increase; the effect of this would be that the fox or dog would be driven to try to catch more hares: his organization, however, being slightly plastic, those individuals with the lightest forms, longest limbs, and best eyesight, let the difference be ever so small, would be slightly favoured, and would tend to live longer, and to survive during that time of the year when food was scarcest; they would also rear more young, which would tend to inherit these slight peculiarities. The less fleet ones would be rigidly destroyed. I can see no more reason to doubt that these causes in a thousand generations would produce a marked effect, and adapt the form of the fox or dog to the catching of hares instead of rabbits, than that greyhounds can be improved by selection and careful breeding. So would it be with plants under similar circumstances. If the number of individuals of a species with plumed seeds could be increased by greater powers of dissemination within its own area (that is, if the check to increase fell chiefly on the seeds), those seeds which were provided with ever so little more down, would in the long run be most disseminated; hence a greater number of seeds thus formed would germinate, and would tend to produce plants inheriting the slightly better-adapted down*.

Besides this natural means of selection, by which those individuals are preserved, whether in their egg, or larval, or mature state, which are best adapted to the place they fill in nature, there is a second agency at work in most unisexual animals, tending to produce the same effect, namely, the struggle of the males for the females. These struggles are generally decided by the law of battle, but in the case of birds, apparently, by the charms of their song, by their beauty or their power of courtship, as in the dancing rock-thrush of Guiana. The most vigorous and healthy males, implying perfect adaptation, must generally gain the victory in their contests. This kind of selection, however, is less rigorous than the other; it does not require the death of the less successful, but gives to them fewer descendants. The struggle falls, moreover, at a time of year when food is generally abundant, and perhaps the effect chiefly produced would be the modification of the secondary sexual characters, which are not related to the power of obtaining food, or to defence from enemies, but to fighting with or rivalling other males. The result of this struggle amongst the males may be compared in some respects to that produced by those agriculturists who pay less attention to the careful selection of all their young animals, and more to the occasional use of a choice mate.

II. *Abstract of a Letter from* C. DARWIN, Esq., *to* Prof. ASA GRAY, *Boston, U.S., dated Down, September 5th*, 1857.

1. It is wonderful what the principle of selection by man, that is the picking out of individuals with any desired quality, and breeding from them, and again picking out, can do. Even breeders have been astounded at their own results. They can act on differences inappreciable to an uneducated eye. Selection has been *methodically* followed in *Europe* for only the last half century; but it was occasionally, and even in some degree methodically, followed in the most ancient times. There must have been also a kind of unconscious selection from a remote period, namely in

* I can see no more difficulty in this, than in the planter improving his varieties of the cotton plant.—C. D. 1858.

the preservation of the individual animals (without any thought of their offspring) most useful to each race of man in his particular circumstances. The "roguing," as nurserymen call the destroying of varieties which depart from their type, is a kind of selection. I am convinced that intentional and occasional selection has been the main agent in the production of our domestic races; but however this may be, its great power of modification has been indisputably shown in later times. Selection acts only by the accumulation of slight or greater variations, caused by external conditions, or by the mere fact that in generation the child is not absolutely similar to its parent. Man, by this power of accumulating variations, adapts living beings to his wants—may be said to make the wool of one sheep good for carpets, of another for cloth, &c.

2. Now suppose there were a being who did not judge by mere external appearances, but who could study the whole internal organization, who was never capricious, and should go on selecting for one object during millions of generations; who will say what he might not effect? In nature we have some *slight* variation occasionally in all parts; and I think it can be shown that changed conditions of existence is the main cause of the child not exactly resembling its parents; and in nature geology shows us what changes have taken place, and are taking place. We have almost unlimited time; no one but a practical geologist can fully appreciate this. Think of the Glacial period, during the whole of which the same species at least of shells have existed; there must have been during this period millions on millions of generations.

3. I think it can be shown that there is such an unerring power at work in *Natural Selection* (the title of my book), which selects exclusively for the good of each organic being. The elder De Candolle, W. Herbert, and Lyell have written excellently on the struggle for life; but even they have not written strongly enough. Reflect that every being (even the elephant) breeds at such a rate, that in a few years, or at most a few centuries, the surface of the earth would not hold the progeny of one pair. I have found it hard constantly to bear in mind that the increase of every single species is checked during some part of its life, or during some shortly recurrent generation. Only a few of those annually born can live to propagate their kind. What a trifling difference must often determine which shall survive, and which perish!

4. Now take the case of a country undergoing some change. This will tend to cause some of its inhabitants to vary slightly—

not but that I believe most beings vary at all times enough for selection to act on them. Some of its inhabitants will be exterminated; and the remainder will be exposed to the mutual action of a different set of inhabitants, which I believe to be far more important to the life of each being than mere climate. Considering the infinitely various methods which living beings follow to obtain food by struggling with other organisms, to escape danger at various times of life, to have their eggs or seeds disseminated, &c. &c., I cannot doubt that during millions of generations individuals of a species will be occasionally born with some slight variation, profitable to some part of their economy. Such individuals will have a better chance of surviving, and of propagating their new and slightly different structure; and the modification may be slowly increased by the accumulative action of natural selection to any profitable extent. The variety thus formed will either coexist with, or, more commonly, will exterminate its parent form. An organic being, like the woodpecker or misseltoe, may thus come to be adapted to a score of contingences—natural selection accumulating those slight variations in all parts of its structure, which are in any way useful to it during any part of its life.

5. Multiform difficulties will occur to every one, with respect to this theory. Many can, I think, be satisfactorily answered. *Natura non facit saltum* answers some of the most obvious. The slowness of the change, and only a very few individuals undergoing change at any one time, answers others. The extreme imperfection of our geological records answers others.

6. Another principle, which may be called the principle of divergence, plays, I believe, an important part in the origin of species. The same spot will support more life if occupied by very diverse forms. We see this in the many generic forms in a square yard of turf, and in the plants or insects on any little uniform islet, belonging almost invariably to as many genera and families as species. We can understand the meaning of this fact amongst the higher animals, whose habits we understand. We know that it has been experimentally shown that a plot of land will yield a greater weight if sown with several species and genera of grasses, than if sown with only two or three species. Now, every organic being, by propagating so rapidly, may be said to be striving its utmost to increase in numbers. So it will be with the offspring of any species after it has become diversified into varieties, or subspecies, or true species. And it follows, I think, from the foregoing facts, that the varying offspring of each species will try

(only few will succeed) to seize on as many and as diverse places in the economy of nature as possible. Each new variety or species, when formed, will generally take the place of, and thus exterminate its less well-fitted parent. This I believe to be the origin of the classification and affinities of organic beings at all times; for organic beings always *seem* to branch and sub-branch like the limbs of a tree from a common trunk, the flourishing and diverging twigs destroying the less vigorous—the dead and lost branches rudely representing extinct genera and families.

This sketch is *most* imperfect; but in so short a space I cannot make it better. Your imagination must fill up very wide blanks.

C. DARWIN.

III. *On the Tendency of Varieties to depart indefinitely from the Original Type.* By ALFRED RUSSEL WALLACE.

One of the strongest arguments which have been adduced to prove the original and permanent distinctness of species is, that *varieties* produced in a state of domesticity are more or less unstable, and often have a tendency, if left to themselves, to return to the normal form of the parent species; and this instability is considered to be a distinctive peculiarity of all varieties, even of those occurring among wild animals in a state of nature, and to constitute a provision for preserving unchanged the originally created distinct species.

In the absence or scarcity of facts and observations as to *varieties* occurring among wild animals, this argument has had great weight with naturalists, and has led to a very general and somewhat prejudiced belief in the stability of species. Equally general, however, is the belief in what are called "permanent or true varieties,"—races of animals which continually propagate their like, but which differ so slightly (although constantly) from some other race, that the one is considered to be a *variety* of the other. Which is the *variety* and which the original *species*, there is generally no means of determining, except in those rare cases in which the one race has been known to produce an offspring unlike itself and resembling the other. This, however, would seem quite incompatible with the "permanent invariability of species," but the difficulty is overcome by assuming that such varieties have strict limits, and can never again vary further from the original type, although they may return to it, which, from the analogy of the domesticated animals, is considered to be highly probable, if not certainly proved.

It will be observed that this argument rests entirely on the assumption, that *varieties* occurring in a state of nature are in all respects analogous to or even identical with those of domestic animals, and are governed by the same laws as regards their permanence or further variation. But it is the object of the present paper to show that this assumption is altogether false, that there is a general principle in nature which will cause many *varieties* to survive the parent species, and to give rise to successive variations departing further and further from the original type, and which also produces, in domesticated animals, the tendency of varieties to return to the parent form.

The life of wild animals is a struggle for existence. The full exertion of all their faculties and all their energies is required to preserve their own existence and provide for that of their infant offspring. The possibility of procuring food during the least favourable seasons, and of escaping the attacks of their most dangerous enemies, are the primary conditions which determine the existence both of individuals and of entire species. These conditions will also determine the population of a species; and by a careful consideration of all the circumstances we may be enabled to comprehend, and in some degree to explain, what at first sight appears so inexplicable—the excessive abundance of some species, while others closely allied to them are very rare.

The general proportion that must obtain between certain groups of animals is readily seen. Large animals cannot be so abundant as small ones; the carnivora must be less numerous than the herbivora; eagles and lions can never be so plentiful as pigeons and antelopes; the wild asses of the Tartarian deserts cannot equal in numbers the horses of the more luxuriant prairies and pampas of America. The greater or less fecundity of an animal is often considered to be one of the chief causes of its abundance or scarcity; but a consideration of the facts will show us that it really has little or nothing to do with the matter. Even the least prolific of animals would increase rapidly if unchecked, whereas it is evident that the animal population of the globe must be stationary, or perhaps, through the influence of man, decreasing. Fluctuations there may be; but permanent increase, except in restricted localities, is almost impossible. For example, our own observation must convince us that birds do not go on increasing every year in a geometrical ratio, as they would do, were there not some powerful check to their natural increase. Very few birds produce less than two young ones each year, while many have six, eight, or ten; four will certainly be below the average; and if we suppose that each pair produce young only four times in their life, that will also be below the average, supposing them not to die either by violence or want of food. Yet at this rate how tremendous would be the increase in a few years from a single pair! A simple calculation will show that in fifteen years each pair of birds would have increased to nearly ten millions! whereas we have no reason to believe that the number of the birds of any country increases at all in fifteen or in one hundred and fifty years. With such powers of increase the population must have reached its limits, and have become stationary, in a very few years after the origin of each species. It is evident, therefore, that each year an immense number of birds must perish—as many in fact as are born; and as on the lowest calculation the progeny are each year twice as numerous as their parents, it follows that, whatever be the average number of individuals existing in any given country, *twice that number must perish annually*,—a striking result, but one which seems at least highly probable, and is perhaps under rather than over the truth. It would therefore appear that, as far as the continuance of the species and the keeping up the average number of individuals are concerned, large broods are superfluous. On the average all above *one* become food for hawks and kites, wild cats and weasels, or perish of cold and hunger as winter comes on. This is strikingly proved by the case of particular species; for we find that their abundance in individuals bears no relation whatever to their fertility in producing offspring. Perhaps the most remarkable instance of an immense bird population is that of the passenger pigeon of the United States, which lays only one, or at most two eggs, and is said to rear generally but one young one. Why is this bird so extraordinarily abundant, while others producing two or three times as many young are much less plentiful? The explanation is not difficult. The food most congenial to this species, and on which it thrives best, is abundantly distributed over a very extensive region, offering such differences of soil and climate, that in one part or another of the area the supply never fails. The bird is capable of a very rapid and long-continued flight, so that it can pass without fatigue over the whole of the district it inhabits, and as soon as the supply of food begins to fail in one place is able to discover a fresh feeding-ground. This example strikingly shows us that the procuring a constant supply of wholesome food is almost the sole condition requisite for ensuring the rapid increase of a given species, since neither the limited fecundity, nor the unrestrained attacks of birds of prey and of man are here sufficient to check it. In no other birds are these peculiar circumstances so strikingly combined. Either their food is more liable to failure, or they have not sufficient power of wing to search for it over an extensive area, or during some season of the year it becomes very scarce, and less wholesome substitutes have to be found; and thus, though more fertile in offspring, they can never increase beyond the supply of food in the least favourable seasons. Many birds can only exist by migrating, when their food becomes scarce, to regions possessing a milder, or at least a different climate, though, as these migrating birds are seldom excessively abundant, it is evident that the countries they visit are still deficient in a constant and abundant supply of wholesome food. Those whose organization does not permit them to migrate when their food becomes periodically scarce, can never attain a large population. This is probably the reason why woodpeckers are scarce with us, while in the tropics they are among the most abundant of solitary birds. Thus the house sparrow is more abundant than the redbreast, because its food is more constant and plentiful,—seeds of grasses being preserved during the winter, and our farm-yards and stubble-fields furnishing an almost inexhaustible supply. Why, as a general rule, are aquatic, and especially sea birds, very numerous in individuals? Not because they are more prolific than others, generally the contrary; but because their food never fails, the sea-shores and river-banks daily swarming with a fresh supply of small mollusca and crustacea. Exactly the same laws will apply to mammals. Wild cats are prolific and have few enemies; why then are they never as abundant as rabbits? The only intelligible answer is, that their supply of food is more precarious. It appears evident, therefore, that so long as a country remains physically unchanged, the numbers of its animal population cannot materially increase. If one species does so, some others requiring the same kind of food must diminish in proportion. The numbers that die annually must be immense; and as the individual existence of each animal depends upon itself, those that die must be the weakest—the very young, the aged, and the diseased,—while those that prolong their existence can only be the most perfect in health and vigour—those who are best able to obtain food regularly, and avoid their numerous enemies. It is, as we commenced by remarking, "a struggle for existence," in which the weakest and least perfectly organized must always succumb.

Now it is clear that what takes place among the individuals of a species must also occur among the several allied species of a group,—viz. that those which are best adapted to obtain a regular supply of food, and to defend themselves against the attacks of their enemies and the vicissitudes of the seasons, must necessarily obtain and preserve a superiority in population; while those species which from some defect of power or organization are the least capable of counteracting the vicissitudes of food, supply, &c., must diminish in numbers, and, in extreme cases, become altogether extinct. Between these extremes the species will present various degrees of capacity for ensuring the means of preserving life; and it is thus we account for the abundance or rarity of species. Our ignorance will generally prevent us from accurately tracing the effects to their causes; but could we become perfectly acquainted with the organization and habits of the various species of animals, and could we measure the capacity of each for performing the different acts necessary to its safety and existence under all the varying circumstances by which it is surrounded, we might be able even to calculate the proportionate abundance of individuals which is the necessary result.

If now we have succeeded in establishing these two points—1st, *that the animal population of a country is generally stationary, being kept down by a periodical deficiency of food, and other checks*; and, 2nd, *that the comparative abundance or scarcity of the individuals of the several species is entirely due to their organization and resulting habits, which, rendering it more difficult to procure a regular supply of food and to provide for their personal safety in some cases than in others, can only be balanced by a difference in the population which have to exist in a given area*—we shall be in a condition to proceed to the consideration of *varieties*, to which the preceding remarks have a direct and very important application.

Most or perhaps all the variations from the typical form of a species must have some definite effect, however slight, on the habits or capacities of the individuals. Even a change of colour might, by rendering them more or less distinguishable, affect their safety; a greater or less development of hair might modify their habits. More important changes, such as an increase in the power or dimensions of the limbs or any of the external organs, would more or less affect their mode of procuring food or the range of country which they inhabit. It is also evident that most changes would affect, either favourably or adversely, the powers of prolonging existence. An antelope with shorter or weaker legs must necessarily suffer more from the attacks of the feline carnivora; the passenger pigeon with less powerful wings would sooner or later be affected in its powers of procuring a regular supply of food; and in both cases the result must necessarily be a diminution of the population of the modified species. If, on the other hand, any species should produce a variety having slightly increased powers of preserving existence, that variety must inevitably in time acquire a superiority in numbers. These results must follow as surely as old age, intemperance, or scarcity of food produce an increased mortality. In both cases there may be many individual exceptions; but on the average the rule will invariably be found to hold good. All varieties will therefore fall into two classes—those which under the same conditions would never reach the population of the parent species, and those which would in time obtain and keep a numerical superiority. Now, let some alteration of physical conditions occur in the district—a long period of drought, a destruction of vegetation by locusts, the irruption of some new carnivorous animal seeking "pastures new"—any change in fact tending to render existence more difficult to the species in question, and tasking its utmost powers to avoid complete extermination; it is evident that, of all the individuals composing the species, those forming the least numerous and most feebly organized variety would suffer first, and, were the pressure severe, must soon become extinct. The same causes continuing in action, the parent species would next suffer, would gradually diminish in numbers, and with a recurrence of similar unfavourable conditions might also become extinct. The superior variety would then alone remain, and on a return to favourable circumstances would rapidly increase in numbers and occupy the place of the extinct species and variety.

The *variety* would now have replaced the *species*, of which it would be a more perfectly developed and more highly organized form. It would be in all respects better adapted to secure its safety, and to prolong its individual existence and that of the race. Such a variety *could not* return to the original form; for that form is an inferior one, and could never compete with it for existence. Granted, therefore, a "tendency" to reproduce the original type of the species, still the variety must ever remain preponderant in numbers, and under adverse physical conditions *again alone survive*.

But this new, improved, and populous race might itself, in course of time, give rise to new varieties, exhibiting several diverging modifications of form, any of which, tending to increase the facilities for preserving existence, must, by the same general law, in their turn become predominant. Here, then, we have *progression and continued divergence* deduced from the general laws which regulate the existence of animals in a state of nature, and from the undisputed fact that varieties do frequently occur. It is not, however, contended that this result would be invariable; a change of physical conditions in the district might at times materially modify it, rendering the race which had been the most capable of supporting existence under the former conditions now the least so, and even causing the extinction of the newer and, for a time, superior race, while the old or parent species and its first inferior varieties continued to flourish. Variations in unimportant parts might also occur, having no perceptible effect on the life-preserving powers; and the varieties so furnished might run a course parallel with the parent species, either giving rise to further variations or returning to the former type. All we argue for is, that certain varieties have a tendency to maintain their existence longer than the original species, and this tendency must make itself felt; for though the doctrine of chances or averages can never be trusted to on a limited scale, yet, if applied to high numbers, the results come nearer to what theory demands, and, as we approach to an infinity of examples, become strictly accurate. Now the scale on which nature works is so vast—the numbers of individuals and periods of time with which she deals approach so near to infinity, that any cause, however slight, and however liable to be veiled and counteracted by accidental circumstances, must in the end produce its full legitimate results.

Let us now turn to domesticated animals, and inquire how varieties produced among them are affected by the principles here enunciated. The essential difference in the condition of wild and domestic animals is this,—that among the former, their well-being and very existence depend upon the full exercise and healthy condition of all their senses and physical powers, whereas, among the latter, these are only partially exercised, and in some cases are absolutely unused. A wild animal has to search, and often to labour, for every mouthful of food—to exercise sight, hearing, and smell in seeking it, and in avoiding dangers, in procuring shelter from the inclemency of the seasons, and in providing for the subsistence and safety of its offspring. There is no muscle of its body that is not called into daily and hourly activity; there is no sense or faculty that is not strengthened by continual exercise. The domestic animal, on the other hand, has food provided for it, is sheltered, and often confined, to guard it against the vicissitudes of the seasons, is carefully secured from the attacks of its natural enemies, and seldom even rears its young without human assistance. Half of its senses and faculties are quite useless; and the other half are but occasionally called into feeble exercise, while even its muscular system is only irregularly called into action.

Now when a variety of such an animal occurs, having increased power or capacity in any organ or sense, such increase is totally useless, is never called into action, and may even exist without the animal ever becoming aware of it. In the wild animal, on the contrary, all its faculties and powers being brought into full action for the necessities of existence, any increase becomes immediately available, is strengthened by exercise, and must even slightly modify the food, the habits, and the whole economy of the race. It creates as it were a new animal, one of superior powers, and which will necessarily increase in numbers and outlive those inferior to it.

Again, in the domesticated animal all variations have an equal chance of continuance; and those which would decidedly render a wild animal unable to compete with its fellows and continue its existence are no disadvantage whatever in a state of domesticity. Our quickly fattening pigs, short-legged sheep, pouter pigeons, and poodle dogs could never have come into existence in a state of nature, because the very first step towards such inferior forms would have led to the rapid extinction of the race; still less could they now exist in competition with their wild allies. The great speed but slight endurance of the race horse, the unwieldy strength of the ploughman's team, would both be useless in a state of nature. If turned wild on the pampas, such animals would probably soon become extinct, or under favourable circumstances might each lose those extreme qualities which would never be called into action, and in a few generations would revert to a common type, which must be that in which the various powers and faculties are so proportioned to each other as to be best adapted to procure food and secure safety,—that in which by the full exercise of every part of his organization the animal can alone continue to live. Domestic varieties, when turned wild, *must* return to something near the type of the original wild stock, *or become altogether extinct.*

We see, then, that no inferences as to varieties in a state of nature can be deduced from the observation of those occurring among domestic animals. The two are so much opposed to each other in every circumstance of their existence, that what applies to the one is almost sure not to apply to the other. Domestic animals are abnormal, irregular, artificial; they are subject to varieties which never occur and never can occur in a state of nature: their very existence depends altogether on human care; so far are many of them removed from that just proportion of faculties, that true balance of organization, by means of which alone an animal left to its own resources can preserve its existence and continue its race.

The hypothesis of Lamarck—that progressive changes in species have been produced by the attempts of animals to increase the development of their own organs, and thus modify their structure and habits—has been repeatedly and easily refuted by all writers on the subject of varieties and species, and it seems to have been considered that when this was done the whole question has been finally settled; but the view here developed renders such an hypothesis quite unnecessary, by showing that similar results must be produced by the action of principles constantly at work in nature. The powerful retractile talons of the falcon- and the cat-tribes have not been produced or increased by the volition of those animals; but among the different varieties which occurred in the earlier and less highly organized forms of these groups, *those always survived longest which had the greatest facilities for seizing their prey*. Neither did the giraffe acquire its long neck by desiring to reach the foliage of the more lofty shrubs, and constantly stretching its neck for the purpose, but because any varieties which occurred among its antitypes with a longer neck than usual *at once secured a fresh range of pasture over the same ground as their shorter-necked companions, and on the first scarcity of food were thereby enabled to outlive them.* Even the peculiar colours of many animals, especially insects, so closely resembling the soil or the leaves or the trunks on which they habitually reside, are explained on the same principle; for though in the course of ages varieties of many tints may have occurred, *yet those races having colours best adapted to concealment from their enemies would inevitably survive the longest.* We have also here an acting cause to account for that balance so often observed in nature,—a deficiency in one set of organs always being compensated by an increased development of some others—powerful wings accompanying weak feet, or great velocity making up for the absence of defensive weapons; for it has been shown that all varieties in which an unbalanced deficiency occurred could not long continue their existence. The action of this principle is exactly like that of the centrifugal governor of the steam engine, which checks and corrects any irregularities almost before they become evident; and in like manner no unbalanced deficiency in the animal kingdom can ever reach any conspicuous magnitude, because it would make itself felt at the very first step, by rendering existence difficult and extinction almost sure soon to follow. An origin such as is here advocated will also agree with the peculiar character of the modifications of form and structure which obtain in organized beings—the many lines of divergence from a central type, the increasing efficiency and power of a particular organ through a succession of allied species, and the remarkable persistence of unimportant parts such as colour, texture of plumage and hair, form of horns or crests, through a series of species differing considerably in more essential characters. It also furnishes us with a reason for that "more specialized structure" which Professor Owen states to be a characteristic of recent compared with extinct forms, and which would evidently be the result of the progressive modification of any organ applied to a special purpose in the animal economy.

We believe we have now shown that there is a tendency in nature to the continued progression of certain classes of *varieties* further and further from the original type—a progression to which there appears no reason to assign any definite limits—and that the same principle which produces this result in a state of nature will also explain why domestic varieties have a tendency to revert to the original type. This progression, by minute steps, in various directions, but always checked and balanced by the necessary conditions, subject to which alone existence can be preserved, may, it is believed, be followed out so as to agree with all the phenomena presented by organized beings, their extinction and succession in past ages, and all the extraordinary modifications of form, instinct, and habits which they exhibit.

Ternate, February, 1858.

2

Reprinted from *Roy. Soc. (Lond.) Proc.* **61**:401–413 (1897)

THE AVERAGE CONTRIBUTION OF EACH SEVERAL ANCESTOR TO THE TOTAL HERITAGE OF THE OFFSPRING

Francis Galton

In the following memoir the truth will be verified in a particular instance, of a statistical law of heredity that appears to be universally applicable to bisexual descent. I stated it briefly and with hesitation in my book 'Natural Inheritance' (Macmillan, 1889; page 134), because it was then unsupported by sufficient evidence. Its existence was originally suggested by general considerations, and it might, as will be shown, have been inferred from them with considerable assurance. Consequently, as it is now found to hold good in a special case, there are strong grounds for believing it to be a general law of heredity.

I have had great difficulty in obtaining a sufficient amount of suitable evidence for the purpose of verification. A somewhat extensive series of experiments with moths were carried on, in order to supply it, but they unfortunately failed, partly owing to the diminishing fertility of successive broods and partly to the large disturbing effects of differences in food and environment on different broods and in different places and years. No statistical results of any consistence or value could be obtained from them. Latterly, while engaged in planning another extensive experiment with small, fast-breeding mammals, I became acquainted with the existence of a long series of records, preserved by Sir Everett Millais, of the colours during many successive generations of a large pedigree stock of Basset hounds, that he originated some twenty years ago, having purchased ninety-three of them on the Continent, for the purpose. These records afford the foundation upon which this memoir rests.

The law to be verified may seem at first sight too artificial to be true, but a closer examination shows that prejudice arising from the cursory impression is unfounded. This subject will be alluded to again, in the meantime the law shall be stated. It is that the two parents contribute between them on the average one-half, or (0·5) of the total heritage of the offspring; the four grandparents, one-quarter, or $(0 \cdot 5)^2$; the eight great-grandparents, one-eighth, or $(0 \cdot 5)^3$, and so on. Thus the sum of the ancestral contributions is expressed by the series $\{(0 \cdot 5)+(0 \cdot 5)^2+(0 \cdot 5)^3$, &c.$\}$, which, being equal to 1, accounts for the whole heritage.

The same statement may be put into a different form, in which a parent, grandparent, &c., is spoken of without reference to sex, by saying that each parent contributes on an average one-quarter, or $(0 \cdot 5)^2$, each grandparent one-sixteenth, or $(0 \cdot 5)^4$, and so on, and that generally the occupier of each ancestral place in the nth degree, whatever be the value of n, contribntes $(0 \cdot 5)^{2n}$ of the heritage.

In interbred stock there are always fewer, and usually far fewer, different individuals among the ancestry than ancestral places for them to fill. A pedigree stock descended from a single couple, m generations back, will have 2^m ancestral places of the mth order, but only two individuals to fill them; therefore if $m = 10$ there are 1024 such places; if $m = 20$ there are more than a million. Whenever the same individual occupies many places he will be separately rated for each of them.

The neglect of individual prepotencies is justified in a law that avowedly relates to average results; they must of course be taken into account when applying the general law to individual cases. No difficulty arises in dealing with characters that are limited by sex, when their equivalents in the opposite sex are known, for instance in the statures of men and women.

The law may be applied *either* to total values or to deviations, as will be gathered from the following equation. Let M be the mean value from which all deviations are reckoned, and let D_1, D_2, &c., be the means of all the deviations, including their signs, of the ancestors in the 1st, 2nd, &c., degrees respectively; then

$$\tfrac{1}{2}(M+D_1)+\tfrac{1}{4}(M+D_2)+\&c. = M+(\tfrac{1}{2}D_1+\tfrac{1}{4}D_2+\&c.)$$

It should noted that nothing in this statistical law contradicts the generally accepted view that the chief, if not the sole, line of descent runs from germ to germ and not from person to person. The person may be accepted on the whole as a fair representative of the germ, and, being so, the statistical laws which apply to the persons would apply to the germs also, though with less precision in individual cases. Now this law is strictly consonant with the observed binary subdivisions of the germ cells, and the concomitant extrusion and loss of one-half of the several contributions from each of the two parents to the germ-cell of the offspring. The apparent artificiality of the law ceases on these grounds to afford cause for doubt; its close agreement with physiological phenomena ought to give a prejudice in *favour* of its truth rather than the contrary.

Again, a wide though limited range of observation assures us that the occupier of each ancestral place *may* contribute something of his own personal peculiarity, apart from all others, to the heritage of the offspring. Therefore there is such a thing as an average contribution appropriate to each ancestral place, which admits of statistical valuation, however minute it may be. It is also well known that the more remote stages of ancestry contribute considerably less than the nearer ones. Further, it is reasonable to believe that the contributions of parents to children are in the same proportion as those of the grandparents to the parents, of the great-grandparents to the grandparents, and so on; in short, that their total amount is to be expressed by the sum of the terms in an infinite geometric series diminishing to zero. Lastly, it is an essential condition that their total amount should be equal to 1, in order to account for the whole of the heritage. All these conditions are fulfilled by the series of $\frac{1}{2}+\frac{1}{2}^2+\frac{1}{2}^3+$ &c., and by no other. These and the foregoing considerations were referred to when saying that the law might be inferred with considerable assurance *à priori*; consequently, being found true in the particular case about to be stated, there is good reason to accept the law in a general sense.

The Bassets are dwarf blood-hounds, of two, and only two, recognised varieties of colour. Excluding, as I have done, a solitary exception of black and tan, they are either white, with large blotches ranging between red and yellow, or they may in addition be marked with more or less black. In the former case they are technically known and registered as "lemon and white," in the latter case as "tricolour." Tricolour is, in fact, the introduction of melanism, so I shall treat the colours simply as being "tricolour" or "non-tricolour;" more briefly, as T. or N. I am assured that transitional cases between T. and N. are very rare, and that experts would hardly ever disagree about the class to which any particular hound should be assigned. A stud-book is published from time to time containing the pedigrees, dates of birth, and the names of the breeders of these valuable animals. The one I have used bears the title 'The Basset Hound Club Rules and Stud-Book,' compiled by Everett Millais, 1874–1896. It contains the names of nearly 1000 hounds, to which Sir Everett Millais has very obligingly, at my request, appended their colours so far as they have been registered, which during later years has almost invariably been done. The upshot is that I have had the good fortune to discuss a total of 817 hounds of known colour, all descended from parents of known colour. In 567 out of these 817, the colours of all four grandparents were also known. These two sets are summarised in Table I and discussed in Table V, and they afford the data for Tables II, III, and IV. In 188 of the above cases the colours of all the eight great-grandparents were known as well; this third set is discussed in Table VI.

Partly owing to inequality in the numbers of the tricolours and non-tricolours, and partly owing to a selective mating in favour of the former, the different possible combinations of T. and N. ancestry are by no means equally common. The effect of this is conspicuous in Table I, where the entries are huddled together in some parts and absent in others. Still, though the data are not distributed as evenly as could be wished, they will serve our purpose if we are justified in grouping them without regard to sex; or, more generally, if we treat the 2^n components of each several A_n, whatever be the value of n, as equally efficient contributors.

Our first inquiry then must be, "Is or is not one sex so markedly prepotent over the other, in transmitting colour, that a disregard of sex would introduce statistical error?" In answering this, we should bear in mind a common experience, that statistical questions relating to sex are very difficult to deal with. Large and unknown disturbing causes appear commonly to exist, that make data which are seemingly homogeneous, very heterogeneous in reality. Some of these are undoubtedly present here, especially such as may be due to individual prepotencies combined with close interbreeding. For although this pedigree stock originated in as many as ninety-three different hounds, presumably more or less distant relations to one another, some of them proved of so much greater value than the rest that very close interbreeding has subsequently been resorted to in numerous instances. In order to show the danger of trusting blindly to averages of sex, even when the numbers are large, I have compared the results derived from different sets of data, namely from those contained in the last two columns of Table I, where they are distinguished by the letters A and B, and have treated them both separately and together in Table II. They will be seen to disagree widely, concurring only in showing that the dam is prepotent over the sire in transmitting colour. According to the A data, their relative efficacy in this respect is as 58 to 51, say 114 to 100; according to the B data, it is as 47 to 32, say 147 to 100. Taking all the data together, it is as 54 to 45, say 120 to 100, or as 6 to 5.

It does not seem to me that this ratio of efficacy of 6 to 5 is sufficient to overbear the statistical advantages of grouping the sexes as if they were equally efficient, the error in one case being more or less balanced by an opposite error in the other. It is true that the reciprocal forms of mating are by no means equally numerous, the prevailing tendency to use tricolours as sires being conspicuous. Still, as will be found later, on the application of a general test, the error feared is too insignificant to be observed. Should, however, a much larger collection of these data be obtained hereafter, minutiæ ought to be taken into account which may now be disregarded, and the neglect of female prepotency would cease to be justified.

The law to be verified supposes all the ancestors to be known, or to be known for so many generations back that the effects of the unknown residue are too small for consideration. The amount of the residual effect, beyond any given generation, is easily determined by the fact that in the series $\frac{1}{2}+\frac{1}{4}+\frac{1}{8}$, &c., each term is equal to the sum of all its successors. Now in the two sets of cases to be dealt with the larger refers to only two generations, therefore as the effect of the second generation is $\frac{1}{4}$, that of the unknown residue is $\frac{1}{4}$ also. The smaller set refers to three generations, leaving an unknown residual effect of $\frac{1}{8}$. These large residues cannot be ignored, amounting, as they do, to 25 and 12·5 per cent. respectively. We have, therefore, to determine fixed and reasonable rules by which they should be apportioned.

The requisite data for doing this are given in Table III, which shows that 79 per cent. of the parents of tricolour hounds are tricolour also, and that 56 per cent. of the parents of non-tricolour hounds are tricolour. It is not to be supposed that the trustworthiness of these results reaches to 1 per cent., but they are the best available data, so I adopt them.

It will be convenient to use the following nomenclature in calculation:—

a_0 stands for a single member of the offspring.

a_1 for a single parent; a_2 for a single grandparent, and so on, the suffix denoting the number of the generation. A parallel nomenclature, using capital letters, is:—

A_0 stands for all the offspring of the same ancestry.

A_1 for the two parents; A_2 for all the four grandparents, and so on. Consequently A_n contains 2^n individuals, each of the form a_n, and A_n contributes $(0{\cdot}5)^n$ to the heritage of each a_0; while each a_n contributes $(0{\cdot}5)^{2n}$ to it.

In the upper part of Table IV the ratios are entered of the average contributions of T. supplied by *known* ancestors. Nothing further need be said about these, except that they are styled coefficients because they must be multiplied into the total number of offspring, in order to calculate the number of them that will, on their separate and independent accounts, be probably tricolours.

We have next to explain how the coefficients for the *unknown* ancestry have been calculated, namely, those which are entered in the lower part of Table IV. Suppose all the four grandparents, A_2, to be tricolour, then only 0·79 of A_3 will be tricolour also, $(0{\cdot}79)^2$ of A_4, and so on. These several orders of ancestry will respectively contribute an average of tricolour to each a_0 of the amounts of $(0{\cdot}5)^3 \times (0{\cdot}79)$, $(0{\cdot}5)^4 \times (0{\cdot}79)^2$, &c. Consequently the sum of their tricolour contributions is

$$(0{\cdot}5)^3 \times (0{\cdot}79)\,\{1 + (0{\cdot}5) \times (0{\cdot}79) + (0{\cdot}5)^2 \times (0{\cdot}79)^2 + \&c.\},$$

which equals 0·1632. The average tricolour contribution from the ancestry of *each* of the four tricolour grandparents must be reckoned as the quarter of this, namely, 0·0408.

By a similar process, the average tricolour contribution from the ancestry of *each* non-tricolour grandparent is found to be 0·0243.

When the furthest known generation is that of the great-grandparents, the formula differs from the foregoing only by substituting $(0{\cdot}5)^4 \times (0{\cdot}79)$ for $(0{\cdot}5)^3 \times (0{\cdot}79)$. This makes the average tricolour contribution from the ancestry of the whole eight tricolour great-grandparents equal to 0·08160, and that from the ancestry of *each* of them to be one-eighth of this, or 0·0102.

In a similar way the tricolour contribution from the ancestry of *each* non-tricolour great-grandparent is found to be 0·0061.

The following example shows how the coefficients in Table IV were utilised in calculating the general coefficients entered in Table V.

2 Parents, T_1 (personal)	0·5000
3 Grandparents, T_2 (personal)	0·1875
1 Grandparent, N_2 (personal)	—
3 Grandparents, T_2 (ancestral)......	0·1224
1 Grandparent, N_2 (ancestral)	0·0243
Total tricolour contribution ..	0·8342

The coefficient 0·83 will consequently be found under the appropriate head in Table V, where the total number of offspring ("all cases") is recorded as 119. By multiplying these together, viz., 0·83 × 119, the "calculated" number of 99 is obtained. It will be seen that the observed number was 101, a difference of only 2 per cent.

The extraordinarily close coincidence throughout the two tables,

V and VI, between calculation and observation, proves that the law is correct in the present instance, and that the principle by which the unknown ancestry was apportioned, is practically exact also. It is not so strictly exact as it might have been, because the whole of the available knowledge has not been utilised. The 0·79 applied to A_4, &c., requires some small correction according to the known colours of the *offspring* of A_3. If they had been all tricolour the 0·79 would have to be increased; if all non-tricolour, it would have to be diminished. Having insufficient data to check a theoretical emendation, I note its omission, but shall not discuss the matter further.

It will be easily understood from these remarks how *collateral* data are to be brought into calculation, for if the collaterals were more tricolour than the average of hounds, the 0·79 would have to be somewhat increased (but not beyond the limiting value of 1·00); if less tricolour than the average, the 0·79 would have to be diminished. The knowledge of collaterals would be superfluous, if that of the direct ancestry were complete, but this important prolongation of the present subject must not be considered further on this occasion.

There are three stages in Tables V and VI at which comparisons may be made between calculated results and observed facts.

(1) *The Grand Totals.*—In Table V the sum of all the calculated values amounts to 391; that of all the observed ones to 387, which are closely alike. In Table VI they amount to 180 and 181 respectively, which is a still closer resemblance. Consequently the calculations are practically exact *on the whole*, and the error occasioned by neglect of sex, &c., is insignificant.

(2) *The Subordinate Pairs of Totals.*—These are entered at the sides of the tables, and are nine in number, namely, 236, 239; 149, 139; 6, 9; 53, 56; 52, 56: 9, 9; 8, 6; 49, 46; 9, 8. The coincidences are striking, in comparison with such results as statisticians have usually to be contented with; the second pair, 149, 139, is the least good, and will be considered in the next paragraph.

(3) *Individual Pairs of Entries.*—There are 32 of these; here also calculation compares excellently well with observation, excepting in the line that furnishes the "subordinate totals" of 149, 139, where the "all cases" of 37, 158, 60 yield the tricolour contingents of 20, 79, 36. Dividing each tricolour by the corresponding "all cases," we obtain what may be called "Coefficients from Observation," to compare with the calculated coefficients. They are as follows:—

		Diff.		Diff.	
Coefficients from observation	54	(−4)	50	(+10)	60
" " calculation	66	(−8)	58	(−7)	51

The great irregularity of the entries in the upper line shows that the observed values cannot be accepted as true representatives of the normal condition. I have not unravelled the causes of this error, and it is not urgent to do so, since its ill effects are swamped by the large number of successes elsewhere.

In order to satisfy myself that the correspondence between calculated and observed values was a sharp test of the correctness of the coefficients, I made many experiments by altering them slightly, and recalculating. In every case there was a notable diminution in the accuracy of the results. The test that the theory has successfully undergone appeared on that account, to be even more searching and severe than I had anticipated.

It is hardly necessary to insist on the importance of possessing a correct law of heredity. Vast sums of money are spent in rearing pedigree stock of the most varied kinds, as horses, cattle, sheep, pigs, dogs, and other animals, besides flowers and fruits. The current views of breeders and horticulturists on heredity are contradictory in important respects, and therefore *must* be more or less erroneous. Certainly no popular view at all resembles that which is justified by the present memoir. A correct law of heredity would also be of service in discussing actuarial problems relating to hereditary longevity and disease, and it might throw light on many questions connected with the theory of evolution.

Table I.—Pedigrees of Parental and Grand Parental Colours (Tricolour (T) or Non-Tricolour (N)).

The Sex of the Ancestors is taken into Account, but not that of the Offspring.

Sire.	Dam.	Offspring.	a	b	c	d	e	f	g	h	i	j	k	l	m	n	o	p	Totals. A.	Others of which all G. P.'s are not known. B.
Sire's sire			T	T	T	T	T	T	T	T	N	N	N	N	N	N	N	N		
Sire's dam			T	T	N	N	T	T	N	N	T	T	N	N	T	T	N	N		
Dam's sire			T	T	T	T	N	N	N	N	T	T	T	T	N	N	N	N		
Dam's dam			T	N	T	N	T	N	T	N	T	N	T	N	T	N	T	N		
Tric.	Tric. ..	Tric.	106	38	47	12	12	7	2	4	3	..	3	4	..	..	..	..	239	87
		Non-T.	13	4	12	2	2	1	1	..	..	..	..	3	..	..	..	..	38	20
Tric.	Non-T.	Tric.	..	..	16	3	..	..	1	..	3	..	..	..	1	1	..	..	25	15
		Non-T.	..	..	6	1	..	..	3	..	2	..	..	4	1	1	..	..	18	17
Non-T. ..	Tric. ..	Tric.	20	45	4	20	11	4	4	..	..	2	..	3	1	..	..	..	114	30
		Non-T.	17	50	16	9	5	..	1	1	..	8	.	1	1	..	..	..	109	64
Non-T. ..	Non-T.	Tric.	..	..	..	2	..	..	..	..	..	5	..	1	..	1	..	..	9	11
		Non-T.	..	..	..	8	..	..	..	4	..	3	..	..	..	..	..	..	15	6
Totals			156	138	101	57	30	12	12	9	8	18	3	16	4	3	..	..	567	250

Table II.—Offspring of one parent Tricolour (T) and of one Non-tricolour (N). The sex of the parents is not regarded.

From Table I.		Observed.			Per cents.	
		Tricolour.	Non-tricolour.		Tricolour.	Non-tricolour.
Sire T, dam N	A	114	109	223	51	49
	B	30	64	94	32	68
	Sum	144	173	317	45	55
Dam T, sire N	A	25	18	43	58	42
	B	15	17	32	47	53
	Sum	40	35	75	54	46

Table III.—Distribution of T and N colour in Parents, when the Offspring are T and N respectively.

From Table I.		No. of T offspring.			Parents* of T offspring.	
Sires.	Dams.	A.	B.	Total.	T.	N.
T	T	239	87	326	652	0
T	N	25	15	40	40	40
N	T	114	30	144	144	144
N	N	9	11	20	0	40
Totals				530	836	224
Per cent. of parents*				50	79	21

From Table I.		No. of N offspring.			Parents* of N offspring.	
Sires.	Dams.	A.	B.	Total.	T.	N.
T	T	38	20	58	116	0
T	N	18	17	35	35	35
N	T	109	64	173	173	173
N	N	15	6	21	0	42
Totals				287	324	250
Per cent. of Parents*				50	56	44

* More properly "Parental Places"; the number of these, though not that of the individual parents, being always double the number of any group of offspring.

Table IV.—Tricolour coefficients.

	Ancestry known up to and inclusive of	
	Grandparents.	Great-grandparents.
Personal allowance of T for each		
Tricolour parent	0·2500	0·2500
„ grandparent	0·0625	0·0625
„ great-grandparent	—	0·0156
(No allowance for Non-tricolours.)	—	—
Ancestral allowance of T for each		
Tricolour grandparent	0·0408	—
Non-tricolour „	0·0243	—
Tricolour great-grandparent	—	0·0102
Non-tricolour „ „	—	0·0061

Table V.—Calculation and Observation Compared.

The pedigrees are utilised up to the second ascending generation. Sex not taken into account.

No. of tricolours in parents.		Number of tricolours in grand-parents.				Total tricolour offspring.	
		4	3	2	1	Calculated.	Observed.
		a	*bcei*	*dfgjkm*	*hlno*		
2	All cases	119	119	28	11		
	Coefficient......	0·91	0·83	0·76	0·68		
	Tricolour calc'd.	*108*	*99*	*21*	*8*	*236*	
	„ observed	106	101	24	8		239
1	All cases	37	158	60	6		
	Coefficient......	0·66	0·58	0·51	0·43		
	Tricolour calc'd.	*24*	*92*	*30*	*3*	*149*	
	„ observed	20	79	36	4		139
0	All cases	..	..	18	6		
	Coefficient......	..	..	0·26	0·18		
	Tricolour calc'd.	..	..	*5*	*1*	*6*	
	„ observed	..	..	7	2		9
				Grand totals		*391*	387

Table VI.—Calculation and Observation Compared.

The pedigrees are utilised up to the third ascending generation. Sex not taken into account.

No. of tricolours in parents.	No. of tricolours in grand-parents.		Number of tricolours in great-grandparents: 8	7	6	5	4	Total tricolour offspring: Calcu-lated.	Observed
2	4	All cases	2	25	14	16			
		Coefficient......	0·96	0·94	0·92	0·90			
		Tricolours calc'd.	*2*	*24*	*13*	*14*	..	*53*	
		,, observed	2	25	14	15	..		56
	3	All cases	..	18	21	16	6		
		Coefficient......	..	0·87	0·85	0·83	0·81		
		Tricolours calc'd.	..	*16*	*18*	*13*	*5*	*52*	
		,, observed	..	17	19	14	6		56
	2	All cases	..	3	2	3	3		
		Coefficient......	..	0·81	0·79	0·77	0·75		
		Tricolours calc'd.	..	*2*	*3*	*2*	*2*	*9*	
		,, observed	..	2	2	3	2		9
1	4	All cases	..	2	1	9			
		Coefficient......	..	0·69	0·67	0·65			
		Tricolours calc'd.	..	*1*	*1*	*6*	..	*8*	
		,, observed	..	1	..	5	..		6
	3	All cases	1	28	14	31	9		
		Coefficient......	0·64	0·62	0·60	0·58	0·56		
		Tricolours calc'd.	*1*	*17*	*8*	*18*	*5*	*49*	
		,, observed.	1	16	12	8	9		46
	2	All cases	..	..	4	13			
		Coefficient......	..	..	0·54	0·52			
		Tricolours calc'd.	..	..	*2*	7	..	*9*	
		,, observed	..	..	1	7	..		8
							Grand totals ..	*180*	181

The summed data derived from Table IV, form the coefficients entered in Tables V and VI. These are multiplied into the corresponding number of "all cases," and the result gives the "calculated" number of tricolour hounds among them.

The entries of "all cases" and of "tricolours observed" in Table V are deduced from Table I, by combining the appropriate columns. The letters at the top show which columns are combined.

Seven other observed cases, disposed in three groups, are scattered beyond the limits of Table VI; two of these seven cases are tricolour.

ERRATUM.

Vol. 61, p. 410.

In the memoir read June 3, 1897, entitled "The average Contribution of each several Ancestor to the total Heritage of the Offspring" the words *sire* and *dam* have been accidentally transposed in Table II, and consequently, in the deduction therefrom at the bottom of page 404, the latter should be that, in the present case, the sire is more potent in transmitting colour than the dam, in the ratio of 6 to 5. This error does not affect the general conclusions of the memoir, because the ratio of 6 to 5 was treated as an insignificant disproportion, and the two sexes were dealt with on equal terms.

FRANCIS GALTON.

3

Reprinted from *Amer. Naturalist* **30**:441-445 (1896)

A NEW FACTOR IN EVOLUTION.

BY J. MARK BALDWIN.

In several recent publications I have developed, from different points of view, some considerations which tend to bring out a certain influence at work in organic evolution which I venture to call "a new factor." I give below a list of references[1] to these publications and shall refer to them by number as this paper proceeds. The object of the present paper is to

[1] References:

(1). *Imitation: a Chapter in the Natural History of Consciousness, Mind* (London), Jan., 1894. Citations from earlier papers will be found in this article and in the next reference.

(2). *Mental Development in the Child and the Race* (1st. ed., April, 1895; 2nd. ed., Oct., 1895; Macmillan & Co. The present paper expands an additional chapter (Chap. XVII) added in the German and French editions and to be incorporated in the third English edition.

(3). *Consciousness and Evolution, Science*, N. Y., August, 23, 1895; reprinted printed in the AMERICAN NATURALIST, April, 1896.

(4). *Heredity and Instinct* (I), *Science*, March 20, 1896. Discussion before N. Y. Acad. of Sci., Jan. 31, 1896.

(5). *Heredity and Instinct* (II), *Science*, April 10, 1896.

(6). *Physical and Social Heredity, Amer. Naturalist*, May, 1896.

(7). *Consciousness and Evolution, Psychol. Review*, May, 1896. Discussion before Amer. Psychol. Association, Dec. 28, 1895.

gather into one sketch an outline of the view of the process of development which these different publications have hinged upon.

The problems involved in a theory of organic development may be gathered up under three great heads: Ontogeny, Phylogeny, Heredity. The general consideration, the "factor" which I propose to bring out, is operative in the first instance, in the field of *Ontogeny*; I shall consequently speak first of the problem of Ontogeny, then of that of Phylogeny, in so far as the topic dealt with makes it necessary, then of that of Heredity, under the same limitation, and finally, give some definitions and conclusions.

I.

Ontogeny: "Organic Selection" (see ref. 2, chap. vii).—The series of facts which investigation in this field has to deal with are those of the individual creature's development; and two sorts of facts may be distinguished from the point of view of the *functions which an organism performs in the course of his life history.* There is, in the first place, the development of his heredity impulse, the unfolding of his heredity in the forms and functions which characterize his kind, together with the congenital variations which characterize the particular indiual—the phylogenetic variations, which are constitutional to him; and there is, in the second place, the series of functions, acts, etc., *which he learns to do himself in the course of his life.* All of these latter, the *special modifications which an organism undergoes during its ontogeny*, thrown together, have been called "acquired characters," and we may use that expression or adopt one recently suggested by Osborn,[2] "ontogenic variations" (except that I should prefer the form "ontogenetic variations"), if the word variations seems appropriate at all.

[2] Reported in *Science*, April 3rd.; also used by him before N. Y. Acad. of Sci., April 13th. There is some confusion between the two terminations "genic" and "genetic." I think the proper distinction is that which reserves the former, "genic," for application in cases in which the word to which it is affixed qualifies a term used *actively*, while the other, "genetic" conveys similarly a *passive* signification; thus agencies, causes, influences, etc., and "ontogenic phylogenic, etc.," while effects, consequences, etc, and "ontogenetic, phylogenetic, etc."

Assuming that there are such new or modified functions, in the first instance, and such "acquired characters," arising by the law of "use and disuse" from these new functions, our farther question is about them. And the question is this: How does an organism come to be modified during its life history?

In answer to this question we find that there are three different sorts of ontogenic agencies which should be distinguished—each of which works to produce ontogenetic modifications, adaptations, or variations. These are: first, the physical agencies and influences in the environment which work upon the organism to produce modifications of its form and functions. They include all chemical agents, strains, contacts, hindrances to growth, temperature changes, etc. As far as these forces work changes in the organism, the changes may be considered largely "fortuitous" or accidental. Considering the forces which produce them I propose to call them "physico-genetic." Spencer's theory of ontogenetic development rests largely upon the occurrence of lucky movements brought out by such accidental influences. Second, there is a class of modifications which arise from the spontaneous activities of the organism itself in the carrying out of its normal congenital functions. These variations and adaptations are seen in a remarkable way in plants, in unicellular creatures, in very young children. There seems to be a readiness and capacity on the part of the organism to "rise to the occasion," as it were, and make gain out of the circumstances of its life. The facts have been put in evidence (for plants) by Henslow, Pfeffer, Sachs; (for micro-organisms) by Binet, Bunge; (in human pathology) by Bernheim, Janet; (in children) by Baldwin (ref. 2, chap. vi.) (See citations in ref. 2, chap. ix, and in Orr, *Theory of Development*, chap. iv). These changes I propose to call "neuro-genetic," laying emphasis on what is called by Romanes, Morgan and others, the "selective property" of the nervous system, and of life generally. Third, there is the great series of adaptations secured by conscious agency, which we may throw together as "psycho-genetic." The processes involved here are all classed broadly under the term "intelligent," i. e., imitation, gregarious influences, maternal instruction, the lessons of pleasure and pain, and of experience generally, and reasoning from means to ends, etc.

We reach, therefore, the following scheme:

Ontogenetic Modifications.	*Ontogenic Agencies.*
1. Physico-genetic.	1. Mechanical.
2. Neuro-genetic.	2. Nervous.
3. Psycho-genetic.	3. Intelligent. Imitation. Pleasure and pain. Reasoning.

Now it is evident that there are two very distinct questions which come up as soon as we admit modifications of function and of structure in ontogenetic development: first, there is the question as to how these modifications can come to be adaptive in the life of the individual creature. Or in other words: What is the method of the individual's growth and adaptation as shown in the well known law of ".use and disuse?" Looked at functionally, we see that the organism manages somehow to accommodate itself to conditions which are favorable, to repeat movements which are adaptive, and so to grow by the principle of use. This involves some sort of selection, from the actual ontogenetic variations, of certain ones—certain functions, etc. Certain other possible and actual functions and structures decay from disuse. Whatever the method of doing this may be, we may simply, at this point, claim the law of use and disuse, as applicable in ontogenetic development, and apply the phrase, "Organic Selection," to the organism's behavior in acquiring new modes or modifications of adaptive function with its influence of structure. The question of the method of "Organic Selection" is taken up below (IV); here, I may repeat, we simply assume what every one admits in some form, that such adaptations of function—"accommodations" the psychologist calls them, the processes of learning new movements, etc.—*do occur.* We then reach another question, second; what place these adaptations have in the general theory of development.

Effects of Organic Selection.—First, we may note the results of this principle in the creature's own private life.

1. *By securing adaptations, accommodations, in special circumstances the creature is kept alive* (ref. 2, 1st ed., pp. 172 ff.). This is true in all the three spheres of ontogenetic variation distinguished in the table above. The creatures which can stand the "storm and stress" of the physical influences of the environment, and of the changes which occur in the environment, *by undergoing modifications of their congenital functions or of the structures which they get congenitally—these creatures will live; while those which cannot, will not.* In the sphere of neurogenetic variations we find a superb series of adaptations by lower as well as higher organisms during the course of ontogenetic development (ref. 2, chap. ix). And in the highest sphere, that of intelligence (including the phenomena of consciousness of all kinds, experience of pleasure and pain, imitation, etc.), we find individual accommodations on the tremendous scale which culminates in the skilful performances of human volition, invention, etc. The progress of the child in all the learning processes which lead him on to be a man, just illustrates this higher form of ontogenetic adaptation (ref. 2, chap. x–xiii).

All these instances are associated in the higher organisms, and all of them unite to *keep the creature alive.*

2. By this means *those congenital or phylogenetic variations are kept in existence, which lend themselves to intelligent, imitative, adaptive, and mechanical modification during the lifetime of the creatures which have them.* Other congenital variations are not thus kept in existence. So there arises a more or less wide-spread series of *determinate variations in each generation's ontogenesis* (ref. 3, 4, 5).[3]

[3] "It is necessary to consider further how certain reactions of one single organism can be selected so as to adapt the organism better and give it a life history. Let us at the outset call this process "Organic Selection" in contrast with the Natural Selection of whole organisms. . . . If this (natural selection) worked alone, every change in the environment would weed out all life except those organisms, which by accidental variation reacted already in the way demanded by the changed conditions—in every case new organisms showing variations, not, in any case, new elements of life-history in the old organisms. In order to the latter we would have to conceive . . . some modification of the old reactions in an organism through the influence of new conditions. . . . We are, accordingly, left to the view that the new stimulations brought by changes in the environment

themselves modify the reactions of an organism. . . . The facts show that individual organisms do acquire new adaptations in their lifetime, and that is our first problem. If in solving it we find a principle which may also serve as a principle of race-development, then we may possibly use it against the 'all sufficiency of natural selection' or in its support" (ref. 2, 1st. ed., pp. 175–6.)

4

Reprinted from *J. Roy. Horticultural Soc.* **26**:1-32 (1901)

EXPERIMENTS IN PLANT HYBRIDISATION.

By Gregor Mendel.

With an Introductory Note by W. Bateson, M.A., F.R.S.

The original paper, of which the following pages are a translation, was published by Gregor Mendel in the year **1865** *in the "Abhandlungen des naturforschenden Vereines in Brünn," Bd. iv. That periodical is little known, and probably there are not half a dozen copies in the libraries of this country. It will consequently be a matter for satisfaction that the Royal Horticultural Society has undertaken to publish a translation of this extraordinarily valuable contribution to biological science.*

The conclusion which stands out as the chief result of Mendel's admirable experiments is of course the proof that in respect of certain pairs of differentiating characters the germ-cells of a hybrid, or cross-bred, are pure, *being carriers and transmitters of either the one character or the other, not both. That he succeeded in demonstrating this law for the simple cases with which he worked it is scarcely possible to doubt.*

In so far as Mendel's law applies, therefore, the conclusion is forced upon us that a living organism is a complex of characters, of which some, at least, are dissociable and are capable of being replaced by others. We thus reach the conception of unit-characters, *which may be rearranged in the formation of the reproductive cells. It is hardly too much to say that the experiments which led to this advance in knowledge are worthy to rank with those that laid the foundation of the Atomic laws of Chemistry.*

To what extent Mendel's conclusions will be found to apply to other characters, and to other plants and animals, further experiment alone can show. Though little has yet been done, we already know a considerable group of cases in which the law holds, but we also have tolerably clear evidence that many phenomena of cross-breeding point to the coexistence of other laws of a much higher order of complexity. When the paper before us was written Mendel apparently inclined to the view that, with modifications, *his law might be found to include all the phenomena of hybridisation, but in a brief subsequent paper on hybrids of the genus* Hieracium* *he clearly recognised the existence of unconformable cases.*

Nevertheless, however much it may be found possible to limit or to extend the principle discovered by Mendel, there can be no doubt that we have in his work not only a model for future experiments of the same kind, but also a solid foundation from which the problem of Heredity may be attacked in the future.

It may seem surprising that a work of such importance should so long have failed to find recognition and to become current in the world of science. It is true that the journal in which it appeared is scarce, but this circumstance has seldom long delayed general recognition. The cause is unquestionably to be found in the neglect of the experimental study of the problem of Species which supervened on the general acceptance of the Darwinian doctrines. The problem of Species, as Gärtner, Kölreuter, Naudin, Mendel, and the other hybridists of the first half of the nineteenth century conceived it, attracted thenceforth no workers. The question, it was imagined, had been answered and the debate ended. No one felt any interest in the matter. A host of other lines of work were suddenly opened up, and in 1865 *the more vigorous investigators naturally found those new methods of research more attractive than the tedious observations of the hybridisers, whose inquiries were supposed, moreover, to have led to no definite result. But if we are to make progress with the study of Heredity, and to proceed further with the problem "What is a Species?" as distinct from the other problem "How do Species survive?" we must go back and take up the thread of the inquiry exactly where Mendel dropped it.*

As was stated in a lecture to the Royal Horticultural Society in 1900 *it is to De Vries, Correns, and Tschermak that we owe the simultaneous rediscovery, confirmation and extension of Mendel's work. References*† *are there given to the chief recent publications relating to the subject, of which the number is rapidly increasing.*

The whole paper abounds with matters for comment and criticism, which could only be profitable if undertaken at some length. There are also many deductions and lines of inquiry to which Mendel's facts point, which we in a fuller knowledge of physiology can perceive. It may, however, be doubted whether in his own day his conclusions could have been extended.

As some biographical particulars respecting this remarkable investigator will be welcome, I subjoin the following brief notice, which was published by Correns ‡ *on the authority of Dr. von Schanz: Gregor Johann Mendel was born on July* 22, 1822, *at Heinzendorf bei Odrau, in Austrian Silesia. He was the son of well-to-do peasants. In* 1843 *he entered as a novice the "Königinkloster," an Augustinian foundation in Altbrünn. In* 1847 *he was ordained priest. From* 1851 *to* 1853 *he studied physics and natural science at Vienna. Thence he returned to his cloister and became a teacher in the Realschule at Brünn. Subsequently he was*

* *Abh. Naturf. Brünn*, viii. 1869, p. 26.

† *Journal Royal Horticultural Society*, 1900, xxv. p. 54.

‡ *Bot. Zeitg.* lviii. 1900, No. 15, p. 229.

made Abbot, and died January 6, 1884. The experiments described in his papers were carried out in the garden of his Convent.

Besides the two papers on hybridisation, dealing respectively with Pisum *and* Hieracium, *Mendel contributed to the Brünn journal observations of a meteorological character, but, so far as I am aware, no others relating to natural history.*—W. BATESON.]

INTRODUCTORY REMARKS.

ARTIFICIAL fertilisation, such as is effected with decorative plants in order to obtain new variations in colour, has led up to the experiments which will here be discussed. The striking regularity with which the same hybrid forms always reappeared whenever fertilisation took place between the same species induced further experiments to be undertaken, the object of which was to follow up the developments of the hybrids in their progeny.

To this object numerous careful observers, such as Kölreuter, Gärtner, Herbert, Lecoq, Wichura and others, have devoted a part of their lives with inexhaustible perseverance. Gärtner especially, in his work "Die Bastarderzeugung im Pflanzenreiche" (The Production of Hybrids in the Vegetable Kingdom), has recorded very valuable observations; and quite recently Wichura published the results of some profound investigations into the hybrids of the Willow. That, so far, no generally applicable law governing the formation and development of hybrids has been successfully formulated can hardly be wondered at by anyone who is acquainted with the extent of the task, and can appreciate the difficulties with which experiments of this class have to contend. A final decision can only be arrived at when we shall have before us the results of detailed experiments made on plants belonging to the most diverse orders.

Those who survey the work done in this department will arrive at the conviction that among all the numerous experiments made, not one has been carried out to such an extent and in such a way as to permit of the possibility of determining the number of different forms under which the offspring of hybrids appear, or so that these forms may be arranged with certainty according to their separate generations, or that their mutual numerical relations can be definitely ascertained.

It requires indeed some courage to undertake a labour of such far-reaching extent; it appears, however, to be the only right way by which we can finally reach the solution of a question the importance of which cannot be overestimated in connection with the history of the evolution of organic forms.

The paper now presented records the results of such a detailed experiment. This experiment was appropriately confined to a small plant group, and is now, after eight years' pursuit, concluded in all essentials. Whether the plan upon which the separate experiments were conducted and carried out was the best suited to attain the desired end is left to the friendly decision of the reader.

SELECTION OF THE TRIAL PLANTS.

The value and utility of any experiment are determined by the fitness of the material to the purpose for which it is used, and thus in the case before us it cannot be immaterial what plants are subjected to experiment and in what manner such experiments are conducted.

The selection of the plant group which shall serve for experiments of this class must be made with all possible care if it be desired to avoid at the outset every risk of questionable results.

The trial plants must necessarily

1. Possess constant differentiating characters.

2. The hybrids of such plants must, during the flowering period, be protected from the influence of all foreign pollen, or be easily capable of such protection.

The hybrids and their offspring should suffer no marked disturbance in their fertility in the successive generations.

Accidental impregnation by foreign pollen, if such occurred during the experiments and were not recognised, would lead to entirely erroneous conclusions. Reduced fertility or entire sterility of certain forms, such as occurs in the offspring of many hybrids, would render the trials very difficult or entirely frustrate them. In order to discover the relations in which the hybrid forms stand towards each other and also towards their progenitors it appears to be necessary that all members of the series developed in each successive generation should be, *without exception*, subjected to observation.

At the very outset special attention was devoted to the *Leguminosæ* on account of their peculiar floral structure. Experiments which were made with several members of this family led to the result that the genus *Pisum* was found to possess the necessary conditions.

Some thoroughly distinct forms of this genus possess characters which are constant, and easily and certainly recognisable, and when their hybrids are mutually crossed they yield perfectly fertile progeny. Furthermore, a disturbance through foreign pollen cannot easily occur, since the fertilising organs are closely packed within the keel and the anther bursts within the bud, so that the stigma becomes covered with pollen even before the flower opens. This circumstance is of especial importance. As additional advantages worth mentioning, there may be cited the easy culture of these plants in the open ground and in pots, and also their relatively short period of growth. Artificial fertilisation is certainly a somewhat elaborate process, but nearly always succeeds. For this purpose the bud is opened before it is perfectly developed, the keel is removed, and each stamen carefully extracted by means of forceps, after which the stigma can at once be dusted over with the foreign pollen.

In all, thirty-four more or less different varieties of Peas were obtained from several seedsmen and subjected to a two years' trial. In the case of one variety there were remarked, among a larger number of plants all alike, a few forms which were markedly different. These, however, did not vary in the following year, and agreed entirely with another variety obtained from the same seedsmen; the seeds were therefore doubtless merely accidentally mixed. All the other varieties yielded perfectly constant and similar offspring; at any rate, no essential difference was observed during the two trial years. For fertilisation twenty-two of these were selected and cultivated during the whole period of the experiments. They remained constant without any exception.

Their systematic classification is difficult and uncertain. If we adopt the strictest definition of a species, according to which only those individuals belong to a species which under precisely the same circumstances display precisely similar characters, no two of them could be imputed to one species. According to the opinion of experts, however, the majority belong to the species *Pisum sativum*; while the rest are regarded and classed, some as sub-species of *P. sativum*, and some as independent species, such as *P. quadratum*, *P. saccharatum*, and *P. umbellatum*. The positions, however, which may be assigned to them in a classificatory system are quite immaterial for the purposes of the experiments in question. It has so far been found to be just as impossible to draw a sharp line between the hybrids of species and varieties as between species and varieties themselves.

Division and Arrangement of the Experiments.

If two plants which differ constantly in one or several characters be crossed, numerous experiments have demonstrated that the common characters are transmitted unchanged to the hybrids and their progeny; but each pair of differentiating characters, on the other hand, unite in the hybrid to form a new character, which in the progeny of the hybrid is usually variable. The object of the trial was to observe these variations in the case of each pair of differentiating characters, and to deduce the law according to which they appear in the successive generations. The trial resolves itself therefore into just as many separate experiments as there are constantly differentiating characters presented in the trial plants.

The various forms of Peas selected for crossing showed differences in the length and colour of the stem; in the size and form of the leaves; in the position, colour, and size of the flowers; in the length of the flower stalk; in the colour, form, and size of the pods; in the form and size of the seeds; and in the colour of the seed-coats and the albumen [cotyledons]. Some of the characters noted do not permit of a sharp and certain separation, since the difference is of a "more or less" nature, which is often difficult to define. Such characters could not be utilised for the separate trials; these could only be confined to characters which stand out clearly and definitely in the plants. Lastly, the result must show whether they, in their entirety, observe a regular relation in their hybrid unions, and whether from these facts any conclusion can be come to regarding those characters which possess a subordinate significance in the type.

The characters which were selected for the trials relate:

1. To the *difference in the form of the ripe seeds*. These are either round or roundish, the wrinkling, when such occurs on the surface, being always only shallow; or they are irregularly angular and deeply wrinkled (*P. quadratum*).

2. To the *difference in the colour of the seed albumen* (endosperm).* The albumen of the ripe seeds is either pale yellow, bright yellow and orange coloured, or it possesses a more or less intense green tint. This difference of colour is easily seen in the seeds, as their coats are transparent.

* [Mendel uses the terms "albumen" and "endosperm" somewhat loosely to denote the cotyledons, containing food-material, within the seed. —W. B.

3. To the *difference in the colour of the seed-coat.* This is either white, with which character white flowers are constantly correlated; or it is grey, grey-brown, leather-brown, with or without violet spotting, in which case the colour of the standards is violet, that of the wings purple, and the stem in the axils of the leaves is of a reddish tint. The grey seed-coats become dark brown in boiling water.

4. To the *difference in the form of the ripe pods.* These are either simply inflated, never contracted in places; or they are deeply constricted between the seeds and more or less wrinkled (*P. saccharatum*).

5. To the *difference in the colour of the unripe pods.* They are either light to dark green, or vividly yellow, in which colouring the stalks, leaf-veins, and blossom participate.*

6. To the *difference in the position of the flowers.* They are either axial, that is, distributed along the main stem; or they are terminal, that is, bunched at the top of the stem and arranged almost in a false umbel; in this case the upper part of the stem is more or less widened in section (*P. umbellatum*).

7. To the *difference in the length of the stem.* The length of the stem† is very various in some forms; it is, however, a constant character for each, in so far that in healthy plants, grown in the same soil, it is only subject to unimportant variations.

In trials with this character, in order to be able to discriminate with certainty, the long axis of 6 – 7 ft. was always crossed with the short one of ¾ ft. to 1½ ft.

Each two of the differentiating characters enumerated above were united by cross-fertilisation. There were made for the

1st	trial	60	fertilisations	on	15	plants.
2nd	„	58	„	„	10	„
3rd	„	35	„	„	10	„
4th	„	40	„	„	10	„
5th	„	23	„	„	5	„
6th	„	34	„	„	10	„
7th	„	37	„	„	10	„

From a larger number of plants of the same variety only the most vigorous were chosen for fertilisation. Weakly plants afford always uncertain results, because even in the first generation of hybrids, and still more so in the subsequent ones, many of the offspring either entirely fail to flower or only form a few and inferior seeds.

Furthermore, in all the trials reciprocal crossings were effected in such a way, that is, that each of the two varieties which in one set of fertilisations served as seed-bearers in the other set were used as pollen plants.

The plants were grown in garden beds, a few also in pots, and were maintained in their natural upright position by means of sticks, branches

* One pecies possesses a beautifully brownish-red coloured pod, which when ripening turns to violet and blue. Trials with this character were only begun last year. Of these further experiments it seems no account was published.

† In my account of these experiments (*R.H.S. Journal*, vol. xxv. p. 54) I misunderstood this paragraph and took "axis" to mean the *floral* axis, instead of the main axis of the plant. The unit of measurement, being indicated in the original by a dash, I thus took to have been an *inch*, but the translation here given is evidently correct.—W. B.

of trees, and strings stretched between. For each trial a number of pot plants were placed during the blooming period in a greenhouse, to serve as control plants with respect to the main trial in the open as regards possible disturbance by insects. Among the insects* which visit Peas the beetle *Bruchus pisi* might be detrimental to the trials should they appear in numbers. The female of this species is known to lay the eggs in the flower, and in so doing opens the keel; upon the tarsi of one specimen, which was caught in a flower, some pollen grains could clearly be seen under a lens. Mention must also be made of a circumstance which possibly might lead to the introduction of foreign pollen. It occurs, for instance, in some rare cases that certain parts of an otherwise quite normally developed flower wither, which results in a partial exposure of the fertilising organs. A defective development of the keel has also been observed, owing to which the stigma and anthers remained partially uncovered. It also sometimes happens that the pollen does not reach full perfection. In this event there occurs a gradual lengthening of the stigma during the blooming period, until the stigmatic tip protrudes at the point of the keel. This remarkable appearance has also been observed in hybrids of *Phaseolus* and *Lathyrus*.

The risk of false impregnation by foreign pollen is, however, a very slight one with *Pisum*, and is quite incapable of disturbing the general result. Among more than 10,000 plants which were carefully examined there were only a very few cases where an indubitable false impregnation had occurred. Since in the greenhouse such a case was never remarked, it may well be supposed that *Bruchus pisi*, and possibly also the described abnormalities in the floral structure, were to blame.

The Forms of the Hybrids.†

Experiments which in previous years were made with decorative plants have already afforded evidence that the hybrids, as a rule, are not exactly intermediate between the parental species. With some of the more striking characters, those, for instance, which relate to the form and size of the leaves, the pubescence of the several parts, &c., the intermediate, indeed, was nearly always to be seen; in other cases, however, one of the two parental characters was so preponderant that it was difficult, or quite impossible, to detect the other in the hybrid.

This is precisely the case with Pea hybrids. In the case of each of the seven crosses the hybrid character resembles that of one of the parental forms so closely that the other either escapes observation completely or cannot be detected with certainty. This circumstance is of great importance in the determination and classification of the forms under which the offspring of the hybrids appear. Henceforth in this paper those characters which are transmitted entirely, or almost unchanged in the hybridisation, and therefore in themselves represent the hybrid characters, are termed the *dominant*, and those which become latent in the process *recessive*. The expression "recessive" has been chosen

* [It is somewhat surprising that no mention is made of Thrips, which swarm in Pea flowers.]

† [Mendel throughout speaks of his cross-bred Peas as "hybrids," a term which many restrict to the offspring of two distinct *species*. He, as he explains, held this to be only a question of degree.—W. B.]

because the characters thereby designated withdraw or entirely disappear in the hybrids, but nevertheless reappear unchanged in their progeny, as will be demonstrated later on.

It was furthermore shown by the whole of the experiments that it is perfectly immaterial whether the dominant character belong to the seed-bearer or to the pollen parent; the form of the hybrid remains identical in both cases. This interesting fact was also emphasised by Gärtner, with the remark that even the most practised expert is not in a position to determine in a hybrid which of the two parental species was the seed or the pollen plant.

Of the differentiating characters which were used in the experiments the following are dominant:—

1. The round or roundish form of the seed with or without shallow depressions.

2. The yellow colouring of the seed albumen [cotyledons].

3. The grey, grey-brown, or leather-brown colour of the seed-coat, in connection with violet-red blossoms and reddish spots in the leaf axils.

4. The simply inflated form of the pod.

5. The green colouring of the unripe pod in connection with the same colour in the stems, the leaf-veins and the calyx.

6. The distribution of the flowers along the stem.

7. The greater length of stem.

With regard to this last character it must be stated that the longer of the two parental stems is usually exceeded by the hybrid, which is possibly only attributable to the greater luxuriance which appears in all parts of plants when stems of very different length are crossed. Thus, for instance, in repeated experiments, stems of 1 ft. and 6 ft. in length yielded without exception hybrids which varied in length between 6 ft. and 7½ ft.

The seeds of hybrids in the experiments with seed-coat are often more spotted, and the spots sometimes coalesce into small bluish-violet patches. The spotting also frequently appears even when it is absent as a parental character.

The hybrid forms of the seed-shape and of the albumen are developed immediately after the artificial fertilisation by the mere influence of the foreign pollen. They can, therefore, be observed even in the first trial year, whilst all the other characters naturally only appear in the following year in such plants as have been raised from the crossed seed.

The First Generation from the Hybrids.

In this generation there reappear, together with the dominant characters, also the recessive ones with their full peculiarities, and this occurs in the definitely expressed average proportion of three to one, so that among each four plants of this generation three receive the dominant character and one the recessive. This relates without exception to all the characters which were embraced in the trials. The angular wrinkled form of the seed, the green colour of the albumen, the white colour of the seed-coats and the flowers, the constrictions of the pods, the yellow colour of the unripe pod, of the stalk, the calyx, and the leaf venation, the umbel-like form of the inflorescence, and the dwarfed stem, all reappear in the numerical proportion given without any essential alteration. *Transitional forms were not observed in any experiment.*

Once the hybrids resulting from reciprocal crosses are fully formed, they present no appreciable difference in their subsequent development, and consequently the results [of the reciprocal crosses] can be reckoned together in each experiment. The relative numbers which were obtained for each pair of different characters are as follows :—

Trial 1. Form of seed.—From 253 hybrids 7,324 seeds were obtained in the second trial year. Among them were 5,474 round or roundish ones and 1,850 angular wrinkled ones. Therefrom the relation is deduced of 2·96 to 1.

Trial 2. Colour of albumen.—258 plants yielded 8,023 seeds, 6,022 yellow, and 2,001 green; their relation, therefore, is as 3·01 to 1.

In these two trials each pod yielded usually both kinds of seed. In well-developed pods which contained on the average six to nine seeds, it often occurred that all the seeds were round (Trial 1) or all yellow (Trial 2); on the other hand there were never observed more than five angular or five green ones in one pod. It appears to make no difference whether the pods are developed early or later in the hybrid or whether they spring from the main axis or from a lateral one. In some few plants only a few seeds developed in the first formed pods, and these possessed exclusively one of the two characters, but in the subsequently developed pods the normal proportions were maintained nevertheless.

As in separate pods, so did the distribution of the characters vary in separate plants. By way of illustration the first ten individuals from both series of trials may serve.

	Trial 1. Form of Seed.		Trial 2. Colour of Albumen.	
Plants.	Round.	Angular.	Yellow.	Green.
1	45	12	25	11
2	27	8	32	7
3	24	7	14	5
4	19	10	70	27
5	32	11	24	13
6	26	6	20	6
7	88	24	32	13
8	22	10	44	9
9	28	6	50	14
10	25	7	44	18

As extremes in the distribution of the two seed characters in one plant, there were observed in Trial 1 an instance of 43 round and only 2 angular, and another of 14 round and 15 angular seeds. In Trial 2 there was a case of 32 yellow and only 1 green seed, but also one of 20 yellow and 19 green.

These two trials are important for the determination of the average relative figures, because with a smaller number of trial plants they show that very considerable fluctuations may occur. In counting the seeds, also, especially in Trial 2, some care is requisite, since in some of the seeds of many plants the green colour of the albumen is less developed, and at first may be easily overlooked. The cause of the partial disappearance of the green colouring has no connection with the hybrid character of the plants, as it likewise occurs in the parental variety. This peculiarity is also confined to the individual and is not inherited by the offspring. In luxuriant plants this appearance was frequently noted. Seeds which are damaged by insects during their development often vary in colour and form, but with a little practice in sorting errors are easily avoided. It is almost superfluous to mention that the pods must remain on the plants until they are thoroughly ripened and have become dried, since it is only then that the shape and colour of the seed are fully developed.

Trial 3. Colour of the seed-coats.—Among 929 plants 705 bore violet-red flowers and grey-brown seed-coats; 224 had white flowers and white seed-coats. Thence results the proportion 3·15 to 1.

Trial 4. Form of pods.—Of 1,181 plants 882 had them simply inflated, and in 299 they were constricted. Resulting proportion, 2·95 to 1.

Trial 5. Colour of the unripe pods.—The number of trial plants was 580, of which 428 had green pods and 152 yellow ones. Consequently these stand in proportion as 2·82 to 1.

Trial 6. Position of flowers.—Among 858 cases 651 blossoms were axial and 207 terminal. Proportion, 3·14 to 1.

Trial 7. Length of stem.—Out of 1,064 plants, in 787 cases the stem was long, and in 277 short. Hence a mutual proportion of 2·84 to 1. In this trial the dwarfed plants were carefully lifted and transferred to a special bed. This precaution was necessary, as otherwise they would have perished through being overgrown by their tall relatives. Even in their quite young state they can be easily picked out by their compact growth and thick dark-green foliage.

If now the results of the whole of the trials be brought together, there is found, as between the number of forms with the dominant and recessive characters, an average proportion of 2·98 to 1, or 3 to 1.

The dominant character can have here a *double significance*—viz. that of the parental character, or the character of the hybrid. In which of the two significations it appears in each separate case can only be determined by the following generation. As a parental character it must be transmitted unchanged to the whole of the offspring; as a hybrid character, on the other hand, it must observe the same proportion as in the first generation.

The Second Generation from the Hybrids.

Those forms which in the first generation maintain the recessive character do not further vary in the second generation as regards this character; they remain constant in their offspring.

It is otherwise with those which possess the dominant character in the first generation. Of these *two*-thirds yield offspring which display the dominant and recessive characters in the proportion of 3 to 1, and thereby show exactly the same ratio as the hybrid forms, while only *one*-third remains with the dominant character constant.

The separate trials yielded the following results:—

Trial 1.—Among 565 plants which were raised from round seeds of the first generation, 193 yielded round seeds only, and remained therefore constant in this character; 372, however, gave both round and angular seeds, in the proportion of 3 to 1. The number of the hybrids, therefore, as compared with the constants is 1·93 to 1.

Trial 2.—Of 519 plants which were raised from seeds whose albumen was of yellow colour in the first generation, 166 yielded exclusively yellow, while 353, however, yielded yellow and green seeds in the proportion of 3 to 1. There resulted, therefore, a splitting into hybrid and constant forms in the proportion of 2·13 to 1.

For each separate trial in the following experiments 100 plants were selected which displayed the dominant character in the first generation, and in order to ascertain the significance of this, ten seeds of each were cultivated.

Trial 3.—The offspring of 36 plants yielded exclusively grey-brown seed-coats, while 64 plants yielded partly grey-brown and partly white.

Trial 4.—The offspring of 29 plants had only simply inflated pods; of the offspring of 71, on the other hand, some had inflated and some constricted.

Trial 5.—The offspring of 40 plants had only green pods; of the offspring of 60 plants some had green, some yellow ones.

Trial 6.—The offspring of 33 plants had only axial flowers; of the offspring of 67, on the other hand, some had axial and some terminal flowers.

Trial 7.—The offspring of 28 plants inherited the long axis, and those of 72 plants some the long and some the short axis.

In each of these trials a certain number of the plants came constant with the dominant character. For the determination of the proportion in which the separation of the forms with the constantly persistent character results, the two first trials are of especial importance, since in these a larger number of plants can be compared. The ratios 1·93 to 1 and 2·13 to 1 gave together almost exactly the average ratio of 2 to 1. The sixth trial has a quite concordant result; in the others the ratio varies more or less, as was only to be expected in view of the smaller number of 100 trial plants. Trial 5, which shows the greatest departure, was repeated, and then in lieu of the ratio of 6C and 40 that of 65 and 35 resulted. *The average ratio of 2 to 1 appears, therefore, as fixed with certainty.* It is therefore demonstrated that, of those forms which possess the dominant character in the first generation, in two-thirds the hybrid character is embodied, while one-third remains constant with the dominant character.

The ratio of 3 to 1, in accordance with which the distribution of the dominant and recessive characters results in the first generation, resolves itself therefore in all trials into the ratio of 2 : 1 : 1 if the dominant character be differentiated according to its significance as a hybrid character or a parental one. Since the members of the first generation spring directly from the seed of the hybrids, *it is now clear that the hybrids form seeds having one or other of the two differentiating characters, and of these one-half develop again the hybrid form, while the other half yield plants which remain constant and receive the dominant and recessive characters in equal numbers.*

The Subsequent Generations from the Hybrids.

The proportions in which the descendants of the hybrids develop and split up in the first and second generations presumably hold good for all subsequent progeny. Trials 1 and 2 have already been carried through six generations, 3 and 7 through five, and 4, 5, and 6 through four, these trials being continued from the third generation with a small number of plants, and no departure from the rule has been perceptible. The offspring of the hybrids separated in each generation in the ratio of 2 : 1 : 1 into hybrids and constant forms.

If A be taken as denoting one of the two constant characters, for instance the dominant, a, the recessive, and Aa the hybrid form in which both are conjoined, the formula

$$A + 2Aa + a$$

shows the order of development for the progeny of the hybrids of two differentiating characters.

The observation made by Gärtner, Kölreuter, and others, that hybrids are inclined to revert to the parental forms, is also confirmed by the trials described. It is seen that the number of the hybrids which arise from one fertilisation, as compared with the number of forms which become constant and the progeny of such from generation to generation, is continually diminishing, but that nevertheless they could not entirely disappear. If there be assumed an average equality of fertility in all plants in all generations, and that, furthermore, each hybrid forms seed of which one-half yields hybrids again, while the other half is constant to both characters in equal proportions, the ratio of numbers for the offspring in each generation is seen by the following summary, in which A and a denote again the two parental characters, and Aa the hybrid forms. For brevity's sake it may be assumed that each plant in each generation furnishes only 4 seeds.

				Ratios.		
Generation	A	Aa	a	A	Aa	a
1	1	2	1	1 :	2 :	1
2	6	4	6	3 :	2 :	3
3	28	8	28	7 :	2 :	7
4	120	16	120	15 :	2 :	15
5	496	32	496	31 :	2 :	31
n				2^n-1 :	2 :	2^n-1

In the tenth generation, for instance, $2^n-1=1023$. There result, therefore, in each 2,048 plants which arise in this generation 1,023 with the constant dominant character, 1,023 with the recessive character, and only two hybrids.

The Offspring of Hybrids in which Several Differentiating Characters are Associated.

In the trials above described plants were used which differed only in one essential character. The next task consisted in ascertaining whether the law of development discovered in these applied to each pair of differentiating characters when several diverse characters are united in the hybrid by crossing. As regards the form of the hybrids in these cases, the trials showed throughout that this invariably more nearly approaches to that one of the two parental plants which possesses the greater number of dominant characters. If, for instance, the seed plant has a short stem, terminal white flowers, and simply inflated pods; the pollen plant, on the other hand, a long stem, violet-red flowers distributed along the stem, and constricted pods, the hybrid resembles the seed parent only in the form of the pod; in the other characters it agrees with the pollen parent. Should one of the two parental types possess only dominant characters, then the hybrid is scarcely or not at all distinguishable from it.

Two trials were made with a larger number of plants. In the first trial the parental plants differed in the form of the seed and in the colour of the albumen; in the second in the form of the seed, in the colour of the albumen, and in the colour of the seed-coats. Trials with seed characters give the result in the simplest and most certain way.

In order to facilitate study of the data in these trials, the different characters of the seed plant will be indicated by A, B, C, those of the pollen plant by a, b, c, and the hybrid forms of these characters by Aa, Bb, and Cc.

Trial 1.—AB, seed parents; ab, pollen parents;
A, form round; a, form angular;
B, albumen yellow. b, albumen green.

The fertilised seeds appeared round and yellow like those of the seed parents. The plants raised therefrom yielded seeds of four sorts, which frequently presented themselves in one pod. In all 556 seeds were yielded by 15 plants, and of these there were:—

315 round and yellow,
101 angular and yellow,
108 round and green,
32 angular and green.

All were sown the following year. Eleven of the round yellow seeds did not yield plants, and three plants did not form seeds. Among the rest:

38 had round yellow seeds . . . . . . AB
65 round yellow and green seeds . . . . . . ABb
60 round yellow and angular yellow seeds . . . AaB
138 round yellow and green, angular yellow and green seeds AaBb.

From the angular yellow seeds 96 resulting plants bore seed, of which:

28 had only angular yellow seeds aB
68 angular yellow and green seeds aBb.

From 108 round green seeds 102 resulting plants fruited, of which:

35 had only round green seeds Ab
67 round and angular green seeds Aab.

The angular green seeds yielded 30 plants which bore seeds all of like character; they remained constant ab.

The offspring of the hybrids appeared therefore under nine different forms and partly in very unequal numbers. When these are collected and co-ordinated we find :

38	plants with the sign	AB
35	,, ,, ,,	Ab
28	,, ,, ,,	aB
30	,, ,, ,,	ab
65	,, ,, ,,	ABb
68	,, ,, ,,	aBb
60	,, ,, ,,	AaB
67	,, ,, ,,	Aab
138	,, ,, ,,	AaBb

The whole of the forms may be classed into three essentially different groups. The first embraces those with the signs AB, Ab, aB, and ab: they possess only constant characters and do not vary again in the next generation. Each of these forms is represented on the average thirty-three times. The second group embraces the signs ABb, aBb, AaB, Aab: these are constant in one character and hybrid in another, and vary in the next generation only as regards the hybrid character. Each of these appears on an average sixty-five times. The form AaBb occurs 138 times : it is hybrid in both characters, and behaves exactly as do the hybrids from which it is derived.

If the numbers in which the forms of these sections appear be compared, the ratios of 1, 2, 4 are unmistakably evident. The numbers 32, 65, 138 present very favourable approximations to the ratio numbers of 33, 66, 132.

The developmental series consists, therefore, of nine classes, of which four appear therein always once and are constant in both characters ; the forms AB, ab resemble the parental forms, the two others present combinations between the conjoined characters A, a, B, b, which combinations are likewise possibly constant. Four classes appear always twice, and are constant in one character and hybrid in the other. One class appears four times, and is hybrid in both characters. Consequently the offspring of the hybrids, if two kinds of differentiating characters are combined therein, are developed according to the formula

$$AB + Ab + aB + ab + 2\ ABb + 2\ aBb + 2\ AaB + 2\ Aab + 4\ AaBb.$$

This developmental series is incontestably a combination series in which the two developmental series for the characters A and a, B and b, are combined. We arrive at the full number of the classes of the series by the combination of the formulæ :

$$A + 2\ Aa + a$$
$$B + 2\ Bb + b.$$

Second Trial.—ABC, seed parents ;	abc, pollen parents ;
A, form round ;	a, form angular ;
B, albumen yellow ;	b, albumen green ;
C, seed-coat grey-brown.	c, seed-coat white.

This trial was made in precisely the same way as the previous one. Among all the trials it demanded the most time and trouble. From 24 hybrids 687 seeds were obtained in all: these were all either spotted, grey-brown or grey-green, round or angular. From these in the following year 639 plants fruited, and, as further investigation showed, there were among them :

8 plants	ABC.	22 plants	ABCc.	45 plants	ABbCc.
14 „	ABc.	17 „	AbCc.	36 „	aBbCc.
9 „	AbC.	25 „	aBCc.	38 „	AaBCc.
11 „	Abc.	20 „	abCc.	40 „	AabCc.
8 „	aBC.	15 „	ABbC.	49 „	AabbC.
10 „	aBc.	18 „	ABbc.	48 „	AaBbc.
10 „	abC.	19 „	aBbC.		
7 „	abc.	24 „	aBbc.		
		14 „	AaBC.	78 „	AaBbCc.
		18 „	AaBc.		
		20 „	AabC.		
		16 „	Aabc.		

The developmental series embraced 27 classes. Of these 8 are constant in all characters, and each appears on the average 10 times; 12 are constant in two characters, and hybrid in the third, each appears on the average 19 times; 6 are constant in one character and hybrid in the other two; each appears on the average 43 times. One form appears 78 times and is hybrid in all of the characters. The ratios 10, 19, 43, 78 agree so closely with the ratios 10, 20, 40, 80, or 1, 2, 4, 8, that this last undoubtedy represents the true value.

The development of the hybrids when the original parents differ in three characters results therefore according to the following formula :

ABC + ABc + AbC + Abc + aBC + aBc + abC + abc + 2 ABCc + 2 AbCc +2 aBCc +2 abCc +2 ABbC + 2 ABbc +2 aBbC + 2 aBbc + 2 AaBC + 2 AaBc + 2 AabC + 2 Aabc + 4 ABbCc + 4 aBbCc + 4 AaBCc + 4 AabCc + 4 AaBbC + 4 AaBbc + 8 AaBbCc.

Here also is involved a combination series in which the developmental series for the characters A and a, B and b, C and c, are united. The formulæ

$$A+2\,Aa+a$$
$$B+2\,Bb+b$$
$$C+2\,Cc+c$$

give all the classes of the series. The constant combinations which occur therein agree with all combinations which are possible between the characters A, B, C, a, b, c; two thereof, ABC and abc, resemble the two original parental stocks.

In addition, further experiments were made with a smaller number of trial plants in which the remaining characters by twos and threes were united as hybrids: all yielded approximately the same results. There is therefore no doubt that for the whole of the characters involved in the trials the principle applies that *the offspring of the hybrids in which several essentially different characters are combined represent the components of a series of combinations, in which the developmental series for each two different characters are associated.* It is demonstrated at the same time that *the relation of each two different characters in hybrid connection is independent of the other differences in the two original parental stocks.*

If n represent the number of the characteristic differences in the two original stocks, 3^n gives the number of components of the combination series, 4^n the number of individuals which belong to the series, and 2^n the number of connections which remain constant. The series therefore embraces, if the original stocks differ in four characters, $3^4=81$ component classes, $4^4=256$ individuals, and $2^4=16$ constant forms; or, which is the same, among each 256 offspring of the hybrids there are 81 different combinations, 16 of which are constant.

All constant combinations which in Peas are possible by the combination of the said seven characteristic features were actually obtained by repeated crossing. Their number is given by $2^7=128$. Thereby is simultaneously given the practical proof *that the constant characters which appear in various forms of a plant group may be obtained in all the associations which are possible according to the laws of combination by means of repeated artificial fertilisation.*

As regards the flowering time of the hybrids, the trials are not yet concluded. It can, however, already be stated that the period stands almost exactly between those of the seed and pollen parents, and that the development of the hybrids with respect to this character probably happens in the same way as in the case of the other characters. The forms which are selected for trials of this class must have a difference of at least twenty days from the middle flowering period of one to that of the other; furthermore, the seeds when sown must all be placed at the same depth in the earth, so that they may germinate simultaneously. Also, during the whole flowering period, the more important variations in temperature must be taken into account, and the partial hastening or delaying of the flowering which may result therefrom. It is clear that this experiment presents many difficulties to be overcome and necessitates great attention.

If we endeavour to collate in a brief form the results arrived at, we find that those differentiating characters which admit of easy and certain recognition in the trial plants all behave exactly alike in their hybrid associations. The offspring of the hybrids of each pair of differentiating characters are, one-half, hybrid again, while the other half are constant in equal proportions with the characters of the seed and pollen parents respectively. If several differentiating characters are combined by cross-fertilisation in a hybrid, the resulting offspring form the components of a combination series in which the developmental series for each pair of differentiating characters are united.

The uniformity of behaviour shown by the whole of the characters submitted to trial permits, and fully justifies, the acceptance of the principle that a similar relation exists in the other characters which appear less sharply defined in plants, and therefore could not be included in the separate experiments. An experiment with peduncles of different lengths gave on the whole a fairly satisfactory result, although the differentiation and serial arrangement of the forms could not be effected with that certainty which is indispensable for correct experiment.

The Reproductive Cells of Hybrids.

The results of the previously described experiments induced further experiments, the results of which appear fitted to afford some conclusions as regards the composition of the egg and pollen cells of hybrids. An important basis for argument is afforded in *Pisum* by the circumstance that among the progeny of the hybrids constant forms appear, and that this occurs, too, in all combinations of the associated characters. So far as experience goes, we find it in every case confirmed that constant progeny can only be formed when the egg cells and the fertilising pollen are of like character, so that both are provided with the material for vitalising quite similar individuals, as is the case with the normal fertilisation of pure species. We must therefore regard it as essential that exactly similar factors are at work also in the production of the constant forms in the hybrid plants. Since the various constant forms are produced in *one* plant, or even in *one* flower of a plant, the conclusion appears logical that in the ovaries of the hybrids there are formed as many sorts of egg cells, and in the anthers as many sorts of pollen cells, as there are possible constant combination forms, and that these egg and pollen cells agree in their internal composition with those of the separate forms.

In point of fact it is possible to demonstrate theoretically that this hypothesis would fully suffice to account for the development of the hybrids in the separate generations, if we might at the same time assume that the various kinds of egg and pollen cells were formed in the hybrids on the average in equal numbers.

In order to bring these assumptions experimentally to the proof, the following trials were selected. Two forms which were constantly different in the form of the seed and the colour of the albumen were united by fertilisation.

If the differentiating characters are again indicated as A, B, a, b, we have:

AB, seed parent;	ab, pollen parent;
A, form round;	a, form angular;
B, albumen yellow.	b, albumen green.

The artificially fertilised seeds were sown together with several seeds of both original stocks, and the most vigorous examples were chosen for the reciprocal crossing. There were fertilised:—

1. The hybrids	with the	pollen of	AB.
2. The hybrids	,,	,,	ab.
3. AB	,,	,,	the hybrids.
4. ab	,,	,,	the hybrids.

For each of these four trials the whole of the flowers on three plants were fertilised. If the above theory be correct, there must be developed on the hybrids egg and pollen cells of the forms AB, Ab, aB, ab, and there would be combined:—

1. The egg cells	AB, Ab, aB, ab	with the	pollen cells	AB.
2. ,,	AB, Ab, aB, ab	,,	,,	ab.
3. ,,	AB	,,	,,	AB, Ab, aB, ab.
4 ,,	ab	,,	,,	AB, Ab, aB, ab.

C

From each of these trials there could then result only the following forms :—

1. AB, ABb, AaB, AaBb.
2. AaBb, Aab, aBb, ab.
3. AB, ABb, AaB, AaBb.
4. AaBb, Aab, aBb, ab.

If, furthermore, the several forms of the egg and pollen cells of the hybrids were produced on an average in equal numbers, then in each trial the said four combinations should stand in the same numerical relation. A perfect agreement in the numerical relations was, however, not to be expected, since in each fertilisation, even in normal cases, some egg cells remain undeveloped or subsequently die, and many even of the well-formed seeds fail to germinate when sown. The above assumption is also limited in so far that, while it demands the formation of an equal number of the various sorts of egg and pollen cells, it does not require that this should apply to each separate hybrid with mathematical exactness.

The first and second trials had pre-eminently the object of proving the composition of the hybrid egg cells, while the third and fourth trials were to decide that of the pollen cells. As is shown by the above demonstration the first and second trials and the third and fourth trials should produce precisely the same combinations, and even in the second year the result should be partially visible in the form and colour of the artificially fertilised seed. In the first and third trials the dominant characters of form and colour, A and B, appear in each union, and are also partly constant and partly in hybrid union with the recessive characters a and b, for which reason they must impress their peculiarity upon the whole of the seeds. All seeds should therefore appear round and yellow, if the theory be justified. In the second and fourth trials, on the other hand, one union is hybrid in form and in colour, and consequently the seeds are round and yellow; another is hybrid in form, but constant in the recessive character of colour, whence the seeds are round and green; the third is constant in the recessive character of form but hybrid in colour, consequently the seeds are angular and yellow; the fourth is constant in both recessive characters, so that the seeds are angular and green. In both these trials there were consequently four sorts of seed to be expected—viz. round and yellow, round and green, angular and yellow, angular and green.

The crop fulfilled these expectations perfectly. There were obtained in the

1st Trial, 98 exclusively round yellow seeds ;
3rd „ 94 „ „ „ „

In the 2nd Trial, 31 round and yellow, 26 round and green, 27 angular and yellow, 26 angular and green seeds.

In the 4th Trial, 24 round and yellow, 25 round and green, 22 angular and yellow, 27 angular and green seeds.

A favourable result could now scarcely be doubted; the next generation must afford the final proof. From the seed sown there resulted for the first trial 90 plants, and for the third 87 plants which fruited: these yielded for the—

1st Experiment	3rd Experiment		
20	25	round yellow seeds	AB.
23	19	round yellow and green seeds . . .	ABb.
25	22	round and angular yellow seeds . .	AaB.
22	21	round and angular green and yellow seeds	AaBb.

In the second and fourth trials the round and yellow seeds yielded plants with round and angular yellow and green seeds, AaBb.

From the round green seeds plants resulted with round and angular green seeds, Aab.

The angular yellow seeds gave plants with angular yellow and green seeds, aBb.

From the angular green seeds plants were raised which yielded again only angular green seeds, ab.

Although in these two trials likewise some seeds did not germinate, the figures arrived at already in the previous year were not affected thereby, since each kind of seed gave plants which, as regards their seed, were like each other and different from the others. There resulted therefore from the

2nd Experiment	4th Experiment		
31	24	plants of the form	AaBb
26	25	" "	Aab.
27	22	" "	aBb.
26	27	" "	ab.

In all the trials, therefore, there appeared all the forms which the proposed theory demands, and also in nearly equal numbers.

In a further trial the characters of floral colour and length of stem were experimented upon, and selection so made that in the third trial-year each character ought to appear in half of all the plants if the above theory were correct. A, B, a, b serve again as indicating the various characters.

A, violet-red flowers.	a, white flowers.
B, axis long.	b, axis short.

The form Ab was fertilised with ab, which produced the hybrid Aab. Furthermore, aB was also fertilised with ab, whence the hybrid aBb. In the second year, for further fertilisation, the hybrid Aab was used as seed parent, and hybrid aBb as pollen parent.

Seed parent, Aab.	Pollen parent, aBb.
Possible egg cells, Abab.	Pollen cells, aBab.

From the fertilisation between the possible egg and pollen cells four combinations should result, viz.:—

$$AaBb + aBb + Aab + ab.$$

From this it is perceived that, according to the above theory, in the third trial-year out of all the plants

The half should have violet-red flowers (Aa), Classes	1, 3
„ „ „ white flowers (a) „	2, 4
„ „ „ a long axis (Bb) „	1, 2
„ „ „ a short axis (b) „	3, 4

From 45 fertilisations of the second year 187 seeds resulted, of which only 166 reached the flowering stage in the third year. Among these the separate classes appeared in the numbers following:—

Class.	Colour of flower.	Stem.	
1	violet-red	long	47 times
2	white	long	40 „
3	violet-red	short	38 „
4	white	short	41 „

There consequently appeared—

The violet-red flower colour	(Aa)	in 85 plants.
„ white „ „	(a)	in 81 „
„ long stem	(Bb)	in 87 „
„ short „	(b)	in 79 „

The theory adduced is therefore satisfactorily confirmed in this trial also.

For the characters of form of pod, colour of pod, and position of flowers experiments were also made on a small scale, and results obtained in perfect agreement. All combinations which were possible through the union of the differentiating characters duly appeared, and in nearly equal numbers.

Experimentally, therefore, the theory is justified that *pea hybrids form egg and pollen cells which, in their constitution, represent in equal numbers all constant forms which result from the combination of the characters when conjoined by fertilisation.*

The difference of the forms among the progeny of the hybrids, as well as the relative ratio of numbers in which they are observed, find a sufficient explanation in the principle above deduced. The simplest case is afforded by the developmental series of each pair of differentiating characters. This series is expressed by the formula A + 2Aa + a, in which A and a signify the forms with constant differentiating characters, and Aa the hybrid form of both. It includes in three different classes four individuals. In the formation of these, pollen and egg cells of the form A and a take part on the average equally in the fertilisation; hence each form twice, since four individuals are formed. There participate consequently in the fertilisation—

The pollen cells A + A + a + a
The egg cells A + A + a + a.

It remains, therefore, purely a matter of chance which of the two sorts of pollen will become united with each separate egg cell. According, however, to the law of probability, it will always happen, on the average of many cases, that each pollen form A and a will unite equally often with each egg cell form A and a, consequently one of the two pollen cells A in the fertilisation will meet with the egg cell A and the other with an egg cell a, and so likewise one pollen cell a, will unite with an egg cell A, and the other with egg cell a.

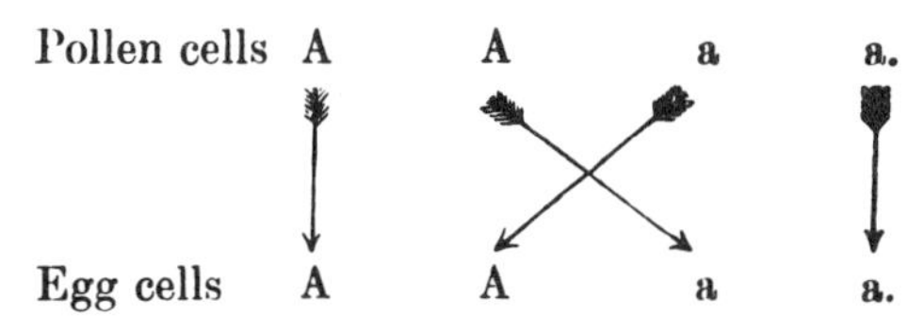

The result of the fertilisation may be made clear by putting the signs for the conjoined egg and pollen cells in the form of fractions, those for the pollen cells above and those for the egg cells below the line. We then have

$$\frac{A}{A}+\frac{A}{a}+\frac{a}{A}+\frac{a}{a}.$$

In the first and fourth factor the egg and pollen cells are of like kind, consequently the product of their union must be constant, viz. A and a; in the second and third, on the other hand, there again results a union of the two differentiating characters of the stocks, consequently the forms resulting from these fertilisations are identical with those of the hybrid from which they sprang. *There occurs accordingly a repeated hybridisation.* This explains the striking fact that the hybrids are able to produce, besides the two parental forms, offspring which are like themselves; $\frac{A}{a}$ and $\frac{a}{A}$ both give the same union Aa, since, as already remarked above, it makes no difference in the result of fertilisation to which of the two characters the pollen or egg cells belong. We may write then—

$$\frac{A}{A}+\frac{A}{a}+\frac{a}{A}+\frac{a}{a}=A+2\ Aa+a.$$

This represents the average result of the self-fertilisation of the hybrids when two differentiating characters are united in them. In solitary flowers and in solitary plants, however, the ratios in which the forms of the series are produced may suffer not inconsiderable fluctuations. Apart from the fact that the numbers in which both sorts of egg cells occur in the seed vessels can only be regarded as equal on the average, it remains purely a matter of chance which of the two sorts of pollen may fertilise each separate egg cell. For this reason the separate values must necessarily be subject to fluctuations, and there are even extreme cases possible, as were described earlier in connection with the experiments with the form of the seed and the colour of the albumen. The true ratios of the numbers can only be ascertained by an average deduced from the sum of as many single values as possible; the greater the number the more are merely chance elements eliminated.

The developmental series for hybrids in which two kinds of differentiating characters are united contains among sixteen individuals nine different forms, viz., AB + Ab + aB + ab + 2 ABb + 2 aBb + 2 AaB + 2 Aab + 4 AaBb. Between the differentiating characters of the original stocks Aa and Bb four constant combinations are possible, and consequently the hybrids produce the corresponding four forms of egg and pollen cells AB, Ab, aB, ab, and each of these will on the average figure four times in the fertilisation, since sixteen individuals are included in the series. Therefore the participators in the fertilisation are—

Pollen cells AB + AB + AB + AB + Ab + Ab + Ab + Ab + aB + aB + aB + aB + ab + ab + ab + ab.

Egg cells AB + AB + AB + AB + Ab + Ab + Ab + Ab + aB + aB + aB + aB + ab + ab + ab + ab.

In the process of fertilisation each pollen form unites on an average equally often with each egg cell form, so that each of the four pollen cells AB unites once with one of the forms of egg cell AB, Ab, aB, ab. In precisely the same way the rest of the pollen cells of the forms Ab, aB, ab unite with all the other egg cells. We obtain therefore—

$$\frac{AB}{AB}+\frac{AB}{Ab}+\frac{AB}{aB}+\frac{AB}{ab}+\frac{Ab}{AB}+\frac{Ab}{Ab}+\frac{Ab}{aB}+\frac{Ab}{ab}+$$

$$\frac{aB}{AB}+\frac{aB}{Ab}+\frac{aB}{aB}+\frac{aB}{ab}+\frac{ab}{AB}+\frac{ab}{Ab}+\frac{ab}{aB}+\frac{ab}{ab},$$

or

AB + ABb + AaB + AaBb + ABb + Ab + AaBb + Aab + AaB + AaBb + aB + aBb + AaBb + Aab + aBb + ab = AB + Ab + aB + ab + 2 ABb + 2 aBb + 2 AaB + 2 Aab + 4 AaBb.*

In precisely similar fashion is the developmental series of hybrids exhibited when three kinds of differentiating characters are conjoined in them. The hybrids form eight various kinds of egg and pollen cells— ABC, ABc, AbC, Abc, aBC, aBc, abC, abc—and each pollen form unites itself again on the average once with each form of egg cell.

The law of combination of different characters which governs the development of the hybrids finds therefore its foundation and explanation in the principle enunciated, that the hybrids produce egg cells and pollen cells which in equal numbers represent all constant forms which result from the combination of characters united by fertilisation.

Experiments with Hybrids of other Species of Plants.

It must be the object of further experiments to ascertain whether the law of development discovered for *Pisum* applies also to the hybrids of other plants. To this end several experiments were recently commenced. Two minor experiments with species of *Phaseolus* have been completed, and may be here mentioned.

A trial with *Phaseolus vulgaris* and *Phaseolus nanus* gave results in perfect agreement. *Ph. nanus* had together with the dwarf axis simply inflated green pods. *Ph. vulgaris* had, on the other hand, an axis 10 feet to 12 feet high, and yellow coloured pods, constricted when ripe. The ratios of the numbers in which the different forms appeared in the separate generations were the same as with *Pisum*. Also the development of the constant combinations resulted according to the law of simple combination of characters, exactly as in the case of *Pisum*. There were obtained—

* [In the original the sign of equality (=) is here represented by +, evidently a misprint. W. B.]

Constant combinations.	Axis.	Colour of the unripe pods.	Form of the ripe pods.
1	long	green	inflated
2	,,	,,	constricted
3	,,	yellow	inflated
4	,,	,,	constricted
5	short	green	inflated
6	,,	,,	constricted
7	,,	yellow	inflated
8	,,	,,	constricted

The green colour of the pod, the inflated forms, and the long axis were, as in *Pisum*, dominant characters.

Another trial with two very different species of *Phaseolus* had only a partial result. *Phaseolus nanus*, L., served as seed parent, a perfectly constant species, with white flowers in short bunches and small white seeds in straight, inflated, smooth pods; as pollen parent was used *Ph. multiflorus*, W., with tall winding stem, purple-red flowers in very long bunches, rough, sickle-shaped crooked pods, and large seeds which bore black flecks and splashes on a peach-blood-red ground.

The hybrids had the greatest similarity to the pollen parent, but the flowers appeared less intensely coloured. Their fertility was very limited; from seventeen plants, which together developed many hundreds of flowers, only forty-nine seeds in all were obtained. These were of medium size, and were flecked and splashed similarly to those of *Ph. multiflorus*, while the ground colour was not materially different. The next year forty-four plants were raised from these seeds, of which only thirty-one reached the flowering stage. The characters of *Ph. nanus*, which had been altogether latent in the hybrids, reappeared in various combinations; their ratio, however, with relation to the dominant characters was necessarily very fluctuating owing to the small number of trial plants. With certain characters, as in those of the axis and the form of pod, it was, however, as in the case of *Pisum*, almost exactly 1 : 3.

Insignificant as the results of this trial may be as regards the determination of the relative numbers in which the various forms appeared, it presents, on the other hand, the phenomenon of a remarkable change of colour in the flowers and seed of the hybrids. In *Pisum* it is known that the characters of the flower- and seed-colour present themselves unchanged in the first and further generations, and that the offspring of the hybrids display exclusively the one or the other of the characters of the original stocks. It is otherwise in the experiment we are considering. The white flowers and the seed-colour of *Ph. nanus* appeared, it is true, at once in the first generation in one fairly fertile example, but the remaining thirty plants developed flower colours which were of various grades of purple-red to pale violet. The colouring of the seed-coat was no less varied than that of the flowers. No plant could rank as fully fertile; many produced no fruit at all; others only yielded fruits from the flowers last produced, and did not ripen. From fifteen plants only were well-developed seeds obtained. The greatest disposition to infertility was seen in the forms with preponderantly red flowers, since out of sixteen of these only four yielded ripe seed. Three of these had a similar seed pattern to

Ph. multiflorus, but with a more or less pale ground colour; the fourth plant yielded only one seed of plain brown tint. The forms with preponderantly violet coloured flowers had dark brown, black-brown, and quite black seeds.

The trial was continued through two more generations under similar unfavourable circumstances, since even among the offspring of fairly fertile plants there were still some which were less fertile or even quite sterile. Other flower—and seed—colours than those cited did not subsequently present themselves. The forms which in the first generation contained one or more of the recessive characters remained, as regards these, constant without exception. Also of those plants which possessed violet flowers and brown or black seed, some did not vary again in these respects in the next generation; the majority, however, yielded, together with offspring exactly like themselves, some which displayed white flowers and white seed-coats. The red flowering plants remained so slightly fertile that nothing can be said with certainty as regards their further development.

Despite the many disturbing factors with which the observations had to contend, it is nevertheless seen by this experiment that the development of the hybrids, with regard to those characters which concern the form of the plants, follows the same laws as does *Pisum*. With regard to the colour characters, it certainly appears difficult to perceive a substantial agreement. Apart from the fact that from the union of a white and a purple-red colouring a whole series of colours results, from purple to pale violet and white, the circumstance is a striking one that among thirty-one flowering plants only one received the recessive character of the white colour, while in *Pisum* this occurs on the average in every fourth plant.

Even these enigmatical results, however, might probably be explained by the law governing *Pisum* if we might assume that the colour of the flowers and seeds of *Ph. multiflorus* is a combination of two or more entirely independent colours, which individually act like any other constant character in the plant. If the flower colour A were a combination of the individual characters $A_1 + A_2 + \ldots$ which produce the total impression of a purple colouration, then by fertilisation with the differentiating character of the white colour a there would be produced the hybrid unions $A_1a + A_2a + \ldots$ and so would it be with the corresponding colouring of the seed-coats. According to the above assumption, each of these hybrid colour unions would be independent, and would consequently develop quite independently from the others. It is then easily seen that from the combination of the separate developmental series a perfect colour-series must result. If, for instance, $A = A_1 + A_2$ then the hybrids A_1a, and A_2a form the developmental series—

$$A_1 + 2A_1a + a$$
$$A_2 + 2A_2a + a$$

The members of this series can enter into nine different combinations and each of these denotes another colour—

1 A_1A_2	2 A_1aA_2	1 A_2a
2 A_1A_2a	4 A_1aA_2a	2 A_2aa
1 A_1a	2 A_1aa	1 aa

The figures prescribed for the separate combinations also indicate how many plants with the corresponding colouring belong to the series. Since the total is sixteen, the whole of the colours are on the average distributed over each sixteen plants, but, as the series itself indicates, in unequal proportions.

Should the colour development really happen in this way, we could offer an explanation of the case above described, viz. that the white flowers and seed-coat colour only appeared once among thirty-one plants of the first generation. This colouring appears only once in the series, and could therefore also only be developed once in the average in each sixteen, and with three colour characters only once even in sixty-four plants.

It must, however, not be forgotten that the explanation here attempted is based on a mere hypothesis, only supported by the very imperfect result of the trial just described. It would, however, be well worth while to follow up the development of colour in hybrids by similar experiments, since it is probable that in this way we might learn the significance of the extraordinary variety in the colouring of our decorative flowers.

So far, little at present is known with certainty beyond the fact that the colour of the flowers in most ornamental plants is an extremely variable character. The opinion has often been expressed that the stability of the species is greatly disturbed or entirely upset by cultivation, and consequently there is an inclination to regard the development of cultivated forms as a matter of chance devoid of rules ; the colouring of decorative plants is indeed usually cited as an example of great instability. It is, however, not clear why the simple transference into garden soil should result in such a thorough and persistent revolution in the plant organism. No one will seriously maintain that the development of plants in the open country is ruled by other laws than in the garden bed. Here, as there, changes of type must take place if the conditions of life be altered, and the species possesses the capacity of fitting itself to its new environment. It is willingly granted that by cultivation the origination of new varieties is favoured, and that by man's labour many varieties are acquired which, under natural conditions, would be lost; but nothing justifies the assumption that the tendency to the formation of varieties is so extraordinarily increased that the species speedily lose all stability, and their offspring diverge into an endless series of extremely variable forms. Were the change in the conditions of vegetation the sole cause of variability we might expect that those cultivated plants which are grown for centuries under almost identical conditions would again attain constancy. That, as is well known, is not the case, since it is precisely under such circumstances that not only the most varied but also the most variable forms are found. It is only the *Leguminosæ*, like *Pisum*, *Phaseolus*, *Lens*, whose organs of fertilisation are protected by the keel, which constitute a noteworthy exception. Even here there have arisen numerous varieties during a cultural period of more than 1,000 years; these maintain, however, under unchanging environments a stability as great as that of species growing wild.

It is more than probable that as regards the variability of cultivated plants there exists a factor which so far has received little attention. Various experiments force us to the conclusion that our cultivated plants, with few exceptions, are *members of various hybrid series*, whose further development in conformity with law is changed and hindered by frequent crossings *inter se.* The circumstance must not be overlooked that cultivated plants are mostly grown in great numbers and close together, which affords the most favourable conditions for reciprocal fertilisation between the varieties present and the species itself. The probability of this is supported by the fact that among the great array of variable forms solitary examples are always found, which in one character or another remain constant, if only foreign influence be carefully excluded. These forms develop precisely as do those which are known to be members of the compound hybrid series. Also with the most susceptible of all characters, that of colour, it cannot escape the careful observer that in the separate forms the inclination to vary is displayed in very different degrees. Among plants which arise from *one* spontaneous fertilisation there are often some whose offspring vary widely in the constitution and arrangement of the colours, while others furnish forms of little deviation, and among a greater number solitary examples occur which transmit the colour of the flowers unchanged to their offspring. The cultivated species of *Dianthus* afford an instructive example of this. A white-flowered example of *Dianthus caryophyllus*, which itself was derived from a white-flowered variety, was shut up during its blooming period in a greenhouse; the numerous seeds obtained therefrom yielded plants entirely white-flowered like itself. A similar result was obtained from a subspecies, with red flowers somewhat flushed with violet, and one with flowers white, striped with red. Many others, on the other hand, which were similarly protected, yielded progeny which were more or less variously coloured and marked.

Whoever studies the colouration which results in decorative plants from similar fertilisation can hardly escape the conviction that here also the development follows a definite law which possibly finds its expression *in the combination of several independent colour characters.*

Concluding Remarks.

It can hardly fail to be of interest to compare the observations made regarding *Pisum* with the results arrived at by the two authorities in this branch of knowledge, Kölreuter and Gärtner, in their investigations. According to the opinion of both, the hybrids in outer appearance present either a form intermediate between the original species, or they closely resemble either the one or the other type, and sometimes can hardly be discriminated from it. From their seeds usually arise, if the fertilisation was effected by their own pollen, various forms which differ from the normal type. As a rule, the majority of individuals obtained by one fertilisation maintain the hybrid form, while some few others come more like the seed parent, and one or the other individual approaches the pollen parent. This, however, is not the case with all hybrids without exception. With some the offspring have more nearly approached, some the one and some the other, original stock, or they all incline more to one or the other side; while with others *they remain perfectly like the hybrid* and continue constant in their offspring. The hybrids of varieties behave like hybrids of species, but they possess greater variability of form and a more pronounced tendency to revert to the original type.

With regard to the form of the hybrids and their development, as a rule an agreement with the observations made in *Pisum* is unmistakable. It is otherwise with the exceptional cases cited. Gärtner confesses even that the exact determination whether a form bears a greater resemblance to one or to the other of the two original species often involved great difficulty, so much depending upon the subjective point of view of the observer. Another circumstance could, however, contribute to render the results fluctuating and uncertain, despite the most careful observation and differentiation; for the experiments plants were mostly used which rank as good species and are differentiated by a large number of characters. In addition to the sharply defined characters, where it is a question of greater or less similarity, those characters must also be taken into account which are often difficult to define in words, but yet suffice, as every plant connoisseur knows, to give the forms a strange appearance. If it be accepted that the development of hybrids follows the law which is valid for *Pisum*, the series in each separate trial must embrace very many forms, since the number of the components, as is known, increases with the number of the differentiating characters in *cubic ratio*. With a relatively small number of trial-plants the result therefore could only be approximately right, and in single cases might fluctuate considerably. If, for instance, the two original stocks differ in seven characters, and 100 and 200 plants were raised from the seeds of their hybrids to determine the grade of relationship of the offspring, we can easily see how uncertain the decision must become, since for seven differentiating characters the developmental series contains 16,384 individuals under 2,187 various forms; now one and then another relationship could assert its predominance, just according as chance presented this or that form to the observer in a majority of instances.

If, furthermore, there appear among the differentiating characters at the same time dominant characters, which are transferred entire or nearly unchanged to the hybrids, then in the components of the developmental series that one of the two original stocks which possesses the majority of dominant characters must always be predominant. In the experiment described relative to *Pisum*, in which three kinds of differentiating characters were concerned, all the dominant characters belonged to the seed parent. Although the components of the series in their internal composition approach both original stock plants equally, in this trial the type of the seed parent obtained so great a preponderance that out of each sixty-four plants of the first generation fifty-four exactly resembled it, or only differed in one character. It is seen how rash it may be under such circumstances to draw from the external resemblances of hybrids conclusions as to their internal relations.

Gärtner mentions that in those cases where the development was regular among the offspring of the hybrids the two original species were not reproduced, but only a few closely approximating individuals. With very extended developmental series it could not in fact be otherwise. For seven differentiating characters, for instance, among more than 16,000 individuals—offspring of the hybrids—each of the two original species would occur only once. It is therefore hardly possible that such should appear at all among a small number of trial plants; with some probability, however, we might reckon upon the appearance of a few forms which approach them in the series.

We meet with an essential difference in those hybrids which remain constant in their progeny and propagate themselves as truly as the pure species. According to Gärtner, to this class belong the remarkably fertile hybrids *Aquilegia atropurpurea canadensis*, *Lavatera pseudolbia thuringiaca*, *Geum urbano-rivale*, and some *Dianthus* hybrids; and, according to Wichura, the hybrids of the Willow species. For the history of the evolution of plants this circumstance is of special importance, since constant hybrids acquire the status of new species. The correctness of this is evidenced by most excellent observers, and cannot be doubted. Gärtner had opportunity to follow up *Dianthus Armeria deltoides* to the tenth generation, since it regularly propagated itself in the garden.

With *Pisum* it was shown by trials that the hybrids form egg and pollen cells of different kinds, and that herein lies the reason of the variability of their offspring. In other hybrids, likewise, whose offspring behave similarly we may assume a like cause; for those, on the other hand, which remain constant the assumption appears justifiable that their fertilising cells are all alike and agree with the foundation-cell of the hybrid. In the opinion of renowned physiologists, for the purpose of propagation one pollen cell and one egg cell unite in Phanerogams * into a single cell, which is capable by assimilation and formation of new cells to develop an independent organism. This development follows a constant law, which is founded on the material composition and arrangement of the elements which meet in the cell in a vivifying union. If the reproductive cells be of the same kind and agree with the foundation cell of the mother plant, then the development of the new individual will follow the same law which rules the mother plant. If it chance that an egg cell unites with a dissimilar pollen cell, we must then assume that between those elements of both cells, which determine the mutual differences, some sort of compromise is effected. The resulting compound cell becomes the foundation of the hybrid organism, the development of which necessarily follows a different law from that obtaining in each of the two original species. If the compromise be taken to be a complete one, in the sense, namely, that the hybrid embryo is formed from cells of like kind, in which the differences are entirely and permanently accommodated together, the further result follows that the hybrids, like any other stable plant species, remain true to themselves in their offspring. The reproductive cells which are formed

* In *Pisum* it is placed beyond doubt that for the formation of the new embryo a perfect union of the elements of both fertilising cells must take place. How could we otherwise explain that among the offspring of the hybrids both original types reappear in equal numbers and with all their peculiarities? If the influence of the egg cell upon the pollen cell were only external, if it fulfilled the *rôle* of a nurse only, then the result of each artificial fertilisation could be no other than that the developed hybrid should exactly resemble the pollen parent, or at any rate do so very closely. This the experiments so far have in no wise confirmed. An evident proof of the complete union of the contents of both cells is afforded by the experience gained on all sides that it is immaterial, as regards the form of the hybrid, which of the original species is the seed parent or which the pollen parent.

in their seed vessels and anthers are of one kind, and agree with the fundamental compound cell.

With regard to those hybrids whose progeny is variable we may perhaps assume that between the differentiating elements of the egg and pollen cells there also occurs a compromise, in so far that the formation of a cell as foundation of the hybrid becomes possible; but, nevertheless, the arrangement between the conflicting elements is only temporary and does not endure throughout the life of the hybrid plant. Since in the habit of the plant no changes are perceptible during the whole period of vegetation, we must further assume that it is only possible for the differentiating elements to liberate themselves from the enforced union when the fertilising cells are developed. In the formation of these cells all existing elements participate in an entirely free and equal arrangement, in which it is only the differentiating ones which mutually separate themselves. In this way the production would be rendered possible of as many sorts of egg and pollen cells as there are combinations possible of the formative elements.

The attribution attempted here of the essential difference in the development of hybrids to *a permanent or temporary union* of the differing cell elements can, of course, only claim the value of an hypothesis for which the lack of definite data offers a wide field. Some justification of the opinion expressed lies in the evidence afforded by *Pisum* that the behaviour of each pair of differentiating characters in hybrid union is independent of the other differences between the two original plants, and, further, that the hybrid produces just so many kinds of egg and pollen cells as there are possible constant combination forms. The differentiating characters of two plants can finally, however, only depend upon differences in the composition and grouping of the elements which exist in the fundamental cells of the same in vital interaction.

Even the validity of the law formulated for *Pisum* requires still to be confirmed, and a repetition of the more important experiments is consequently much to be desired, that, for instance, relating to the composition of the hybrid fertilising cells. A differential [factor] may easily escape the single observer, which although at the outset may appear to be unimportant, may yet accumulate to such an extent that it must not be ignored in the total result. Whether the variable hybrids of other plant species observe an entire agreement must also be first decided experimentally. In the meantime we may assume that in material points a difference in principle can scarcely occur, since the unity in the developmental plan of organic life is beyond question.

In conclusion, the experiments carried out by Kölreuter, Gärtner, and others with respect to *the transformation of one species into another by artificial fertilisation* merit special mention. A special importance has been attached to these experiments, and Gärtner reckons them among "the most difficult of all in hybridisation."

Should a species A be transformed into a species B, both would be united by fertilisation and the resulting hybrids then be fertilised with the pollen of B; then out of the various offspring resulting that form would be selected which stood in nearest relation to B and once more be fertilised with B pollen, and so continuously until finally a form was arrived at which was like B and constant in its progeny. By this process the species A would change into the species B. Gärtner alone has effected thirty such trials with plants of genera *Aquilegia*, *Dianthus*, *Geum*, *Lavatera*, *Lychnis*, *Malva*, *Nicotiana*, and *Œnothera*. The period of transformation was not alike for all species. While with some a triple fertilisation sufficed, with others this had to be repeated five or six times, and even in the same species fluctuations were observed in various experiments. Gärtner ascribes this difference to the circumstance that "the typical force by which a species, during reproduction, effects the change and transformation of the maternal type varies considerably in different plants, and that, consequently, the periods must also vary within which the one species is changed into the other, as also the number of generations, so that the transformation in some species is perfected in more, and in others in fewer generations." Further, the same observer remarks "that in these transformation trials a good deal depends upon which type and which individual be chosen for further transformation."

If it may be assumed that in these trials the development of the forms resulted in a similar way to that of *Pisum*, the entire process of transformation would find a fairly simple explanation. The hybrid forms as many kinds of egg cells as there are constant combinations possible of the characters conjoined therein, and one of these is always of the same kind as the fertilising pollen cells. Consequently there always exists the possibility with all such trials that even from the second fertilisation there may result a constant form identical with that of the pollen parent. Whether this really be obtained depends in each separate case upon the number of the trial plants, as well as upon the number of differentiating characters which are united by the fertilisation. Let us, for instance, assume that the plants selected for trial differed in three characters, and the species ABC is to be transformed into the other species abc by repeated fertilisation with the pollen of the latter; the hybrids resulting from the first cross form eight different kinds of egg cells, viz.:

ABC, ABc, AbC, aBC, Abc, aBc, abC, abc.

These in the second trial year are united again with the pollen cells abc, and we obtain the series

AaBbCc + AaBbc + AabCc + aBbCc + Aabc + aBbc + abCc + abc.

Since the form abc occurs once in the series of eight components, it is consequently little likely that it would be missing among the trial plants, even were these raised in a smaller number, and the transformation would be perfected already by a second fertilisation. If by chance it did not appear, then the fertilisation must be repeated with one of those forms nearest akin, Aabc, aBbc, abCc. It is perceived that such an experiment must extend the farther *the smaller the number of trial plants and the larger the number of differentiating characters* in the two original species; and that, furthermore, in the same species there can easily occur a delay of one or even of two generations such as Gärtner observed. The transformation of widely divergent species could generally only be completed in five or six trial years, since the number of different egg cells which are formed in the hybrid increases in square ratio with the number of differentiating characters.

Gärtner found by repeated trials that the respective period of transformation varies in many species, so that frequently a species A can be transformed into a species B a generation sooner than can species B into species A. He deduces thereform that Kölreuter's opinion can hardly be maintained that "the two natures in hybrids are perfectly in equilibrium." It appears, however, that Kölreuter does not merit this criticism, but that Gärtner rather has overlooked a material point, to which he himself elsewhere draws attention, viz. that "it depends which individual is chosen for further transformation." Experiments which in this connection were carried out with two species of *Pisum* demonstrated that as regards the choice of the fittest individuals for the purpose of further fertilisation it may make a great difference which of two species is transformed into the other. The two trial plants differed in five characters, while at the same time those of species A were all dominant and those of species B all recessive. For mutual transformation A was fertilised with pollen of B, and B with pollen of A, and this was repeated with both hybrids the following year. With the first trial $\frac{B}{A}$ there were eighty-seven plants available in the third trial year for the selections of individuals for further crossing, and these were of the possible thirty-two forms; with the second trial $\frac{A}{B}$ seventy-three plants resulted, which *agreed throughout perfectly in habit with the pollen parent*; in their internal composition, however, they must have been just as varied as the forms of the other trial. A definite selection was consequently only possible with the first trial; with the second some plants selected at random had to be excluded. Of the latter only a portion of the flowers were crossed with the A pollen, the others were left to fertilise themselves. Among each five plants which were selected in both trials for fertilisation there agreed, as the following year's culture showed, with the pollen parent:—

First Trial.	Second Trial.	
2 Plants	—	in all characters
3 ,,	—	,, 4 ,,
—	2 plants	,, 3 ,,
—	2 ,,	,, 2 ,,
—	1 plant	,, 1 character

In the first trial, therefore, the transformation was completed; in the second, which was not continued further, two more fertilisations would probably have been required.

Although the case may not frequently occur that the dominant characters belong exclusively to one or the other of the original parent plants, it will always make a difference which of the two possesses the majority. If the pollen parent shows the majority, then the selection of forms for further crossing will afford a less degree of security than in the reverse case, which must imply a delay in the period of transformation, provided that the trial is only considered as completed when a form is arrived at which not only exactly resembles the pollen p'ant in form, but also remains as constant in its progeny.

Gärtner, by the results of these transformation experiments, was led to oppose the opinion of those naturalists who dispute the stability of plant species and believe in a continuous evolution of vegetation. He perceives in the complete transformation of one species into another an indubitable proof that species are fixed within limits beyond which they cannot change. Although this opinion cannot be unconditionally accepted, we find on the other hand in Gärtner's experiments a noteworthy confirmation of the supposition regarding variability of cultivated plants which has already been expressed.

Among the trial species there were cultivated plants, such as *Aquilegia atropurpurea* and *canadensis*, *Dianthus caryophyllus*, *chinensis*, and *japonicus*, *Nicotiana rustica* and *paniculata*, and hybrids between these species lost none of their stability after four or five generations.

Part II

CONFLICTS AND RESOLUTIONS

Editor's Comments on Papers 5 Through 11

5 CASTLE
The Laws of Heredity of Galton and Mendel and Some Laws Governing Race Improvement by Selection

6 YULE
On the Theory of Inheritance of Quantitative Compound Characters on the Basis of Mendel's Laws—A Preliminary Note

7 HARDY
Mendelian Proportions in a Mixed Population

8 WEINBERG
On the Demonstration of Inheritance in Man

9 PEARSON
The Theory of Ancestral Contributions in Heredity

10 PEARSON
On the Ancestral Gametic Correlations of a Mendelian Population Mating at Random

11 EAST
A Mendelian Interpretation of Variation That is Apparently Continuous

In 1893, at the instigation of Galton, Weldon, and others, the Royal Society formed a committee for conducting "Statistical Inquiries into the Measurable Characteristics of Plants and Animals." The basic problem with the function of the committee was the difference of opinion between Galton and Weldon on the nature of evolution. Galton took the position that evolution occurred by jumps (saltations), Weldon believed in the gradual change championed by Darwin. The great difficulties of defining terms and in determining what was a character, what evolved (individuals, populations, varieties, races, species), and exactly how they actually evolved arose early in the study. In 1894 Bateson's *Materials for the Study of Variation, Treated with Especial Regard to Discontinuity in the Origin of Species* was published. By mid-1895 Bateson and Weldon, formerly friends, had established unyielding conflicting positions and each felt the other had reduced the arguments to personal attacks; they had. In early 1897 the Royal Committee was expanded to include Bateson and others and the name was changed to the Evolution Committee. In 1900 Mendel's principles were rediscovered by de Vries, Correns, and von Tschermak; Zirkle (1964) and Olby (1966) provide details. In 1902 Bateson and Saunders reported to the Evolution Committee their experiments and Bateson published a small volume in defense of Mendel's principles. By this time inflammatory words were being published and personal attacks were no longer hidden. Late that year Yule (1902) published a review of the position of the two sides in which he rebuked Bateson for his words and the University Press for allowing publication. He suggested that much of the differences resulted from differences in use and definition of the terms "heredity, variation, variable, variability." Clearly strong personalities were involved. Yule, a competent mathematician, attempted to understand both positions and bridge the gap between the statisticians and the Mendelians. His article (1902) summarized many points of each side and concluded: "It is however essential, if progress is to be made, that biologists—statistical or otherwise—should recognize that Mendel's Laws and the Law of Ancestral Heredity are not necessarily contradictory statements, one or the other of which must be mythical in character, but are perfectly consistent the one with the other and may quite well form parts of one homogeneous theory of heredity" (p. 236). And finally Yule says, "That an ideally complete theory cannot come yet, may be conceded at once; that it is impossible in the present state of biology to form a quantitative theory, founded on clear and definite physical conception, which will carry one some steps on the way, I do not believe" (p. 237). A preliminary step in this direction was made in 1906 (Paper 6).

Unfortunately, the two theories, and particularly the various derivations required to account for the exceptions to the first look at the two theories, were not "perfectly consistent" and the emotions of the time often emphasized these inconsistencies and efforts to prove one's own position rather than to discover the workings of nature. Therefore, the papers selected for this volume are designed to emphasize the attempts by various workers to resolve the conflicts. The details of the position of Bateson and Weldon, and the considerable rhetoric which accompanied their analyses, may be obtained by reading the papers listed in the bibliography.

In 1903 (Paper 5) Castle analyzed some data on crosses between variously pigmented mice some of which breed true, others giving various ratios. He compared the results with those expected from the Law of Ancestral Heredity and with those expected from the Law of Mendel. He appears to have misunderstood Yule's (1902) position. He found that these alternative characters follow Mendel's Law but not the Law of Ancestral Heredity. He was the first to calculate the progress to be obtained by selective procedures for the desirable or by the elimination of undesirable individuals, providing some interesting tables and a particularly interesting figure. Even more striking is his recognition that "as soon as selection is arrested the race remains stable at the degree of purity then attained" (p. 237). He even identified the roles of differential fertility and mortality.

William E. Castle was born in 1867 in Ohio, attended Denison and Harvard, where he completed his doctorate in 1895. He taught at the University of Wisconsin and at Knox College before he returned to Harvard, where he remained until he retired in 1936. Fron then until his death in 1962 he held a research appointment at the University of California in Berkeley. His early work on genetics was with the coat color of rats, mice, and guinea pigs, and he and S. Wright reported the first demonstration of linkage in mammals. He studied size in mammals and coat color in horses and trained many students of genetics. Castle was a confirmed proponent of the gradual process of natural selection and early believed that Mendelian units were subject to continuous variation.

The next year (1904) Pearson's long and detailed contribution "On a Generalized Theory of Alternative Inheritance, with Special Reference to Mendel's Laws" appeared. The paper was stimulated by Galton's work and by extensive interaction with Weldon, and was one of a number of papers Pearson wrote on the subject. In this paper Pearson developed the theory of the product moment correlation used previously by Galton. Pearson, still in doubt as to the universality of Mendel's laws, worked out the theoretical results of a number of Mendelian crosses and noted the advantages of a generalized theory of the pure gamete but felt "Mendelian principles remained in a state of flux." The most interesting passage for our purposes he italicized:

> *However many couplets we suppose the character under investigation to depend upon, the offspring of the hybrids—or the segregating generation—if they breed at random inter se, will not segregate further, but continue to reproduce themselves in the same proportions as a stable population* (p. 60).

This passage was later referred to by Hardy (Paper 7). Pearson also noted:

> Toss two pennies, and the result of 4n tossings will closely approximate to the distribution n (HH+2HT+TT). Load one or both coins, and the

possible variations will still be HH, HT or TT, but their proportions will be far from n:2n:n (p. 86).

His discussions here do not suggest that he perceived the importance of the principle proposed by Hardy and Weinberg as applied to biological questions, and he certainly failed to develop the loading factors of selection, migration, and mutation.

Karl Pearson (1857-1936) was educated as an attorney and in mathematics. He was admitted to the bar in 1882 and lectured part time until 1884, when he was appointed professor of mathematics at University College, London. The *Journal Biometrika* was started in 1901 by Pearson, Weldon, and Galton. In 1911 he became the Galton Professor of National Eugenics at the University of London. The *Annals of Eugenics* was initiated in 1925. In addition to being a biographer of Galton (1924, 1930), he wrote *Tables for Statisticians* and many other books on philosophy, mathematics, sciences, eugenics, and evolution and served as editor of *Biometrika* until his death.

In 1906 (Paper 6) Yule made a preliminary attempt to use Mendel's laws to propose a "general though still quite limited" theory of the inheritance of quantitative compound characters. This paper shows how close, yet how far, was the integrative theory. Fisher (1918) noted that Yule's effort was almost general and that he showed the similarity of the effects of dominance and of environment in reducing the correlations between relatives. Fisher expanded on the approach. Yule (1871-1951) was trained as a statistician at London, Bonn, and Cambridge. He taught at London and Cambridge and contributed books on theoretical statistics and a number of papers on evolutionary theory.

In 1908 a brief note in *Science* by Hardy responded to a statement by Yule which suggested that dominant genes would be more frequent than recessives in natural populations. I was once told that the mentioned conversation between Punnett and G. H. Hardy took place on a golf course. I have been unable to confirm this story and Punnett's published account (1950) suggests the cricket fields. Whether it is true or not, I can account from personal experience that the position taken by Yule regarding brachydactyly was still being held, and taught, by some zoologists as recently as 1965. In summary, the position of Hardy is that gene frequency, genotype frequency, and phenotype frequency (terms not defined at the time) will remain constant from generation to generation except for "casual deviations," mating not "purely random," or the character "has an influence on fertility." G. H. Hardy (1877-1947) was a mathematician who trained at Cambridge and taught at Oxford, Cambridge, Princeton, and the California Institute of Technology. He worked on number theory, Fourier series, and other aspects of pure mathematics, and was considered the leading mathematician of his day.

Wilhelm Weinberg (1862–1937), a few months prior to Hardy, published a more expanded version of the equilibrium principle in German. Unfortunately the paper did not attract attention for some time. A series of papers (1909, 1910) followed in which he introduced genic and environmental variance, provided a Mendelian basis for the correlation between relatives and expanded the equilibrium concept to multiple alleles. Stern (1962) has pointed out that Weinberg, like Mendel, made his discoveries at a time when science was not ready, yet his contributions to the problems of multiple births, population genetics, ascertainment theory, and medical statistics were original and pioneering. Weinberg trained in medicine at Tubingen and Munich, completing his medical degree in 1886. Clinical studies in Berlin, Vienna, and Frankfurt preceded his return to his birthplace of Stuttgart. He published more than 160 papers during a seventy-five-year life which included forty-two years of active private and public practice. Truly an intellectial giant and "one of the foremost creators and pioneers in the genetics of the twentieth century" (Stern, 1962:5).

In 1909 two short papers by K. Pearson (Papers 9 and 10) concluded that there would be little left of contradiction between biometric and Mendelian results if the absolute law of dominance were not a stumbling block. Weinberg wrote that the quarrel between the biometricians and the Mendelians was completely baseless. Johannsen (1909) produced the operational definitions of phenotype, genotype, and gene which were necessary to clarify the argument. East (Paper 11) showed that a color variation in maize depends on two independent loci, and this and later papers (East and Hayes, 1912) along with the study of Nilsson-Ehle (1909), established the multiple factor hypothesis of quantitative inheritance.

Edward M. East was born in 1879. He attended Case and the University of Illinois, obtaining the doctorate in 1907. He went to Harvard in 1909, where he remained until his death in 1938. He was a principal contributor, along with Shull (1908) and D. F. Jones, to the development of the multiple factor hypothesis and to the development of the theory and procedures of inbreeding and outbreeding which led to a revolution in seed corn production. He was interested and involved in the application of genetics to human welfare, producing several books in the field. D. F. Jones (1939) has provided a biography.

In the first decade of the twentieth century Mendel's principles had been affirmed in a variety of plants and animals including man (Bateson and Saunders, 1902; Weinberg, (Paper 8). Theoretical models had been developed by Yule (1902), Castle (Paper 5), Pearson (Papers 9 and 10), and Weinberg (1909, 1910), which provided the basis for a Mendelian analysis of quantitative characters. A series of studies by Shull (1908), Nilsson-Ehle (1909), East (Paper 11), and others had demonstrated that quantitative characters were determined by multiple factors and multiple alleles. While later papers by Fisher, Wright, Haldane, and others provided the coup de grace, continued battles over the evolutionary role of major saltations and of small continuous changes were unnecessary. The task at hand was to determine where and how saltations occurred and where and how small continuous changes operated. Thus, a decade after the rediscovery of Mendel, the war was over, but the battles went on and on, and its vestiges can still be discerned. Well-educated scholars hold a position long after the position is fruitful and sometimes after the position is untenable.

5

Reprinted by permission from *Amer. Acad. Arts Sci. Proc.* **39**:223-242 (1903)

THE LAWS OF HEREDITY OF GALTON AND MENDEL, AND SOME LAWS GOVERNING RACE IMPROVEMENT BY SELECTION.

By W. E. Castle.

Presented October 14, 1903. Received September 26, 1903.

Contents.

I. The "Law of Ancestral Heredity."

In the year 1889, the eminent English statistician, Francis Galton, attempted to give precise mathematical expression to the well-known fact that the child resembles in varying degree its ancestors near and remote. From a study of family statistics of stature, he found that children resemble their parents, on the average, more closely than their grandparents, and the latter more closely than their great-grandparents, and so on to ancestors still more remote. He tentatively advanced the hypothesis that the resemblance to each earlier generation of ancestors is just half that to the next later.

Galton subsequently tested this hypothesis in the case of a domesticated animal, by applying it to an extensive series of records of the inheritance of black spots in Basset hounds. Satisfied with the result, Galton ('97, p. 502) then formulated as follows the general "Law of Ancestral Heredity": — "The two parents contribute between them, on the average one-half, or (0.5) of the total heritage of the offspring; the four grandparents, one-quarter, or $(0.5)^2$; the eight great-grandparents, one-eighth, or $(0.5)^3$, and so on. Thus the sum of the ancestral contributions is expressed by the series $[(0.5) + (0.5)^2 + (0.5)^3$, etc.$]$, which, being equal to 1, accounts for the whole heritage." Galton found that, allowance being made for male prepotency, the theoretical values calculated in accordance with this "law" conform very closely to the values actually observed in the series of generations of Basset hounds. He, therefore, put his law forward as a general law of ancestral heredity.

But subsequent examination by Pearson ('98) of the material studied by Galton, and of other material similar in nature, has failed to substantiate Galton's conclusion, except in a much modified form. In the most recent statement of his views, Pearson (:03) holds with Galton that the best prediction as to the character of the offspring must be based upon the character of the ancestors, and that the influence of the various ancestors diminishes as they become more remote. He believes that "the contributions of the ancestry follow a geometrical series, although not that originally proposed by Mr. Galton." From a study of the inheritance of eye-color in man and coat-color in thoroughbred horses, he concludes that "as far as the available data at present go, inheritance coefficients for ascending ancestry are within the limits of observational error represented by a geometrical series and by the same series." This series, he observes, approximates those designated I and II below: —

	Pearson's Series I.	Pearson's Series II.	Galton's Series.
Parental influence	.49	.50	.50
Grandparental influence	.32	.33	.25
Great-grandparental influence . . .	.20	.22	.125
Great-great-grandparental influence .	.13	.15	.0625

Comparing Pearson's series with that of Galton, we see that the parental influence is reckoned as substantially the same by both Galton and Pearson, but that Pearson assigns a much greater influence to the more remote ancestors than does Galton.

It should be observed that the "available data" upon which principally Pearson bases his conclusions consist of two cases of pigment inheritance, one in man, the other in the horse. A third well-known series of this sort has not been utilized by Pearson, though our information about it is much more complete and precise than that about either of the other two. I refer to the statistics about color inheritance in mice recorded by von Guaita ('98, :00), of which an analysis has been made by Davenport (:00). In this series the inheritance of color follows closely neither the law of Galton nor the series suggested by Pearson.

The lack of agreement in this case with Galton's law has been pointed out by Davenport for certain of the color categories. He concludes that in the case of gray alone does the color inheritance among von Guaita's mice conform closely with Galton's law. But in reality, even in the case of gray, close agreement does not occur; Davenport's conclusion that it does occur results from the inclusion by him in a single color category of two sorts of mice which are clearly quite distinct, namely, (1) mice gray all over like the wild house-mice, and (2) gray mice with white markings. Even when these two categories are combined, Davenport's figures show close agreement between the observed and calculated numbers in two only of the five filial generations with which he deals, namely, in the third and sixth generations (of von Guaita's nomenclature), in which he finds observed and calculated to agree perfectly. But in the three remaining generations he finds observed and calculated percentages to be related as follows: —

Generation	II.	IV.	V.
Observed	100%	58%	48%
Calculated	0%	48%	60%

Davenport, moreover, has excluded from the category of "albinos" white mice which possess the dancing character. But this is manifestly an error, for the dancing character has nothing to do with coat-color, and is inherited quite independently of it. Davenport's classification, accordingly, makes the category of albinos appear smaller than it really is. If we include all albinos (whether dancers or not) in one category, and make separate classes for gray, gray-white, black, and black-white mice, the relations between the observed and calculated numbers in each generation are found to be as indicated in Table I, Davenport's method of calculation being followed.

An examination of this table shows no close agreement between calculated and observed conditions throughout any single category or any single generation, although the totals turn out better than the predictions for the generations considered separately. In Generation II the discrepancies are glaring. In the column, white, the grand totals alone agree closely, yet this agreement is clearly without significance; it is a chance agreement in the totals of two series divergent throughout.

The observed numbers, it is evident, agree no better with one of Pearson's series than with that of Galton. The discrepancies noted

TABLE I.

TEST OF GALTON'S LAW BY THE STATISTICS OF VON GUAITA.

B, black; *B-W*, black-white; *G*, gray; *G-W*, gray-white; *W*, white.

Generation.		*G.*	*G-W.*	*B.*	*B-W.*	*W.*	Total.
II.	Calc.	0			14	14	
	Obs.	28			0	0	28
III.	Calc.	22	0	0	11	11	
	Obs.	17	8	4	1	14	44
IV.	Calc.	14	1	0	3.8	12.1	
	Obs.	16	5	4	1	5	31
V.	Calc.	23	45	13.7	16	15.5	
	Obs.	3	55	12	24	19	113
VI.	Calc.	10.5	31.2	18.4	6	20.4	
	Obs.	2	22	9	17	36	86
VII.	Calc.	1.2	1	0.7	0.2	3.7	
	Obs.	0	4	0	0	3	7
Total	Calc.	70.7	78	32 8	51	76.7	
	Obs.	66	94	29	43	77	309
Ratio, Obs. : Calc.		93.3%	120 5%	88.4%	84.3%	100+ %	

between observed and calculated will remain and even be accentuated if we replace Galton's series with one of those suggested by Pearson. For the result will be unchanged in Generation II, but the calculated numbers will in most cases diverge still more from the observed ones, in the later generations, because Pearson attaches more weight to the remoter ancestors than does Galton.

It is evident, then, that some fundamental defect exists in the "law of ancestral heredity," as stated by either Galton or Pearson. It fails in the case just examined not only to account for the observed result, but even to enable one to predict that result with any degree of accuracy, and that too in the very category of cases which it was originally formulated to cover, namely in color inheritance among mammals. Galton himself ('97, p. 403) recognized the existence of such a defect, though he considered it, for practical purposes, of little consequence. Stated in his own words it is as follows: —

"The chief line of descent," it is generally believed, "runs from germ to germ and not from person to person." Yet "the person may be accepted on the whole as a fair representative of the germ, and, being so, the statistical laws which apply to persons would apply to the germs also, though with less precision in individual cases." Failure of Galton's law in the case of von Guaita's statistics is due to the falsity of the assumption here made by Galton that the person is "a fair representative of the germ." In all cases of alternative inheritance the person (or soma) represents only a *part* of the ripe germs produced by the individual, in some cases it may even represent none of them. Hence any theory of heredity which bases its predictions as to the character of the offspring solely upon the character of the soma of the ancestors, is clearly inapplicable to cases of alternative inheritance. The presumption is against its application to any other class of cases until that applicability has been demonstrated.

II. Mendel's Law of Heredity.

Certain facts of alternative inheritance were clearly stated and accounted for many years ago by Gregor Mendel ('66). He thus not only formulated laws of alternative inheritance, whose correctness has been fully confirmed by a number of independent observations, but he also laid the foundation for a general theory of heredity. In the history of the study of heredity his discovery is the most fundamental and far-reaching. Its importance is not lessened by the fact that it was long unrecognized. Only under the fertilizing influence of Weismann's ideas was the rediscovery of Mendel's law accomplished independently by de Vries (:00), Correns (:00), Tschermak (:00), and others. To its further development no one has contributed more than Bateson (:02).

Where Galton's law gives us at best rough approximations based upon averages of heterogeneous material, and with no attempt at an explanation of the results, Mendel's law enables us to make predictions for specific cases as to both the character and the numerical proportions of the offspring to be expected, and furnishes us at the same time with a rational explanation of the outcome. It thus meets the two-fold requirements of a scientific theory, a statement of phenomena and an explanation of them; the "law of ancestral heredity" attempts only the first of these two things, and even here fails lamentably. It will thus be seen that the claims of Mendel's law are much greater than those of Galton's law. If it fails, its failure is as much more signal.

The same test may be applied to Mendel's law as to Galton's. Can we, on the basis of Mendel's law, make predictions concerning the various generations of von Guaita's mice with greater accuracy than has been found possible under Galton's law? Before we can frame an answer to this question, we must know precisely what the Mendelian predictions are.

Mendelian predictions are based, not on the *somatic character of the parents*, but on the character of the *germ-cells* formed by the parents. The simplest way of determining the character of the germ-cells formed by an animal or plant is by experimental breeding tests. In cases where this is not practicable, one can often predict with equal confidence from a knowledge concerning the grandparents, not as to their *somatic* character, but as to the character of their *germ-cells* as evidenced by the nature of the offspring produced by them. Stated in the terminology of present-day biology, the principles which underlie the Mendelian predictions are these: —

1. Every gamete (egg or spermatozoön) bears the determinants of a complete set of somatic characters of the species. Accordingly when two gametes (an egg and a spermatozoön) have met in fertilization, there are present in the fertilized egg the representatives of *two sets* of somatic characters, which may or may not be the same. If they are the same for a given character, as, for example, coat-color in mammals, the individual which develops from the egg must inevitably have that same character. Thus when gametes formed by one white mouse meet in fertilization gametes formed by another white mouse, the offspring are invariably white. Similarly when a wild gray mouse is bred to another wild gray mouse the offspring are invariably gray. And when a pure-bred spotted black-white mouse is bred to a mouse like itself, the offspring are all spotted black-white.

2. But when the two gametes uniting bear each what represents a *different* somatic character, only one of these characters may be manifested by the individual (or zygote) formed. Thus, when wild gray mice are mated with white mice, only gray offspring are produced. The gray character is, in Mendel's terminology, *dominant*, the white character

recessive. Or, when wild gray rats are mated with black-white rats, only gray rats are produced. The wild gray character is, accordingly, dominant not only over white, but also over black-white.

3. Sometimes the zygote formed by the union of two unlike gametes (heterozygote, Bateson, :0) develops the character of neither parent in its purity. It may have a character intermediate between those of its parents, or something entirely different from either. Thus when black-white mice are mated with white mice, the offspring are gray like the wild house mouse.

4. Whatever the somatic character of the zygote is, the germ-cells which it forms will be, in respect to any particular character, like those which united to produce it, — half like the maternal and half like the paternal gamete. Thus, a gray mouse obtained by crossing a wild gray mouse with a white one forms in equal numbers gametes which bear the gray character and those which bear the white character. This is conclusively shown by two simple breeding tests: 1) when a cross-bred (or hybrid) gray mouse is bred to a white mouse, half the offspring are hybrid grays, half are white. This is precisely the result we should expect if the cross-bred gray mouse forms, in equal numbers, as we have supposed, gametes which bear the gray and those which bear the white character. For

The gray mouse will produce gametes	G and W
The white mouse, gametes	W and W
And the possible combinations of these 2 sets are their product	$2\ GW + 2\ WW$

But, as we have already stated, when a zygote contains *both* the gray character and the white character, only the former will be visible. This may be indicated by placing the (invisible) W within a parenthesis. Further, in the expression 2 WW one of the identical letters may be dropped as superfluous. Our formula, representing the outcome of the breeding test described, then reads $2\ G\ (W) + 2\ W$, and signifies that two in every four of the offspring produced will be gray hybrids, and the remaining two white. 2) When two cross-bred (or hybrid) gray mice are bred together, the offspring consist of gray mice and white mice in the ratio of three gray to one white. Moreover, breeding tests show that of the three gray mice thus obtained one is pure, that is, will form only gametes bearing the gray character, while two are hybrid, that is, will form gametes some of which bear the gray character, others the white character. This is precisely the result expected under our hypothesis that each hybrid individual forms gametes G and W in equal numbers.

For the possible combinations of two sets of gametes each G and W are represented by their product $GG + 2GW + WW$, or simplified as already explained, $G + 2G(W) + W$.

The principle illustrated by these examples is, as pointed out by Bateson (:02), the most fundamental and far-reaching of the Mendelian ideas. It is known as the law of segregation, or "splitting" (de Vries, :00) of the parental characters at gamete formation, or as the "principle of gametic purity" (Bateson, :02). Dominance is purely a secondary matter; it may or may not occur along with segregation, though the latter can be more easily demonstrated in cases where it is associated with the former. The principle of gametic purity just stated rests upon the assumption that gamete-formation is the reverse of fertilization. In fertilization, gametes A and B unite to form a zygote AB; when this zygote in turn forms gametes, they will be again A and B. From a knowledge of the *somatic form* alone of pure As and Bs, one can make no trustworthy prediction as to the form of AB. — Here is the fundamental error of the "law of ancestral heredity" as stated by Galton ('97) or Pearson (:03). — AB may have invariably the somatic form of A or of B (cases of simple dominance, as of gray over white in mice); or it may have *sometimes* the form of A, sometimes that of B (cases of alternative dominance — see Tschermak (:02) —); or, finally, the somatic form of AB may be different from both that of A and that of B (cases like that of the gray hybrid formed by the cross of black-white with white mice). But, no matter what the somatic form of AB is, we may with confidence predict that its gametes will be essentially pure As and pure Bs, and the two will be produced in proportions approximately equal. This is the Mendelian expectation in all cases of alternative inheritance. Whether it applies to other cases also, and if so to what extent, is not yet known. For the present we may confine our attention to the case which afforded a basis for the "law of ancestral heredity," namely alternative color-inheritance among mammals.

In Table II are given the Mendelian predictions for the inheritance of complete albinism in the various generations and matings of von Guaita's mice. These predictions are based upon the fact repeatedly observed that complete albinism behaves as a *recessive* character in heredity with reference to a pigmented character of any sort (gray, black, or spotted). Predictions are not made for the other color categories separately, because their relations to each other are not entirely clear from von Guaita's experiments. It seems probable, however, that they bear one toward another relations of alternative dominance. This

TABLE II.

TEST OF MENDEL'S LAW BY VON GUAITA'S STATISTICS.

Abbreviations as in Table I.

Generation	Pair.	Total Young.	Mendel's Law, Calc. No. *W*.	Observed No. *W*.	Galton's Law, Calc. No. *W*.
II.		28	0	0	14
III.		44	11	14	11
IV.	(1)	4	4	4	2 5
	(2)	16	0	0	6
	(3)	2	0	0	0.25
	(4)	7	0	0	2.62
	(5)	2	1	1	0.75
	Total	31	5	5	12.12
V.	(1)	16	0	0	1
	(2)	5	1.2+	2	1
	(3)	32	8	7	6
	(4)	13	0	0	1.6
	(5)	44	11	9	5.5
	(6)	3	0.7+	1	0.4
	Total	113	21	19	15.5
VI.	(1)	32	8	10	2
	(2)	6	6	6	4
	(3)	43	21.5	20	14
	(4)	2	0 ?	0	0.2
	(5)	3	0 ?	0	0.2
	Total	86	35.5	36	20.4
VII.	(1)	4	0	0	1.4
	(2)	3	3	3	2.3
	Total	7	3	3	3.7
Total . . .		309	75 5	77	76.7

matter is now undergoing experimental tests which, when complete, may enable us to make predictions for these color categories not less precise than those given for white.

In the last column of the table are given for comparison the predictions based on Galton's law for the corresponding generations and pairs. If we were to consider the grand totals only, we might conclude that the Galtonian predictions are quite as good as the Mendelian, but if we examine item by item the two series from which these totals are made up, we see that there is no comparison in point of accuracy between the two sets of predictions. The Mendelian predictions are very close to the observed numbers throughout the table, generation by generation and pair by pair. In all cases except four the predictions are either perfect or within one of perfection, and in one only of these cases is the error greater than two. This one case is the total for generation III where the observed number is fourteen, the expectation eleven. Rarely do the Galtonian predictions come within one, or anywhere near one, of prefection. They demand the occurrence of white individuals in every generation and among the offspring of nearly every pair in the series, whereas white individuals are entirely wanting, and according to Mendel's law are not to be expected, among the offspring of *all* pairs in the second generation, and of eight other pairs in later generations of the series. The test is conclusive in favor of Mendel's law and against the "law of ancestral heredity," in the special case of albinism in mice. Elsewhere Castle and Allen (:03) have shown that among organisms in general albinism probably follows the same (Mendelian) law of inheritance.

Numerous other cases of Mendelian inheritance covering a wide range of characters are recorded in recent papers by de Vries (:01–03), Correns (:01, :03), Tschermak (:01, :01[a], :02), Bateson and Saunders (:02), Webber (:00), Spillman (:02, :02[a]), Hurst (:02, :03), and others. These cases show that the Mendelian laws are widely applicable. They are not laws of hybridization merely, as Vernon (:03) and some others assume, but are general laws of alternative inheritance.

III. Yule on Galton's Law and Mendel's Law.

Bateson (:02) has taken the very reasonable position that Mendel's law and the law of "ancestral heredity" cannot both be applicable to the same classes of cases. But Yule (:02) sees no incompatibility between the two, and this view Pearson (:03) endorses. Yule says (p. 226),

"Mendel's Laws, so far from being in any way inconsistent with the Law of Ancestral Heredity, lead then directly to a special case of that law, for the *dominant* attribute at least. For the *recessive* attribute it does not hold." Let us see how Yule reaches this curious conclusion, that certain Mendelian predictions are only a special category of the more general predictions of the law of ancestral heredity.

After a statement of the Galton-Pearson law, whereby it is limited to no particular series, geometrical or otherwise, but is made to include any set of empirical averages of the characters of the ancestors, which can be made the basis of predictions, he proceeds as follows: "The first question to be asked is one that does not seem to have occurred to any of Mendel's followers, viz.: what, exactly, happens if the two races *A* [dominants] and *a* [recessives] are left to themselves to inter-cross freely *as if they were one race?*" In answer to this question, Yule draws the correct conclusion that the first cross-bred (or hybrid) generation will consist exclusively of dominants, but that all subsequent generations will consist of dominant and recessive individuals in the proportions, 3 dominant: 1 recessive, [provided no selection is practised and all individuals are equally fertile]. Yule next inquires, if I understand him rightly, what will be the effect of eliminating in each generation *all the recessive individuals.* Starting with 300 dominant individuals, which are in the Mendelian proportions, 100 pure: 200 hybrid, he finds that the successive generations will contain the following proportions of dominant individuals: —

.83333
.85000
.85294
.85345
.85354

He considers it useless to carry the series farther, as it "tends toward the limiting value .85355339 . . ." Now, what, in plain unmathematical language, does this mean? It means that when a dominant form has once been crossed with a recessive (as a pigmented animal, for example, with an albino), the stock of the former is forever contaminated, and cannot be freed entirely from the albino character by mere elimination of white individuals, however long the process is continued. Ever afterward the cross-bred dominant stock will produce on the average at least fourteen or fifteen white individuals in every hundred born. This conclusion is absurd, as every breeder knows. There is certainly something wrong with Yule's figures, for they do not accord with observation. In reality, an error lies in the very first step of his calculation, which invalidates all that follows. He says, "The 100 pure individuals will give rise to dominant forms in the proportion of 50 pure to 50 hybrids." On the contrary, "pure" dominants bred *inter se* will produce only pure dominant offspring; but if they mate at random with any individuals of the entire 300, there are only two chances out of three that they will mate with *hybrid* dominants, which mating alone could yield "dominant forms in the proportion 50 pure to 50 hybrids." Yule accordingly estimates too low the proportion of dominant individuals in the various generations.

IV. Race Improvement by Selection of Desirable or by Elimination of Undesirable Individuals.

On the hypotheses, which I understand Yule to adopt, of random mating and equal fertility on the part of all individuals, 300 dominant forms, of which 100 are pure and 200 hybrid, will produce more than 88 per cent of dominant individuals, instead of 83 per cent as estimated by Yule. For if we suppose each class to consist of males and females in equal numbers, the chances are just twice as great that an individual will mate with a hybrid dominant, $A\ (B)$, as that it will mate with a pure dominant, A. Or, to put the matter in another way, there are, for each individual of the entire 300, 50 possible A mates, and 100 possible $A\ (B)$ mates. This makes the entire number of different matings possible: —

	A.	$A\ (B)$.	B.
5,000 $A \times A$, yielding offspring	5,000		
20,000 $A \times A\ (B)$, yielding offspring . . .	10,000 +	10,000	
20,000 $A\ (B) \times A\ (B)$, yielding offspring .	5,000 +	10,000 +	5,000
Total	20,000 +	20,000 +	5,000

or $4\,A : 4\,A\ (B) : 1\,B$. It will be observed that $\frac{8}{9}$, or 88.8 per cent, of the offspring have the dominant form, being either A or $A\ (B)$ in character. Eliminating the one recessive individual, B, in each nine offspring, the parents of the next generation will consist of 4 As (pure dominants) and 4 $A\ (B)$s (hybrid dominants); that is, of equal numbers of individuals A and $A\ (B)$. The possible matings * in this case will be: —

* To simplify the calculation, it is well to remember that the numerical proportions of the various matings possible within a population are expressed by the square of that population. Knowing the nature and numerical proportions of the possible matings, one can quickly calculate the numerical proportions of the offspring. Thus, in a population consisting of equal numbers of individuals A and $A\ (B)$, the possible matings are expressed by the square of $A + A\ (B)$, or

	A.	$A(B)$.	B.
$2\,A \times A$, yielding offspring	2		
$4\,A \times A(B)$, yielding offspring	2	+ 2	
$2\,A(B) \times A(B)$, yielding offspring	$\frac{1}{2}$	+ 1	+ $\frac{1}{2}$
Total	$4\frac{1}{2}$	+ 3	+ $\frac{1}{2}$

or $9\,A : 6\,A(B) : 1\,B$. The offspring in this generation are $\frac{15}{16}$, or 93.7 per cent, of the dominant form. Calculating in a similar way for the next four generations, we find that the proportion of dominant individuals steadily increases. The complete series for generations 1–8 following the cross between a pure A and a pure B is shown in Table III.

TABLE III.

RESULTS OF SELECTION FOR THE DOMINANT CHARACTER A IN THE VARIOUS GENERATIONS FOLLOWING A CROSS BETWEEN A PURE A AND A PURE B.

Generation.	Parents.	Offspring.	Per cent A or $A(B)$.
1	$A + B$	$A(B)$	100
2	$A(B)$	$A + 2\,A(B) + B$	75
3	$A + 2\,A(B)$	$4\,A + 4\,A(B) + B$	88.8
4	$A + A(B)$	$9\,A + 6\,A(B) + B$	93.7
5	$3\,A + 2\,A(B)$	$16\,A + 8\,A(B) + B$	96
6	$2\,A + A(B)$	$25\,A + 10\,A(B) + B$	97.2
7	$5\,A + 2\,A(B)$	$36\,A + 12\,A(B) + B$	98
8	$3\,A + A(B)$	$49\,A + 14\,A(B) + B$	98.4

Inspection of the table will allow one to continue it to any desired extent.* Compare Diagram on p. 239, *D*.

$A^2 + 2\,A \cdot A(B) + \overline{A(B)}^2$. Treating the progeny of each mating as equal to four, we have

	A.	$A(B)$.	B.
1 mating $A \times A$, yielding offspring	4		
2 matings $A \times A(B)$, yielding offspring . . .	4	+ 4	
1 mating $A(B) \times A(B)$, yielding offspring .	1	+ 2	+ 1
Total offspring	$9\,A$	$+ 6\,A(B)$	$+ B$.

* The percentage of dominant forms in the various generations may be quickly calculated by observing that it equals the series

$$\frac{1}{1} + \frac{(2)^2 - 1}{(2)^2} + \frac{(3)^2 - 1}{(3)^2} + \frac{(4)^2 - 1}{(4)^2} + \frac{(5)^2 - 1}{(5)^2} \text{ etc.}$$

From the foregoing considerations, we see that it is entirely possible for a breeder, under the conditions stated, practically to eliminate an undesirable recessive character, in a very few generations, *merely by not breeding from individuals which manifest that character.* This accords with experience. There is, however, a much quicker and surer way of accomplishing the desired result, namely, by selection of *pure* dominants only for breeding purposes. If a dominant individual, when bred to a recessive mate, has produced among two or more offspring no recessive individual, it is probable that the dominant is *pure*, and if mated to a similar individual will produce no recessive offspring in subsequent generations.

By means of a few preliminary breeding tests of individual animals or plants the breeder is thus enabled to establish a race of pure dominants as early as the second generation following a cross with recessives. A race of recessives which will breed true, may of course be established at any time by mating two recessive individuals. If the Galton-Pearson law were correct, neither of these things would be possible.

Suppose that the breeder, as is often the case, does not care to take the trouble to establish a perfectly pure race, being anxious to market large numbers of individuals as soon as possible. By merely weeding out the undesirable recessive individuals his race will steadily improve, as indicated by Table III and the diagram on p. 239, *D*. In the second generation following a cross between pure dominant and pure recessive individuals it will, as we have seen, consist of 75 per cent dominant individuals; in the next generation it will consist of 88.8 per cent dominants, and so on.

If the breeder eliminates recessives but once, namely, in the second generation following the cross, the series will be as follows: —

1st generation	100	% dominants
2d generation	75	"
3d generation	88.8	"
4th generation	88.8	"

etc., *ad infinitum.* If he eliminates recessives twice only, namely, in the second and third generations, the race will thereafter continue to contain 93.7 per cent dominant individuals, as follows: —

1st generation	100	% dominants
2d generation	75	"
3d generation	88.8	"
4th generation	93.7	"
5th generation	93.7	"

etc., *ad infinitum*. Similarly, if recessives are eliminated three times only, the race will be stable at 96 per cent dominants; and if four times, at 97.1 per cent dominants. In general, *as soon as selection is arrested the race remains stable at the degree of purity then attained*, provided of course that one form is as fertile as the other, and subject to no greater mortality.

Such is the law governing the transmission of a dimorphic condition within a race, or, to give the matter a practical bearing, we may call it

TABLE IV.

RESULTS OF SELECTION FOR THE CHARACTER A IN THE VARIOUS GENERATIONS FOLLOWING A CROSS BETWEEN A PURE A AND A PURE B.

Generation.	When Dominance is Alternative between A and B.			When A is uniformly Dominant over B.
	Parents.	Offspring.	Per cent A or $A(B)$	Per cent A or $A(B)$.
1	$A + B$	$A(B) + B(A)$	50	100
2	$A(B)$	$A + A(B) + B(A) + B$	50	75
3	$A + A(B)$	$9A + 3A(B) + 3B(A) + B$	75	88.8
4	$3A + A(B)$	$49A + 7A(B) + 7B(A) + B$	87.5	93.7
5	$7A + A(B)$	$(15)^2 A + 15A(B) + 15B(A) + B$	93.7	96
6	$15A + A(B)$	$(31)^2 A + 31A(B) + 31B(A) + B$	96.8	97.2
7	$31A + A(B)$	$(63)^2 A + 63A(B) + 63B(A) + B$	98.4	98
8	$63A + A(B)$	$(127)^2 A + 127A(B) + 127B(A) + B$	99.2	98.4

the *law governing race improvement*, in cases of alternative inheritance, in which one of a pair of characters is uniformly dominant over the other. In cases in which dominance alternates between the two characters A and B (and such cases are probably commoner than is generally suspected) the process of race improvement by elimination of undesirable individuals progresses at first somewhat more slowly, but ultimately even more rapidly than in the case already discussed. A cross between A and B will, when dominance is alternative, yield offspring 50 per cent $A(B)$, 50 per cent $B(A)$. Selecting for A, that is, breeding only from $A(B)$s, the next generation will consist of equal numbers of forms

A, *A*(*B*), *B*(*A*), and *B* respectively, or once more 50 per cent individuals *A* in appearance. See Table IV. Selecting again for *A*, the parents for generation 3 will consist of equal numbers of individuals *A* and *A*(*B*) in character. Continuing the calculation in this way, we get the series of generations indicated in Table IV, and expressed graphically in the diagram on p. 239, *A*. For convenience in comparison, there are also given in the last column of Table IV the percentages of *A* and *A*(*B*) individuals to be expected when *A* is uniformly dominant over *B*. Compare Table III. Inspection of Table IV will allow one to continue it to any desired extent.

TABLE V.

Chances in 100 of isolating a Pure *A* by Random Selection from Individuals manifesting that Character in the various Generations following a Cross between a Pure *A* and a Pure *B*.

Generation.	When Dominance is Alternative between *A* and *B*.*	When *A* is uniformly Dominant over *B*.†
1	0	0
2	50	33.3
3	75	50
4	87.5	60
5	93.7	66.6
6	96.8	71.4
7	98.4	75
8	99.2	77.7

The same law governs arrest of selection in cases of alternative dominance, as in cases of uniform dominance of *A* over *B*. As soon as selection by elimination ceases, the race continues in the condition at that time attained, provided forms *A* and *B* are equally fertile and subject to the same mortality.

On the other hand, if the breeder has the patience to make individual breeding tests, and then to select for *pure* individuals on the basis

* These percentages equal the series 0, $\frac{1}{2}$, $\frac{3}{4}$, $\frac{7}{8}$, $\frac{15}{16}$, $\frac{31}{32}$, etc.

† These percentages equal the series 0, $\frac{1}{3}$, $\frac{2}{4}$, $\frac{3}{5}$, $\frac{4}{6}$, $\frac{5}{7}$, etc.

DIAGRAM SHOWING THE PROGRESS OF SELECTION IN CASES OF ALTERNATIVE INHERITANCE.

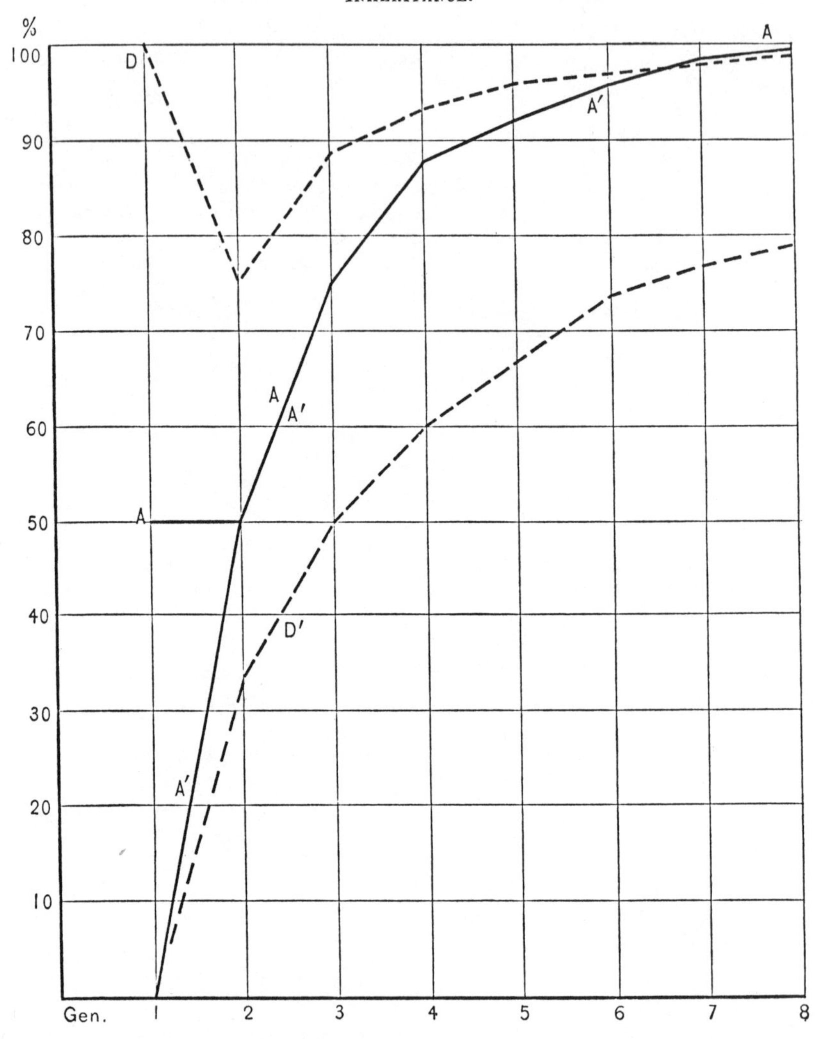

A, Rate of Race Improvement by Elimination, Dominance being Alternative.
D, Same, Dominance being Uniform.
A′, Chances in 100 of obtaining a Pure Individual, Dominance being Alternative.
D′, Same, Dominance being Uniform.

of the tests made, immediate success in obtaining a pure race is assured, whether dominance be alternative or not. But if he selects for pure individuals quite at random, without breeding tests, his chances of success are considerably greater in a case of alternative dominance than in a case of uniform dominance of one character over the other, as will be clear from an examination of Table V. In the second generation following a cross between a pure *A* and a pure *B*, dominance being alternative, the chances are even that any *A* individual selected at random will breed true; and when individuals possessing the character *B* have been eliminated for three successive generations, the chances become approximately 94 in 100 that any *A* individual selected at random will breed true; whereas, when *A* uniformly dominates over *B*, the chances, in the corresponding generations, of securing a pure *A* are only 1 in 3, and 2 in 3 respectively.

A graphic presentation of these facts is made in the diagram on p. 239, *A'*, *D'*.

Bibliography.

Bateson, W.

:02. Mendel's Principles of Heredity, a Defence. With a Translation of Mendel's Original Papers on Hybridisation. xiv + 212 pp. Cambridge. [England. Contains bibliography and portrait of Mendel.]

Bateson, W., and Saunders, E. R.

:02. Experimental Studies in the Physiology of Heredity. Reports to the Evolution Committee of the Royal Society. Report I. 160 pp. London.

Castle, W E., and Allen, G. M.

:03. The Heredity of Albinism. Proc. Amer. Acad. Arts and Sci., Vol. 38, pp. 603–622.

Correns, C.

:01. G. Mendel's Regeln über das Verhalten der Nachkommenschaft der Rassenbastarde. Ber. deutsch. bot. Gesellsch., Bd. 18, pp. 158–168.

Correns, C.

:01. Die Ergebnisse der neuesten Bastardforschungen für die Vererbungslehre. Ber. deutsch. bot. Gesellsch., Bd. 19, Generalversammlungsheft, pp. 72–94.

Correns, C.

:03. Ueber Bastardirungsversuche mit Mirabilis-Sippen. Erste Mittheilung. Ber. deutsch. bot. Gesellsch , Bd. 20, pp. 594–608.

Davenport, C. B.
:00. Review of von Guaita's Experiments in Breeding Mice. Biol. Bull., Vol. 2, pp. 121-128.

Galton, F.
'89. Natural Inheritance. ix. + 259 pp., 16 figs. Macmillan and Co., London and New York.

Galton, F.
'97. The Average Contribution of each Several Ancestor to the Total Heritage of the Offspring. Proc. Roy. Soc., London, Vol. 61, pp. 401-413.

Guaita, G. von.
'98. Versuche mit Kreuzungen von verschiedenen Rassen des Hausmaus. Ber. naturf. Gesellsch. zu Freiburg, Bd. 10, pp. 317-332.

Guaita, G. von.
:00. Zweite Mittheilung über Versuche mit Kreuzungen von verschiedenen Rassen des Hausmaus. Ber. naturf. Gesellsch. zu Freiburg, Bd. 11, pp. 131-138, 3 Taf.

Hurst, C. C.
:02. Mendel's Principles applied to Orchid Hybrids. Reprint from Jour. Roy. Hortic. Soc., Vol. 27, Pts. 2 and 3, 11 pp., 3 fig.

Hurst, C. C.
:03. Mendel's Principles applied to Wheat Hybrids. Reprint from Jour. Roy. Hortic. Soc., Vol. 27, Pt. 4, 18 pp. [Contains Spillman's (:02) account of his experiments in hybridization of wheats.]

Mendel, G.
'66. Versuche über Pflanzenhybriden. Verh. Naturf. Vereins in Brünn, Bd. 4, Abh., pp. 3-47. [Translation in Bateson, :02]

Mendel, G.
'70. Ueber einige aus künstlicher Befruchtung gewonnenen Hieracium-Bastarde. Verh. Naturf. Vereins in Brünn, Bd. 8, Abh., pp. 26-31. [Translation in Bateson, :02.]

Pearson, K.
'98. Mathematical Contributions to the Theory of Evolution. On the Law of Ancestral Heredity. Proc. Roy. Soc., London, Vol. 62, pp. 386-412.

Pearson, K.
:03. The Law of Ancestral Heredity. Biometrika, Vol. 2, Pt. 2, pp. 211-236.

Spillman, W. J.
:02. Quantitative Studies on the Transmission of Parental Characters to Hybrid Offspring. Bull. 115, Office of Exp. Sta., U. S. Dept. Agric., pp. 88-98.

Spillman, W. J.
:02[a]. Exceptions to Mendel's Law. Science, N. S., Vol. 16, pp. 794–796.

Tschermak, E.
:00. Ueber künstliche Kreuzung bei Pisum sativum. Zeitsch. f. landwirths. Versuchswesen in Oester., Jahrg. 3, pp. 465–555.

Tschermak, E.
:01. Weitere Beiträge über Verschiedenwerthigkeit der Merkmale bei Kreuzung von Erbsen und Bohnen. (Vorläufige Mittheilung.) Ber deutsch. bot. Gesellsch., Bd. 19, pp. 35–51.

Tschermak, E.
:01[a]. Ueber Zuchtung neuer Getreiderassen mittlest künstlicher Kreuzung. Zeitsch. f. landwirths. Versuchswesen in Oester., Jahrg. 4, pp. 1029–? .

Tschermak, E.
:02. Ueber die gesetzmässige Gestaltungsweise der Mischlinge. (Fortgesetzte Studie an Erbsen und Bohnen.) Sonderabdruck aus Zeitsch. f. landwirths. Versuchswesen in Oester., Jahrg. 5, 81 pp., 1 Taf.

Tschermak, E.
:02[a]. Der gegenwärtige Stand der Mendel'schen Lehre und die Arbeiten von W. Bateson. Sonderabdruck aus Zeitsch. f. landwirths. Versuchswesen in Oester., Jahrg. 3, 28 pp.

Vernon, H. M.
:03. Variation in Animals and Plants. ix. + 415 pp. Henry Holt and Co., N. Y.

Vries, H. de.
:00. Sur la loi de disjonction des hybrides. Compt. Rend., Paris, Tom. 130, pp. 835–847.

Vries, H. de.
:01–03. Die Mutationstheorie. Bd. 1, xii. + 648 pp. 181 Text fig., 8 Taf.; Bd. 2, xiv + 752 pp., 159 Text fig., 2 Taf. Veit & Co., Leipzig.

Webber, H. J.
:00. Xenia, or the Immediate Effect of Pollen, in Maize. Bull. No. 22, Division of Veg. Physiol. and Pathol., U. S. Dept. Agric., Washington, D. C., 44 pp., 4 pl.

Yule, G. U.
:02. Mendel's Laws and their Probable Relations to Intra-racial Heredity. The New Phytologist, Vol. 1, pp. 193–207 and 222–237.

6

Reprinted from the *Report 3rd Intern. Conf. Genetics* 140–142 (1906)

ON THE THEORY OF INHERITANCE OF QUANTITATIVE COMPOUND CHARACTERS ON THE BASIS OF MENDEL'S LAWS—A PRELIMINARY NOTE.

By G. Udny Yule, University College, London.

In his memoir of 1904 "On a generalised theory of Alternative Inheritance with especial reference to Mendel's laws" (Phil. Trans. Roy. Soc. A, vol. 203) Professor Pearson laid the foundation of the theory of inheritance of a quantitative character determined by n allelomorphic pairs, in a race of which the individuals mate at random. Distinguishing the three types of couplet that can occur as "protogenic," "heterogenic," and "allogenic," he discussed only the theory of inheritance of the number of pairs of the first or last type. Parents containing, say, m allogenic couplets will, he showed, give rise to offspring containing on the average only $\frac{1}{3}$ of m such couplets; that is to say, as the variability of the two successive generations is the same, the coefficient of correlation between parents and offspring is, for this character, $\frac{1}{3}$. The similar coefficients between grandparents and grandchildren, great-grandparents and great-grandchildren were found to be $\frac{1}{6}$, $\frac{1}{12}$ and so on, all these values being quite independent of the total number of couplets by which the character was determined. But the coefficients of correlation between parents and offspring that have been determined from actual data are for the most part greater than $\frac{1}{3}$, and moreover appear to exhibit significant differences as compared with one another. Professor Pearson concluded, accordingly, that the theory was "not sufficiently elastic to cover the observed facts" (p. 73); that "when we come to the actual numerical values for the coefficients of heredity deducible from such a theory of the pure gamete, they do not accord with observation. They diverge in two ways. First, they give a rigid value for these coefficients for all races and characters—a result not in reasonable accordance with observation. Secondly, they give values distinctly too small, as compared with the average values, or with the modal values of large series of population observations." (p. 85).

There does not appear to be any justification in the memoir, however, for the very wide statement in the second passage cited regarding "all races and characters." The only character there dealt with is the number of allogenic or protogenic couplets, and no reason is shown for supposing that this is typical of all characters. There did not appear to me, moreover, to be any obvious reason for making such a supposition, and I accordingly endeavoured to work out a slightly more general, though still quite limited case. Imagine a *length* to be made up of a number of distinct segments, the length of each of which is determined by an independent allelomorphic pair. Let each segment take the length a, b, or c, according as the corresponding protozygote, heterozygote, or allozygote is present; then the total length L is related to the number of proto-, hetero-, and allogenic pairs determining it, m_1, m_2, and m_3 by a relation of the form:

$$L = am_1 + bm_2 + cm_3 \qquad (1)$$

Using methods which are relatively much simpler than those employed by Professor Pearson, the value found for the coefficient of correlation between parent and offspring for such a character was:

$$R = \frac{(a - c)^2}{2(a - c)^2 + (a - 2b + c)^2} \qquad (2)$$

If, now, either $a = b$ or $b = c$, the case reduces to that of dominance, one of the homozygotes giving rise to the same somatic character as the heterozygote: this is virtually the case discussed by Pearson, and accordingly the value of R is the same as that found by him, viz. $\frac{1}{3}$. If, however, the heterozygote give rise in every case to a length exactly intermediate between those due to the respective homozygotes, we must have $b = (a + c)/2$, whence $R = \frac{1}{2}$. This is the greatest value that the above expression for R can attain, and consequently a character of the kind considered may exhibit coefficients of heredity lying anywhere between the limits $\frac{1}{3}$ and $\frac{1}{2}$, for random mating of the parents. With homogamy, higher values could, no doubt, be obtained. There is therefore no difficulty in accounting for a coefficient of 0·5 on the theory of segregation, but such a value probably indicates an absence of the somatic phenomenon of dominance. In the case of characters like stature, span, &c. in man this does not seem very improbable.

As regards the coefficients of correlation with the higher ancestry, the theory leads to results which are still rather limited, for the ratio of successive coefficients appears to be always $\frac{1}{2}$; *i.e.* in the case of dominance or Pearson's case we obtain his series $\frac{1}{3}$, $\frac{1}{6}$, $\frac{1}{12}$ &c., and in the case of perfect blending the series $\frac{1}{2}$, $\frac{1}{4}$, $\frac{1}{8}$ &c. This second series implies, it should be noted, a complete absence of "ancestral inheritance" in the proper sense of the term, the partial coefficients of correlation between the offspring and the higher ancestry being all zero.

A complete theory of heredity should take into account, besides germinal processes, the effect of the environment in modifying the soma obtained from any given type of germ-cell—an effect which is hardly likely to be negligible in the case of such a character as stature. This may be done without much difficulty for the limited case discussed. Let us suppose that the protozygote determines segments which have not all the same length *a*, but, owing to the varying effect of environment, a mean length *a* and a standard deviation *u*. Similarly, let the mean lengths and standard deviations of segments determined by the heterozygote and allozygote be *b* and *v*, *c* and *w* respectively. Then the value of R as given in equation (2) is reduced by the addition of a term

$$3u^2 + 4v^2 + 3w^2$$

to the denominator. The common ratio of the ancestral coefficients remains, however, unaltered at its former value of $\frac{1}{2}$. So far as the coefficients of correlation alone are concerned, it is accordingly impossible to distinguish between the effect of the heterozygote giving rise to forms that are not strictly intermediate, and the effect of the environment in causing somatic variations which are not heritable.

The case taken is a limited one, but the results are sufficient to show that the theory of the pure gamete, as applied to compound characters, is much more flexible than would appear from Professor Pearson's work, and can hardly be summarily dismissed as inapplicable to cases in which the coefficients of correlation approximate to 0·5.

7

Reprinted from *Science* **28**:49-50 (1908)

MENDELIAN PROPORTIONS IN A MIXED POPULATION

G. H. Hardy

To THE EDITOR OF SCIENCE: I am reluctant to intrude in a discussion concerning matters of which I have no expert knowledge, and I should have expected the very simple point which I wish to make to have been familiar to biologists. However, some remarks of Mr. Udny Yule, to which Mr. R. C. Punnett has called my attention, suggest that it may still be worth making.

In the *Proceedings of the Royal Society of Medicine* (Vol. I., p. 165) Mr. Yule is reported to have suggested, as a criticism of the Mendelian position, that if brachydactyly is dominant "in the course of time one would expect, in the absence of counteracting factors, to get three brachydactylous persons to one normal."

It is not difficult to prove, however, that such an expectation would be quite groundless. Suppose that *Aa* is a pair of Mendelian characters, *A* being dominant, and that in any given generation the numbers of pure dominants (*AA*), heterozygotes (*Aa*), and pure recessives (*aa*) are as $p:2q:r$. Finally, suppose that the numbers are fairly large, so that the mating may be regarded as random, that the sexes are evenly distributed among the three varieties, and that all are equally fertile. A little mathematics of the multiplication-table type is enough to show that in the next generation the numbers will be as

$$(p+q)^2 : 2(p+q)(q+r) : (q+r)^2,$$

or as $p_1 : 2q_1 : r_1$, say.

The interesting question is—in what circumstances will this distribution be the same as that in the generation before? It is easy to see that the condition for this is $q^2 = pr$. And since $q_1^2 = p_1 r_1$, whatever the values of p, q and r may be, the distribution will in any case continue unchanged after the second generation.

Suppose, to take a definite instance, that *A* is brachydactyly, and that we start from a population of pure brachydactylous and pure normal persons, say in the ratio of 1:10,000. Then $p=1$, $q=0$, $r=10{,}000$ and $p_1=1$, $q_1=10{,}000$, $r_1=100{,}000{,}000$. If brachydactyly is dominant, the proportion of brachydactylous persons in the second generation is 20,001:100,020,001, or practically 2:10,000, twice that in the first generation; and this proportion will afterwards have no tendency whatever to increase. If, on the other hand, brachydactyly were recessive, the proportion in the second generation would be 1:100,020,001, or practically 1:100,000,000, and this proportion would afterwards have no tendency to decrease.

In a word, there is not the slightest foundation for the idea that a dominant character should show a tendency to spread over a whole population, or that a recessive should tend to die out.

I ought perhaps to add a few words on the effect of the small deviations from the theoretical proportions which will, of course, occur in every generation. Such a distribution as $p_1:2q_1:r_1$, which satisfies the condition $q_1^2=p_1r_1$, we may call a *stable* distribution. In actual fact we shall obtain in the second generation not $p_1:2q_1:r_1$ but a slightly different distribution $p_1':2q_1':r_1'$, which is not "stable." This should, according to theory, give us in the third generation a "stable" distribution $p_2:2q_2:r_2$, also differing slightly from $p_1:2q_1:r_1$; and so on. The sense in which the distribution $p_1:2q_1:r_1$ is "stable" is this, that if we allow for the effect of casual deviations in any subsequent generation, we should, according to theory, obtain at the next generation a new "stable" distribution differing but slightly from the original distribution.

I have, of course, considered only the very simplest hypotheses possible. Hypotheses other that that of purely random mating will give different results, and, of course, if, as appears to be the case sometimes, the character is not independent of that of sex, or has an influence on fertility, the whole question may be greatly complicated. But such complications seem to be irrelevant to the simple issue raised by Mr. Yule's remarks.

G. H. Hardy

Trinity College, Cambridge,
April 5, 1908

P. S. I understand from Mr. Punnett that he has submitted the substance of what I have said above to Mr. Yule, and that the latter would accept it as a satisfactory answer to the difficulty that he raised. The "stability" of the particular ratio 1:2:1 is recognized by Professor Karl Pearson (*Phil. Trans. Roy. Soc.* (A), vol. 203, p. 60).

8

ON THE DEMONSTRATION OF INHERITANCE IN MAN

W. Weinberg

This article was translated expressly for this Benchmark volume by Roy A. Jameson, from "Über den Nachweis der Vererbung beim Menschen," Jahresch. Ver. vaterl. Naturkd. Wuerttemb. Stuttgart *64:368–382 (1908)*

By the term inheritance we understand the fact that through fertilization of the egg by the sperm the developing individual acquires the species and individual characteristics of its parents. The most essential part in this process is attributed to the nucleus, especially the chromosomes of the germ cells. This view, however, has recently again come under strenuous attack. The mature sex cells divide twice before fertilization in the so called reduction division. According to a widespread view, one part of the inheritance derived from each parent is extracted in this process, which is of the utmost importance for the evaluation of both the relationship of the individual

[*Editor's Note:* This translation by Roy A. Jameson places particular emphasis on the science and language of the time the paper was prepared and published.]

to his ancestors and also the laws of inheritance in humans. If we construct an individual's genealogical table - that is, a schematic synopsis of his ancestors - then we will have obtained only a synopsis of those persons who could have influenced the individual's characteristics by means of the germ plasm. However, of these theoretical possibilities only a few fall into consideration, for a continuity of germ plasm does not exist for all ancestors with regard to all characteristics. Most of the ancestors are eliminated in the competition for the determination of the individual's characteristics. We do not know how many ancestors determine the individual with regard to a particular characteristic; we can only say that it must be at least two, one on the paternal, and one on the maternal side. The more ancestors are actually considered, the more gradations or variations of a character will result through the genetic process. From the meaning of the reduction division itself it follows further that a connection with a particular ancestor becomes the more improbable the further the degree of relatedness itself is, for with each degree of removal the number of ancestors doubles, and thus each time the possibility of having influenced the germ plasm halves. Particularly in the case of inheritance under Mendelian Rules it appears as if each characteristic were only derived from two ancestors, which would represent a most sharp selection.

These are the essential viewpoints which we must derive from the results of biological research if we want to investigate and judge the facts of human genetics. By this we have already indicated that the limits of human genetic research are considerably narrower than those of general biology. Essentially human genetics can be concerned only with establishing which cases involve actual inheritance, the measure of influence heredity has with respect to other factors, which ones influence a particular phenomenon, and which special genetic laws must be considered for an individual characteristic. Only general biology itself can provide a cytological understanding of the nature of inheritance; man in particular is no profitable genetic subject since it is not possible to instigate well conceived breeding experiments as is possible with the lower plants and animals and which have lead to such fortuitous results as the discovery of the Mendelian laws of inheritance. With humans we can only evaluate after the fact the experiments which life has thoughtlessly and frequently created without deliberation. For the study of human genetics the experiment must necessarily be replaced by the statistical evaluation of mass phenomena. This leads, however, to uncertainty in the evaluation of the results as to whether a positive or negative selection has taken place, and whether partial inbreeding or choiceless mixing - panmixis - has occurred and to what degree. The consequences of strict inbreeding - that is, sibling mating - under which the classical Mendelian experiments proceeded can not be determined for man and the relatively rare mating of more distant relatives can offer no complete substitute.

Yet these factors contribute only part of the difference between human research and research in plants and animals.

Essential differences lie more in the securing of research material and in the methods of its analysis. With plants and animals it is possible to personally oversee the results of an experiment through several generations. With humans one and the same observer knows at the most only fragments of the history of two generations of a family except for such characteristics as color, which are already established at birth. Even pathological manifestations as well as many normal ones, can be observed only at an age in which the individual lives far from his parents. Many familial traits may therefore be known only through tradition - moreover, a tradition frequently incomplete and which even in the best of cases (which one can not always presuppose) may be deceptive. How many humans do not know the number of births of their mothers, the causes of death of their grandparents, or even completely and correctly their grandparents names?

Thus an entire group of questions can be finally solved only through the help of documentary material whose creation is essentially a matter of the future. One, whose proponents I wish to represent by Goehlert and Ottokar Lorenz, attempt to analyze the history of prominant families, namely those of governing houses and noble families. But these kinds of investigations provide neither a sufficiently large sample, nor assure a uniform reliability and completeness of data. Furthermore, such families represent the product of a selection whose analysis can never offer a picture of average situations.

The other view, in which primarily Ammon and Riffel have been active, attempts to ascertain the anthropological and pathological relations of the total populace of whole regions and communities over long periods. This is also the method I have adopted in my investigations. I was fortunate in being spared the great effort of family compilation from the church and state registers, for the Wurtemberg Family Register provided on one and the same page the demographic history of a household and its relation to ancestors and descendants. I had only to enter the causes of death in the material at my disposal.

Although the unreliability of material obtainable from studies on man of a defect which can be corrected or sometimes even removed by the choice of a suitable method, there remains considerable difficulties. In human studies there arise essential differences in the methods of obtaining and analyzing data such that it is generally impossible to follow the descendants of the personally observed individuals for lengthy periods or even over several generations. In contrast to the investigations which follow the descendants of crosses in plants and animals, one is therefore commonly and well advised to ascertain the ratios of ascendance and collateral relatives. In so far as this entails countless determinations the results must suffer not unimportant displacements, as I shall later show.

The numerous efforts which may without delay explain the decade long stand still in genetics are situated in the domain of method.

One of these errors lay in following the relations of the individual only insofar as pathological conditions could be ascertained, and in the fact that the thus obtained family trees generally were very incomplete and presented an extremely one sided picture. Ottakar Lorenz has correctly pointed out that such family trees are worthless, and pointed to the differences between family trees and genealogical tables. But instead of recommending evaluation of correct, i.e., complete family trees and pedigrees, he strongly preferred the genealogical table and expected from it in particular a reduction of exaggerated views on the importance of pathological inheritance. A comparison with the methods of biological research in plants and animals could have taught him that these methods are successful in examining the results of certain crossings in the descendants. His view that extensive genealogical tables, give a more correct picture of the influence of heredity and that they are minimal for a study of the metnally ill, lies both then in misrecognition of the influence of the reduction division and the partially related variable significance of different degrees of relationship of the inheritance of the individual, but even more in the misrecognition of the most favorable selection which accompanies marriage and which particularly in the case of the mentally ill leads to a high degree of exclusion from the opportunity and possibility of bearing offspring. Thus one seldom finds idiots as children of idiots. For this reason a properly assembled pedigree is to be preferred over the genealogical table because it offers a greater security against one sided selection.

The failure of research on pathological inheritance actually lay in the one sided casuistics and the false statistical views which gradually replaced it without being able to totally suppress it. It was not sufficient that it was recognized that the negative case has the same value as the positive. For in place of the enumeration of typical cases the effort was then made to establish in every case by the greatest possible extension of the investigation of relatives the presence of the trait and thus the greatest possible percentage of the trait. This was also the reason why Riffel's investigation on the inheritance of tuberculosis received such scathing critique from the bacteriological side. This very error, the misrecognition of the varied importance of the different degrees of relatedness of the inheritable mass of an individual, also led Lorenz and Riffel to entirely opposed and false views on the significance of inheritance. Only the comparative method which established the presence of the pathological trait in healthy and sick made it possible to achieve a measure of the influence of heredity. However, the first attempts in this direction by Koller on the mentally ill and by von Kuthry on tuberculious patients do not pay enough attention to the influence of age and the external conditions of life.

The demonstration of a familialy increased trait is not simply and without further investigation identical with inheritance in the genetic, cell-developmental sense. The demonstration may also rely upon the sharing of the outer circumstances of life and daily habits. Thus the family history of diabetics and those suffering from gout (arthritis) does not implicitly demonstrate that these diseases have a genetic foundation. The influence of the external conditions of life on the mentally ill has been too little studied, although outer conditions are probably over valued in the case of tuberculosis. For all these diseases a genetic influence can be recognized only insofar as it involves a hereditary factor whose character is not yet known, and also only insofar as more than a familial burden remains even after excluding the influence of age and external conditions. In order to make this possible for tuberculosis patients I have compared the incidence of tuberculosis in affected persons with those of their spouses. I found familial evidence of tuberculosis among affected individuals to be 50% higher than in their spouses. Furthermore, familial evidence was 100% higher for affected individuals within the group of well off persons.

A certain inbreeding among tuberculosis patients is a necessary result of a definite, if not strong, effort on the part of the healthy to avoid marriage with affected individuals. Thus my figures representing the relative familial burden of tuberculosis may be too low, and for the same reason the incidence of mortality from phthisis among spouses of tuberculosis patients, which I also determined, may be too high. I hope that my remarks have now provided a picture of the difficulties encountered in determining and evaluating the influence of pathological inheritance in man.

Those characteristics whose inheritance is or seem more or less linked to a certain sex are of particular interest for the theory of heredity. Colorblindness and hemophilia belong in this group. According to the literature to date, both diseases are many times more common among men than women, who play an essential role in the hereditary transmission from grandfather to grandson. But recently some doubts have arisen, at least with regard to color blindness, whether it is not significantly more common among women than has been previously believed. It is now particularly recommended that school physicians look for it in their examinations. And one should not exclude the possibility that colorblindness among men is more frequently discovered precisely because it disturbs them in their occupations. Such demonstration would complicate the interpretation of colorblindness to the extent that in this case the question would no longer solely concern the carrier. Hemophilia is also not entirely limited to the male sex. And even here it might be asked whether men - in consequence of their professions and different modes of life - more frequently give the disease opportunity to become manifest.

In contrast to these diseases, the ability to bear twins from two ova represents a trait in which man plays only the role of agent of transmission. According to my investigations (begun over 7 years ago), it is clear that this characteristic is inheritable, just as Darwin assumed, although only on the basis of casuistic interpretation. For at least a portion of the mothers of non identical twins this trait is linked to a particular construction of the ovaries. Consequently the abundance of eggs present in childhood in all females persists in some adult women and allows a more frequent release of eggs from the ovaries. In contrast the majority of adult women have ovaries which are relatively poor in eggs. The ovaries of twin bearing mothers thus resemble those of multiparous animals and so the birth of twins represents an atavism not only physiologically, but also from the standpoint of comparative anatomy. We can see from this, as I stressed then, that no principle difference exists between inheritance and atavism. The fact that children of different sex occur only given several eggs, and the justified assumption that the frequency of unlike sex twins is almost exactly 50% of the total number of dizygotic twins made it possible for me to determine a series of characteristics of dizygotic twins and their mothers from a large mass of population statistics and from specially collected family registers. From this the fact emerges that unlike sex pairs and therefore dizygotic twins occur in very different frequencies among the various peoples of Europe when compared with the total births. Nations of Germanic origin are distinguished in particular by a high frequency of dizygotic twins, while those are relatively rare among Romanic peoples. For this reason it is probably not justified to consider the occurrence of twins in a family as a sign of degeneration as Rosenfield in Vienna tried to do. Such an interpretation shows only once again how thoughtless one can be with this concept. It is a fact that dizygotic twins represent both an inheritable variation and a racial one. The thought thus immediately arises that some racial characteristics are present which are closely linked to a Mendelian form of inheritance. Following a suggestion of Professor Haecker's I investigated my previously collected material for evidence supporting Mendelian inheritance in twins. However, it would have been very difficult to assemble a sufficient number of cases in which children of mothers of twins had married one another and then determined the frequency of twin births among their children. I therefore attempted to construct a formula for the frequency of dominant and recessive traits among the mothers and siblings and daughters of carriers of traits of the same character under the assumption that absolute panmixia is present.

Before I pursue the question of Mendelian inheritance further, I must briefly explain how I was able to demonstrate the fact of inheritance among twins. The proof was a double one.

First I showed that the ability to bear twins is not shared equally by all women. This I was able to demonstrate by investigating the frequency of twins among later or earlier births of mothers of twins. Among dizygotic twins I found a frequency of 1:30; that is, on the average one would have to investigate 30 more births before one would find a further single or dizygotic twin birth. However, this seemingly weak correlation appears significant if one considers that in Würtemberg --- the figures for repeated multiple births rose to 1:18, 1:13, and 1:3, respectively.

Secondly, I was able to demonstrate that there is a direct correlation between the frequency which which multiple births are repeated by a mother, and the frequency of multiple births among mother, sisters, and daughters. Among these individuals the frequency of repeated multiple births rose two times, and for triplets the ratio 1:9 is reached.

From this relatively low figure for the repetition of multiple births (I may perhaps have the opportunity to return to consideration of its cause) I concluded that since the average number of births to a mother of twins reaches only 4 or 5, numerous women with a tendency to twin births do not manifest this trait simply because they have not performed the experiment often enough.

I further showed that among the mothers, sisters, and daughters of twins, triplets, etc., multiple births are significantly more frequent than they are among the total number of births. But here it was not simply the case that the daughters of a woman who in 5 births had once borne twins bore on the average one set of twins in 10 children (because the intensity of the trait among the children represented the mean of that found in their parents). Rather the ratio was such that it could be assumed that the mothers of twins had transmitted their ability with that frequency with which, on the average, she repeated bearing twins, that is, the frequency of 1:30, while the father inherited, on the average, the frequency of 1:90 in the Stuttgart cases. Thus through mixing the actual frequency with which twins were repeated among mothers, sisters, and daughters of twins was 1:30 minus 1:90 equals 1:45. Seven years ago I, like many others, knew nothing of Mendelian genetics. I was then inclined to take these figures as proof that we were dealing with a simple mixture of traits. But I had thereby overseen the fact that the ratios among the relatives of the mothers of twins, quadruplets, and quintuplets did not agree with this --- that is, I had attempted to explain this by the paucity of data. For these one does not obtain among mothers and sisters the simple average of the numbers of repeators of multiple births and the general figures, but rather considerably smaller values, which furthermore are significantly higher among the sisters than among the mothers.

I then asked myself whether these notable results among the triplets could not perhaps be related to the action of Mendel's Rule and reached the conclusion that this is actually the case. However in the presentation of the train of thought which led me to this I must return to the onsetting remarks of my lecture.

As far as my investigations could thus far establish --- and I hope to be able to expand these considerably in the near future --- neither a conscious nor unconscious selection with reference to the hereditary factors for the trait of bearing twins takes place. Theoretical grounds also support the existence of a rather extensive panmixia with regard to this character.

Thus I was confronted with the question: How does the numerical influence of Mendelian Inheritance behave under the influence of panmixia? The typical Mendelian Rule represents only the separation of the inherited factors in the germ cells under the influence of the most absolute inbreeding, which does not occur in man.

If one proceeds with the exclusive crossing of pure types and hybrids for several generations and one counts the hybrids AB among the dominant type AA, while the recessive is signified by BB, then the relative frequency after the first cross in the n the generation is:

$$A=2^{n-1}+1$$

$$B=2^{n-1}-1.$$

The difference then gives the relative numbers of hybrids, which in each case is 2 and is thus relatively less frequent with each generation. One then obtains a different percentage in each generation for A and B.

This situation appears much different when Mendelian inheritance is viewed under the influence of panmixia. I start by assuming that m male and female pure representatives of Type A and likewise n representatives of Type B are present. If these are randomly crossed then, by application of the symbolism of the binomial theorem, the composition of the daughter generation is obtained:

$$(mAA + nBB)^2 = \frac{m^2}{(m+n)^2}AA + \frac{2mn}{(m+n)^2}AB + \frac{n^2B^2}{(m+n)^2}$$

or, if $m + n = 1$

$$m^2AA + 2mnAB + n^2BB$$

If the male and female members of the first generation are randomly crossed, then the following frequencies of different combinations of crosses are obtained:

$$m^2m^2 \ (AA \times AA) = m^4AA$$

$$4m^2mn \ (AA \times AB) = 2m^3nAA + 2m^3nAB$$

$$m^2n^2 \ (AA \times BB) = 2m^2n^2AB$$

$$4(mn)^2 \ (AB \times AB) = m^2n^2AA + 2m^2n^2AB + m^2n^2BB$$

$$4mnn^2 \ (AB \times BB) = 2mn^3AB + 2mn^2BB$$

$$n^2n^2 \ (BB \times BB) = n^4BB$$

or the relative frequencies are:

$$AA : m^2(m+n)^2$$

$$AB : 2m(m+n)^2n$$

$$BB : (m+n)^2n^2$$

and the composition of the second or daughter generation is again:

$$m^2AA + 2mnAB + n^2BB$$

We thus obtain the same distribution of pure types and hybrids for each generation under the influence of panmixia and thereby the possibility of calculating for each generation how these types are represented in panmixia and Mendelian inheritance among the parents, sibling, and children of the various types and hybrids.

If the original distribution of both types and hybrids is represented by

$$m^2AA + 2mnAB + n^2BB$$

and among the relatives the representatives of the dominant types and hybrids are compiled together, in so far as one signifies them with the same letters (A or B), then when A is dominant the frequency of A and B is:

Among the parents of A: $(1+mn)A:n^2B$

Among the siblings of A: $[4\ (1+mn) + mn^2]A:n^2(3+4n)B$

Among the children of A: $(1 + mn)A:n^2B$

But if A is recessive, then one obtains:

For the parents of A: $mA:nB$

For the children of A: $mA:nB$

For the siblings of A: $(2+m)^2A:n(3+4m)B$

For characteristics which are measurable, as in our case, this must lead to different average values for parents and sibling. Assuming a non-Mendelian character, in which the hybrids are represented by several intermediate forms, the average representation of measurable characteristics would be the same in parents and siblings. Thus a real difference exists between Mendelian and non-Mendelian characters. I have found that the trait for triplets results in essentially different values for parents and sibling, which supports a Mendelian trait. Similar small differences occur with twin births.

It may be shown that in the case of dominance by A, type A is always represented in at least half of the parents.

When A is recessive one obtains limiting values for frequencies of A of 1:0 and 0:1 for the parents and 1:0 and 1:3 for the sibling. We can thus quickly note that rare recessive characters can be detected easier in sibling than in parents. The possibility of calculating the expected figures for Mendelian inheritance (not only in the case of absolute inbreeding and in the descendants, but also in panmixia, ascendance, and collateral relatives) allowed me to utilize not only the few cases in which children of mothers of twins married each other but also my entire set of previous data.

There remains only the determination of the value of m. For this the following consideration is necessary. If a dizygotic twin birth occurs once in 35 births among mothers with a tendency towards twin births, but only once in 140 among all mothers, as in Stuttgart, then the former comprise only a fourth of all mothers.

If their frequency is set as $m^2 = 1/4$, then that of the remaining women is $2mn+n^2 = 3/4$, and we obtain $m=n=\frac{1}{2}$. In the case of dominance of the tendency to twin births we would obtain the ratio $m:n = 1:6.5$, if $m^2+2mn:n^2 = 1:3$.

Further, if one triplet birth occurs in approximately 6000 births and in mothers displaying a tendancy towards triplets a triplet occurs once in every 200 single births then the value of

$m = 1/5$ in the case of recessivity and
$n = 1/60$ in the case of dominance.

In each of these cases, however, a twin birth would occur in 84 births to Wurttemberg women lacking the tendancy towards triplets.

If the value for m thus generated is replaced in the above formulas for the hereditary trait through parents, sibling and children, one obtains, from the comparison of the calculated probable number with the actual ratios, the assumption which applies most nearly to the mode of inheritance.

According to whether the trait for multiple births is (I) recessive, (II) dominant, or (III) equivalent with respect to single births, the following expected values are obtained as the frequency of multiple births in the kinships of mothers of twins and triplets in Stuttgart and Wurtemberg:

	I	II	III
(a) among the mothers of mothers of twins	1/52	1/46	1/45
(b) among the daughters of mothers of twins	1/52	1/46.8	1/45
(c) among the sisters of mothers of twins	1/49	1/46.6	1/45
(d) among the mothers of mothers of triplets	1/52	1/29	1/29
(e) among the sisters of mothers of triplets	1/37	1/29	1/29

The observed births in the individual groups were:

(a) 1365
(b) 1464
(c) 1022
(d) 2637
(e) 1666

Thus the absolute number of expected multiple births is:

In group	By Assumption I	II	III	Actual observed
(a)	26	27	30	27
(b)	28	31	33	24
(c)	21	22	23	27
(d)	51	91	91	45
(e)	45	57	57	36
Total	171	228	232	159

The assumption that the tendency for twin births is recessive gives expectable values which are much closer than those expected under any other assumption. The difference in 12 cases lies within the mean error, which is nearly $\sqrt{171} = 13$.

The results found in the inheritance of multiple births therefore are best explained by the assumption that the tendancy to multiple births is inherited according to Mendel's law and is recessive.

This investigation (whose more exhaustive presentation based on a new collection of material must follow elsewhere) may show that one can penetrate into the essence of human heredity by suitable alteration of the method of investigation.

9

Reprinted from *Roy. Soc. (Lond.) Proc.* **81**:219–224 (1909)

The Theory of Ancestral Contributions in Heredity.

By KARL PEARSON, F.R.S.

(Received March 19,—Read April 22, 1909.)

Under the above title a paper has recently appeared by Mr. A. D. Darbishire in the 'Roy. Soc. Proc.,' vol. 81, B, p. 61 *et seq.*, giving further experimental evidence with regard to the inheritance of certain characters in peas. The paper is an interesting one, but the method adopted is not, I venture to think, capable of answering the problem which the author set himself. It has been supposed by some Mendelians that the theory of inheritance summed up in the "law of ancestral heredity" was in some way invalidated by investigations such as Mr. Darbishire's, and that opinion consciously or unconsciously seems to be expressed in the paper just referred to. The law of ancestral heredity is embraced in the following statements:—

(i) In a population breeding without assortative mating the regression line for offspring on any ancestor is linear.

(ii) The correlations between offspring and the successive grades of ancestry form a progression diminishing geometrically as we ascend to distant grades; and

(iii) The general relation of an individual to his ancestry can be closely expressed by the multiple correlation formula.

In a memoir published in the 'Phil. Trans.,' vol. 203, pp. 53–86, I showed that these principles held for material obeying Mendel's laws—in particular (i) and (ii) hold for the simple case of alternative characters such as are said to occur in the case of peas.

The only instance that I am aware of in which ancestry does not matter is that in which the geometrical progression is of the form:

$$\rho, \rho^2, \rho^3, \ldots.$$

I treated this case at length in the 'Phil. Trans.,' vol. 187, A, pp. 304–6 (1896), remarking that the grandparents were quite indifferent, when the parents had been selected. Unfortunately, this is not true when the correlation coefficients are

$$\frac{1}{3}, \frac{1}{2}\times\frac{1}{3}, \frac{1}{2^2}\times\frac{1}{3}, \frac{1}{2^3}\times\frac{1}{3}, \text{ etc.,}$$

as is the case with the somatic correlations on the Mendelian theory. In other words, ancestry does matter in the latter theory. What is the explanation, therefore, of the apparent contradiction between such experiments as those of Mr. Darbishire and the theoretical development of the Mendelism which they profess to establish ?

It does not seem hard to account for the divergence. Experiments such as those of Mr. Darbishire do not deal with a population as a whole, and consider the contributions to the next generation of all its components supposed to be mated at random. I feel quite certain that if Mr. Darbishire makes the requisite crosses in due proportions, and does not weight with differential fertility, he will find that ancestry does matter. That it does matter is just as good a proof of Mendelism as Mr. Darbishire's proof in the simpler case that it has not any effect. If he fails to find its influence, then he will have refuted Mendelian theory.

To illustrate my point, take a population distribution which would follow from crossing two pure races with respectively dominant and recessive characters represented by the letters D and R. Suppose the hybrids to cross at random, then the population will remain absolutely stable with the permanent formula

$$(DD)+2(DR)+(RR).$$

Now suppose this to cross with itself or with

$$(DD)+2(DR)+(RR).$$

Table I gives the scheme of offspring with their parents. This population of 16 individuals of 6 different types of parentage now crosses with itself. The result is a population of 256 individuals showing 15 types of grand-parentage. This is exhibited in Table II. If Mr. Darbishire's principle that ancestry is of no importance were correct, then the differences in type of these grandparents would not be of any significance.

Table I.

Parents.	Offspring.		
	DD.	DR.	RR.
DD, DD	1	—	—
DD, DR	2	2	—
DD, RR	—	2	—
DR, DR	1	2	1
DR, RR	—	2	2
RR, RR	—	—	1

Table II.

Grandparents.					Offspring.		
DD.	DR.	RR.	+	−	DD.	DR.	RR.
4	—	—	4	—	1	—	—
3	1	—	4	—	6	2	—
3	—	1	3	1	2	2	—
2	2	—	4	—	13	10	1
2	1	1	3	1	8	14	2
2	—	2	2	2	1	4	1
1	3	—	4	—	12	16	4
1	2	1	3	1	10	28	10
1	1	2	2	2	2	14	8
1	—	3	1	3	—	2	2
—	4	—	4	—	4	8	4
—	3	1	3	1	4	16	12
—	2	2	2	2	1	10	13
—	1	3	1	3	—	2	6
—	—	4	—	4	—	—	1

Now Table II may be examined from several standpoints. We may first consider the gametic constitutions of the grandparents and of the offspring. Thus we have:—

No. of DD's in grand-parentage.	Percentage of DD's in offspring.
4	100
3	67
2	41
1	22
0	11

In other words, the constitution of the grandparentage substantially modifies the offspring; there exists in this sense an "ancestral contribution" to the heritage.

Of course the gametic constitution RR follows precisely the same system of percentages, and is again influenced by ancestry.

If, however, we take the gametic constitution DR in such a population we find—

No. of DR's in grand-parentage.	Percentage of DR's in offspring.
4	50
3	50
2	50
1	50
0	50

At first sight this seems to indicate that for this case there is no ancestral influence, where we should expect by increasing the number of DR's in the grandparentage to increase the number in the offspring. But this criticism is not valid, for, in the population we are dealing with, it is clear that DR is the modal or mean group, and that, accordingly, it is perfectly neutral in determining the regression or correlation of the *gametic* character. In other words, the deviations of the DR ancestry from the mean population gametic character are all zero and accordingly they have no weight in causing the offspring to deviate from the population norm. They have, in fact, no more effect on the offspring than, in the case of stature, a number of mediocre ancestors have in raising or lowering the average deviation of the offspring from the general population mean.

Lastly, turning from the gametic constitution to the somatic character, I have represented in the fourth and fifth columns of Table II the extent to which the dominant character is present in the ancestry, and in the accompanying table one sees the effect on the offspring:—

No. of grandparents with dominant character.	Percentage of offspring with dominant character.
4	89
3	78
2	59
1	33
0	0

It will thus be obvious that, judging solely by the patent, that is the somatic character of the grandparentage, there is a very marked influence of the ancestry on the heritage; that, if we select ancestry by somatic character only, we shall expect an influence on the offspring varying from 0 to 90 per cent. in intensity, according to the nature of the selection.

I think, therefore, that to deny the influence of ancestry—at any rate that influence in the sense in which the biometrician uses the term—is to deny the application of Mendelism to populations mating at random.

If we start with a population in which the proportions of DD's, DR's, and RR's are not those of a simple hybridisation, but given by

$$p\,(\mathrm{DD}) + 2q\,(\mathrm{DR}) + s\,(\mathrm{RR}),$$

then after the first generation of random mating the population will be

$$(p+2q+s)^2\,(p+q)^2\,(\mathrm{DD}) + 2\,(p+q)\,(s+q)\,(p+2q+s)^2\,(\mathrm{DR}) + (p+2q+s)^2\,(s+q)^2\,(\mathrm{RR}),$$

or its constituents will be proportional to

$$(p+q)^2(\mathrm{DD})+2(p+q)(s+q)(\mathrm{DR})+(s+q)^2(\mathrm{RR}),$$

and this ratio is maintained ever afterwards.*

A little consideration will show that our Table II is obtained by a symbolic process which will not be affected if we replace D by $(p+q)$ D and R by $(s+q)$ R, so that to exhibit the results for a Mendelian population of any constituent proportions we have only to multiply all the numbers in any row of offspring of Table II by $(p+q)^2$ for a DD grandparent, by $(p+q)(s+q)$ for a DR grandparent, and by $(s+q)^2$ for an RR grandparent, starting with the stable population which arises after the first random mating. We then reach the following table for the case of classification by somatic characters, where for brevity I write: $p+q=\pi$, $s+q=\kappa$, and $\pi^2/\kappa^2=n=$ ratio of pure dominants to recessives in the stable population. It will be seen that whatever be the proportions of the Mendelian components in the original population, then a selection of grandparents influences widely the somatic characters of the offspring.

Whether, therefore, Mendelism be or be not the final word as to inheritance (and I personally, especially in the case of human characters, must continue to suspend my judgment), it is clear that ancestral influence cannot be denied in the case of any population mating at random and inheriting on Mendelian lines.

Table III.

No. of grandparents with dominant character.	Percentage of offspring with dominant character.
4	$100\,\frac{(n+1)(n+3)}{(n+2)^2}$
3	$100\,\frac{(n+1)(2n+5)}{2(n+2)^2}$
2	$100\,\frac{(5n+11)(n+1)}{6(n+2)^2}$
1	$100\,\frac{(n+1)}{2(n+2)}$
0	0

* The stability after the first generation is very obvious, but, as far as I know, was first stated in print by G. H. Hardy, 'Science,' vol. 28, p. 49.

We have the following table for various values of n:—

No. of grandparents with dominant character.	Percentages of offspring with dominant character.						
	$n = 10$.	$= 4$.	$= 2$.	$= 1$.	$= \frac{1}{2}$.	$= \frac{1}{4}$.	$= \frac{1}{10}$.
4	99·3	97	94	89	84	80	77
3	95	90	84	78	72	68	65
2	78	72	66	59	54	50	48
1	46	42	37·5	33	30	28	26
0	0	0	0	0	0	0	0

When experimental work is adduced to demonstrate that ancestry has no influence, it will on investigation be found that the writer is:

(i) Confining his attention, as Mr. Darbishire, to isolated lines of inheritance, with restricted matings;

(ii) Asserting that a gametic knowledge of parents is equivalent to a gametic knowledge of ancestry.

In neither case does the argument touch the ancestral position, which is summed up in the assertions that if we measure inheritance by the resemblance of somatic characters between offspring and ancestry, then, in a population mating at random:

The more ancestors of any grade with a given somatic character the more offspring with that character.

For ancestry of different grades the influence is diminished in geometrical progression at each stage.

These principles were first deduced empirically from observations and records without any theory as to the mechanism of heredity. If Mendelism be true for any characters in cross-fertilised plants, then these principles hold also for heredity in that plant-population, for they are essential features of the Mendelian theory (and, as a matter of fact, of a good many other determinantal theories). No proof or disproof of them can be directly deduced from Mr. Darbishire's memoir, but since that memoir brings evidence for the truth of Mendelian theory, it indirectly asserts the truth that ancestry is influential, at least in the field where the biometrician expects and asserts it to play a part. This paper contains only another aspect of the results reached in 1904, but it provides in the simple case—the grandparentage—the actual percentage measures of the influence of ancestry according to Mendel. Its justification is the misinterpretation which is likely to be placed on the statement that "there is nothing like ancestral contributions within the limits of a single unit-character."*

* Darbishire, 'Roy. Soc. Proc.,' B, vol. 81, p. 71.

10

Reprinted from *Roy. Soc. (Lond.) Proc.* **81**:225–229 (1909)

On the Ancestral Gametic Correlations of a Mendelian Population mating at Random.

By KARL PEARSON, F.R.S.

(Received April 2,—Read April 22, 1909.)

(1) The population to be considered in this paper is supposed to be initiated by a group of s_1 individuals with the protogenic constitution (AA), s_2 individuals with the allogenic constitution (aa), and s_3 individuals with the hybrid constitution (Aa), where the mating is given by the simple Mendelian formula: $(\text{AA}) \times (aa) = 4(\text{A}a)$. I do not assume at this stage any relation between the gametic constitution of an individual and its somatic character. I propose first to consider the correlation between any ancestor and the resulting array of offspring, when we regard only their gametic constitutions. I assume that all mating in the population is random, *i.e.* that every possible mating occurs simply in the proportions of the frequency of individuals of given gametic constitution in the population, and that there is no differential fertility or selective death-rate.

In a paper published in the 'Phil. Trans.,' vol. 203, A, 1904, p. 53 *et seq.*, I have dealt with the correlation between the *somatic* characters of the ancestry and the offspring in a population of a Mendelian character, more general in that I supposed the character to depend upon n couplets, and not a single Mendelian couplet, less general in that I supposed the population to have arisen from a series of initial hybridisations, and not from a mixture as in the present case of hybrids and members of two pure races in any proportions. In that paper I showed (a) that there was correlation between any ancestor and the offspring, (b) that the regression for any ancestor and the offspring was linear, and (c) that the correlations decreased in geometrical progression. These are the chief characteristics of the Law of Ancestral Heredity. It was clear that, judged by somatic characters only, ancestry was of importance. The result depended on Mendel's first principle of dominance being absolutely true. The values of the correlations were, however, less than those with which biometric work had made us familiar.

(2) In the present paper I start with a more general population, and investigate the correlation of the gametic not the somatic characters.

The general formula for the population before the first mating is

$$s_1(\text{AA}) + 2s_3(\text{A}a) + s_2(aa). \tag{i}$$

After the first random mating it is

$$(s_1+s_3)^2(\text{AA}) + 2(s_1+s_3)(s_2+s_3)(\text{A}a) + (s_2+s_3)^2(aa).$$

I write this for brevity

$$p^2(\mathrm{AA})+2pq(\mathrm{A}a)+q^2(aa), \qquad \text{(ii)}$$

and this constitution remains permanent in all successive matings. Hence the standard deviations of the gametic constitutions remain the same generation after generation, and the correlation coefficient is in every case equal to the slope of the regression line. I shall determine the slope of this line which will give the correlation and show that the regression is truly linear in each case.

(3) I consider first the effect of individuals of each special type mating with the general population (ii).

(*a*) Type (AA): the array of offspring is $(p+q)\,[p(\mathrm{AA})+q(\mathrm{A}a)]$,
(*b*) Type (aa): „ „ $(p+q)[p(\mathrm{A}a)+q(aa)]$,
(*c*) Type (Aa): „ „ $\frac{1}{2}(p+q)[p(\mathrm{AA})+(p+q)(\mathrm{A}a)+q(aa)]$.

Thus, in seeking what any differentiated group

$$t_1(\mathrm{AA})+t_3(\mathrm{A}a)+t_2(aa)$$

produces when mated with the general population, *i.e.* when mated at random, all we have to do is to replace (AA), (Aa) and (aa) by the above three expressions respectively.

In this manner I obtained the array of offspring due to any parent, any grandparent and any great grandparent. These at once allowed me to reach the general law of distribution, and, assuming this, one multiplication by the general population (ii) demonstrated by induction the validity of the results reached. These are as follows :—

I term nth parent any individual n generations back in the direct ancestry: thus a 1st parent is the father or mother; a 2nd parent, a grandparent; a 3rd parent, a great grandparent, and so on.

(i) If the nth parent be an (AA), then the array of offspring due to random matings is

$$p^2(p+q)^{2(n-1)}(\tfrac{1}{2})^{n-1}\{(2^{n-1}p+q)p(\mathrm{AA})+[(2^n-1)p+q]q(\mathrm{A}a)+(2^{n-1}-1)q^2(aa)\}.$$

(ii) If the nth parent be an (Aa), then the array of offspring is

$$(pq)(p+q)^{2(n-1)}(\tfrac{1}{2})^{n-1}\{[(2^n-1)p+q]p(\mathrm{AA})+[p^2+2(2^n-1)pq+q^2](\mathrm{A}a)+[p+(2^n-1)q]q(aa)\}.$$

(iii) If the nth parent be an (aa), then the array of offspring is

$$q^2(p+q)^{2(n-1)}(\tfrac{1}{2})^{n-1}\{(2^{n-1}-1)p^2(\mathrm{AA})+[(2^n-1)q+p]p(\mathrm{A}a)+(2^{n-1}q+p)q(aa)\}.$$

(4) These distributions correspond to the cases of 2, 1 and 0 A elements in the gametic constitution of the nth parent. And we have at once the following result:—

Number of protogenic elements in nth parent.	Average number of same elements in array of offspring.
2	$\dfrac{2^{n+1}p+2q}{2^n(p+q)} = \bar{y}_2,$
1	$\dfrac{(2^{n+1}-1)p+q}{2^n(p+q)} = \bar{y}_1,$
0	$\dfrac{(2^{n+1}-2)p}{2^n(p+q)} = \bar{y}_0.$

Accordingly, the average number of protogenic elements in the array of offspring decreases uniformly with the decrease in number of the like elements in the nth parent, *i.e.*

$$\bar{y}_2-\bar{y}_1 = (\tfrac{1}{2})^n = \bar{y}_1-\bar{y}_0.$$

Thus the regression between the nth parent and the offspring is linear, and the correlation coefficients form a geometrical series of ratio $\frac{1}{2}$, and first term $\frac{1}{2}$. Further, the exact constitution of the population, as far as the number of protogenic, allogenic or heterogenic individuals is concerned, is of no influence on the result at all. For all mixtures following the simple Mendelian rule: (AA) × (*aa*) = 4 (A*a*), the ancestral correlations for gametic constitution are:

Parental correlation	0·500
Grandparental correlation	0·250
Great grandparental correlation......	0·125 and so on.

It will be seen at once that these correlations are of the type ρ, ρ^2, ρ^3, etc., for which, in my memoir of 1896, I worked out the multiple regression formula, and showed that the ancestors were quite indifferent. "A knowledge of the ancestry beyond the parents in no way alters our judgment as to the size of organ or degree of characteristic probable in the offspring nor its variability."* This remark and the proof apply equally of course to gametic and to somatic characters if the correlation be of the above form.

(5) Accordingly there remains not the least antinomy between the Mendelian theory and the Law of Ancestral Heredity, if we confine our attention to gametic constitution. The Mendelian ancestry is correlated with the offspring in a series descending in a geometrical progression, and the regression is linear. The values of the correlation coefficients are

* "Regression, Heredity, and Panmixia," 'Phil. Trans.,' A, vol. 187, 1896, p. 306.

precisely those which it was pointed out in 1896 would lead to a knowledge of the parental constitution* replacing that of the ancestry.

(6) The striking point, however, of the present investigation is that the values now shown theoretically to exist for the ancestral *gametic* correlations in a simple Mendelian mixture are very close to those determined for *somatic* characters in biometric investigations, whereas the *somatic* correlations for a Mendelian population, if we maintain intact the principle of absolute dominance, appear theoretically to be too low.

Thus the value for parental correlation in man, horse, dog and cattle is about 0·5, and for the grandparental correlation lies between 0·25 and 0·30; but this tendency in the grandparent to some slight excess on the Mendelian gametic value must not be given too much weight.

(7) It seems desirable to consider how far the results in my paper of 1904 for the somatic correlations are modified if we assume for our population

$$p^2(\mathrm{AA})+2pq(\mathrm{A}a)+q^2(aa),$$

and do not make $p = q$.

Assuming the principle of dominance to be absolute, I enquire what is the proportion of offspring possessing the dominant character† (*i.e.* (AA) or (Aa)) supposing the nth parent to possess it (*i.e.* to be (AA) or (Aa)); and again, what is the proportion possessing the dominant character, supposing the nth parent does not possess it (*i.e.* to be aa).

Percentage of dominant offspring.

nth parent dominant in somatic character...$100 \times \dfrac{2^{n-1}p(p+2q)^2+q^3}{2^{n-1}(p+q)^2(p+2q)}$,

nth parent recessive in somatic character ...$100 \times \dfrac{2^{n-1}p(p+2q)^2-pq(p+2q)}{2^{n-1}(p+q)^2(p+2q)}$.

From this it follows that the correlation which is equal to the regression is

$$\frac{1}{2^{n-1}}\frac{q}{p+2q}.$$

If $p = q$, this is $\frac{1}{3}\frac{1}{2^{n-1}}$, in agreement with the conclusion of my memoir of 1904. But unless $q/(p+2q)=\frac{1}{2}$, *i.e.* the number of pure dominants in the population be vanishingly small (as well, of course, as the number of impure dominants !), this is not a series to which the form $\rho, \rho^2, \rho^3 \ldots$ applies, and when we judge (as we must in most instances in man) by the somatic and not the unknown gametic constitution, the ancestry does matter.

* As a matter of fact, a knowledge of the gametic constitution of the ancestry in any generation would be equally sufficient with that of the parents.

† It is assumed that A is dominant over a.

The following table illustrates the percentages of dominant charactered offspring when we selected an ancestor of given character :—

Ancestor.	Percentage of dominants in offspring.					
	$p = 2q$.		$p = q$.		$q = 2p$.	
	Dominant.	Recessive.	Dominant.	Recessive.	Dominant.	Recessive.
Parent	91·7	66·7	83·3	50·0	57·8	33·3
Grandparent	90·3	77·8	79·2	62·5	56·7	44·4
3rd parent	89·6	83·3	77·1	68·7	56·1	50·0
4th parent	89·2	86·1	76·0	71·9	55·8	52·8
5th parent	89·1	87·5	75·5	73·4	55·7	54·2
6th parent	89·0	88·2	75·3	74·2	55·6	54·9
∞th parent..............	88·9	88·9	75·0	75·0	55·6	55·6

It will be clear that the difference of the percentage of dominants in the offspring according as a parent, grandparent or great grandparent was dominant or recessive in somatic character is quite marked ; and only as we approach the higher ancestry, where the correlation is growing very weak, does the percentage difference grow imperceptible.

(8) That ancestry does not matter if we know the gametic constitution of the parents, that it does matter if we only know the somatic character of the parents appears to be the solution of one of the difficulties which some have found between the Mendelian and biometric methods of approaching the subject.

There is, however, I venture to think, another aspect of these results which is worthy of fuller consideration. Namely, the fairly close accordance now shown for the first time to exist between the ancestral gametic correlations in a Mendelian population and the observed ancestral somatic correlations suggests that the accordance between gametic and somatic constitutions is for at least certain characters possibly more intimate than is expressed by an absolute law of dominance. If (A*a*) were a class, or possibly on a wider determinantal theory a group of several classes, marked by an individual somatic character—not invariably identical with the somatic character of (AA)—there would be little left of contradiction between biometric and Mendelian results as judged by populations sensibly mating at random. It is the unqualified assertion of the principle of dominance which appears at present as the stumbling block.

11

Reprinted from *Amer. Naturalist* 44:65–82 (1910)

A MENDELIAN INTERPRETATION OF VARIATION THAT IS APPARENTLY CONTINUOUS[1]

PROFESSOR EDWARD M. EAST

HARVARD UNIVERSITY

THERE are two objects in writing this paper. One is to present some new facts of inheritance obtained from pedigree cultures of maize; the other is to discuss the hypotheses to which an extension of this class of facts naturally leads. This discussion is to be regarded simply as a suggestion toward a working hypothesis, for the facts are not sufficient to support a theory. They do, however, impose certain limitations upon speculation which should receive careful consideration.

The facts which are submitted have to do with independent allelomorphic pairs which cause the formation of like or similar characters in the zygote. Nilsson-Ehle[2] has just published facts of the same character obtained from cultures of oats and of wheat. My own work is largely supplementary to his, but it had been given these interpretations previous to the publication of his paper.

In brief, Nilsson-Ehle's results are as follows: He found that while in most varieties of oats with black

[1] Contributions from the Laboratory of Genetics, Bussey Institution, Harvard University, No. 4. Read before the annual meeting of the American Society of Naturalists, Boston, December 29, 1909.

[2] Nilsson-Ehle, H. Kreuzungsuntersuchungen an Hafer und Weizen. Lunds Universitets Årsskrift, N. F. Afd. 2., Bd. 5, No. 2, 1909.

glumes blackness behaved as a simple Mendelian monohybrid, yet in one case there were two definite independent Mendelian unit characters, each of which was allelomorphic to its absence. Furthermore, in most varieties of oats having a ligule, the character behaved as a mono-hybrid dominant to absence of ligule, but in one case no less than four independent characters for presence of ligule, each being dominant to its absence, were found. In wheat a similar phenomenon occurred. Many crosses were made between varieties having red seeds and those having white seeds. In every case but one the F_2 generation gave the ordinary ratio of three red to one white. In the one exception—a very old red variety from the north of Sweden—the ratio in the F_2 generation was 63 red to 1 white. The reds of the F_2 generation gave in the F_3 generation a very close approximation to the theoretical expectation, which is 37 constant red, 8 red and white separating in the ratio of 63:1, 12 red and white separating in the ratio of 15:1, 6 red and white separating in the ratio of 3:1, and one constant white. He did not happen to obtain the expected constant white, but in the total progeny of 78 F_2 plants his other results are so close to the theoretical calculation that they quite convince one that he was really dealing with three indistinguishable but independent red characters, each allelomorphic to its absence. Nor can the experimental proof of the two colors of the oat glumes be doubted. The evidence of four characters for presence of ligule in the oat is not so conclusive.

In my own work there is sufficient proof to show that in certain cases the endosperm of maize contains two indistinguishable, independent yellow colors, although in most yellow races only one color is present. There is also some evidence that there are three and possibly four independent red colors in the pericarp, and two colors in the aleurone cells. The colors in the aleurone cells when pure are easily distinguished, but when they are together they grade into each other very gradually.

Fully fifteen different yellow varieties of maize have been crossed with various white varieties, in which the crosses have all given a simple mono-hybrid ratio. In the other cases that follow it is seen that there is a di-hybrid ratio.

No. 5–20, a pure white eight-rowed flint, was pollinated by No. 6, a dent pure for yellow endosperm. An eight-rowed ear was obtained containing 159 medium yellow kernels and 145 light yellow kernels. The pollen parent was evidently a hybrid homozygous for one yellow which we will call Y_1 and heterozygous for another yellow Y_2. The gametes Y_1Y_2 and Y_1 fertilized the white in equal quantities, giving a ratio of approximately one medium yellow to one light yellow. The F_2 kernels from the dark yellow were as follows:

TABLE I.[3]

F_2 SEEDS FROM CROSS OF NO. 5–20, WHITE FLINT × NO. 6 YELLOW DENT, HOMOZYGOUS FOR Y_1 AND HETEROZYGOUS FOR Y_2

Dark Seeds Heterozygous for Both Yellows Planted

Ear No.	Dark Y.	Light Y.	Total Y.	No Y.
1	270	56	326	29
2	101	215	316	27
3	261	52	313	28
5	273	284	557	35
10	358	117	475	25
12	296	72	368	19
13	207	156	363	35
14	387	102	489	29
Total	2153	1054	3207	227
Ratio			14.1	1

The ratios of light yellows to dark yellows is very arbitrary, for there was a fine gradation of shades. The ratio of total yellows to white, however, is unmistakably 15:1.

In the next table (Table II) are given the results of F_2 kernels from the light yellows of F_1. Only ear No. 8, which was really planted with the dark yellows, showed yellows dark enough to be mistaken for kernels containing

[3] In these tables only hand pollinated ears are given.

both Y_1 and Y_2. The remaining ears are clearly mono-hybrids with reference to yellow endosperm.

TABLE II.

F_2 SEEDS FROM SAME CROSS AS SHOWN IN TABLE I

Light Yellow Seeds Heterozygous for Y_1 Planted

Ear No.	Dark Y.	Light Y.	No Y.
1		359	117
2		144	54
3		173	63
4		433	136
6		316	120
8	331		109
8a		229	86
9		325	115
10		227	87
11[4]		4	434
12		318	118
13		256	93
Total		3111	1098
Ratio		2.8	1

In a second case the female parent possessed the yellow endosperm. No. 11, a twelve-rowed yellow flint, was crossed with No. 8, a white dent. The F_2 kernels in part showed clearly a mono-hybrid ratio, and in part blended gradually into white. Two of these indefinite ears proved in the F_3 generation to have had the 15:1 ratio in the F_2 generation. Ear 7 of the F_2 generation calculated from the results of the entire F_3 crop must have had about 547 yellow to 52 white kernels, the theoretical number being 561 to 31. The hand-pollinated ears of the F_3 generation (yellow seeds) gave the results shown in Table III.

The F_3 generation grown from the other ear, Ear No. 8, showed that the ratio of yellows to whites in the F_2 generation was about 227 to 47 As the theoretical ratio is 257 to 17, the ratio obtained is somewhat inconclusive. A classification of the open field crop could not be made accurately on account of the light color of the yellows and

[4] Discarded from average. This ear evidently grew from one kernel of the original white mother that was accidentally self-pollinated. The four yellow kernels all show zenia from accidental pollination in the next generation.

Table III.

No. 11 Yellow × No. 8 White

F_3 Generation from Yellow Seeds of F_2 Generation

Ear No.	Dark Y.	Light Y.	Total Y.	No Y.	Ratio They Approximate.
1	116	95	211	19	15Y : 1 no Y
14			88	5	15Y : 1 no Y
5	181	122			$3Y_1Y_2 : 1\ Y_{1\ or\ 2}$
4		253		68	3Y : 1 no Y
6		193		73	"
8		163		79	"
11		108		35	"
9		456			Constant $Y_{1\ or\ 2}$

the presence of many kernels showing zenia. Table IV, however, showing the hand-pollinated kernels of the interbred yellows of the F_2 generation, settles beyond a doubt the fact that the two yellows were present.

Table IV.

Progeny of Ear No. 8 of the Same Cross as shown in Table III

F_3 Generation from Yellow Seeds of F_2 Generation

Ear No.	Dark Y.	Light Y.	Total Y.	No Y.	Ratio They Approximate.
10	101	188	289	25	15Y : 1 no Y
11	89	219	308	23	15Y : 1 no Y
3		233			constant light Y
9	dark and light		331		3 dark : 1 light Y
13	dark and light		350		3 dark : 1 light Y
8		294		108	3 light : 1 no Y
15		221		87	3 light : 1 no Y
1[5]		197		203	

In a third case an eight-rowed yellow flint, No. 22, was crossed with a white dent, No. 8. Only four selfed ears were obtained in the F_2 generation. Ear 1 had 72 yellow to 37 white kernels. This ear was poorly developed and undoubtedly had some yellow kernels which were classed as whites. Ear 4 had 158 yellow and 42 white kernels. It is very likely that both of these ears were mono-hybrids, but the F_3 generation was not grown. Ear 5 had 148 yellow and 15 white kernels. Ear 7 had 78 yellow and 5 white kernels. It seems probable that both of these ears

[5] Kernel from which this ear grew was evidently pollinated by no Y.

were di-hybrids, but only Ear 5 was grown another generation. The kernels classed as white proved to be pure; the open field crop from the yellow kernels gave 14 pure yellow ears and 14 hybrid yellow. Theoretically the ratio should be 7 pure yellows (that is, pure for either one or both yellows) and 8 hybrid yellows (4 giving 15 yellows to 1 white and 4 giving 3 yellows to 1 white). Five hand-pollinated selfed ears were obtained. Three of these gave mono-hybrid ratios, with a total of 607 yellows to 185 white kernels. One ear was a pure dark yellow (probably $Y_1Y_1Y_2Y_2$). The other ear was poorly filled, but had 27 dark yellows (probably Y_1Y_2) and 7 light yellow kernels (Y_1 or Y_2). Unfortunately no 15:1 ratio was obtained in this generation, but this is quite likely to happen when only five selfed ears are counted. The gradation of colors and the general appearance of the open field crop, however, lead me to believe that we were again dealing with a di-hybrid.

Two yellows appeared in still another case, that of white sweet No. 40♀ × yellow dent No. 3♂. Only one selfed ear was obtained in the F_2 generation giving 599 yellow to 43 white kernels. Of these kernels 486 were starchy and 156 sweet, which complicated matters in the F_3 generation because it was very difficult to separate the light yellow sweet from the white sweet kernels. Among the selfed ears were three pure to the starchy character, and in these ears the dark yellows, the light yellows and whites stood out very distinctly. Ear 12 had 156 dark yellow; 47 light yellow; 14 white kernels. Ear 13 had 347 dark yellow; 93 light yellow; 25 white kernels. The third starchy ear, No. 6, had 320 light yellow; 97 white kernels. Two ears, therefore, were di-hybrids, and one ear a mono-hybrid.

The ears which were heterozygous for starch and no starch and those homozygous for no starch, could not all be classified accurately, but it is certain that some pure dark yellows, some pure light yellows, some showing segregation of yellows and whites at the ratio 15:1, and some showing segregation of yellows and whites at the ratio of 3:1, were obtained,

One other case should be mentioned. One ear of a dent variety of unknown parentage obtained for another purpose was found to have some apparently heterozygous yellow kernels. Seven selfed ears were obtained from them, of which two were pure yellow. The other five ears each gave the di-hybrid ratio. There was a total of 1906 yellow seeds to 181 white seeds, which is reasonably close to the expected ratio, 1956 yellow to 131 white.

It is to be regretted that I can present no other case of this class that has been fully worked out, although several other characters which I have under observation in both maize and tobacco seem likely to be included ultimately. Nevertheless, the fact that we have to deal with conditions of this kind in studying inheritance is established; granting only that they will be somewhat numerous, it opens up an entirely new outlook in the field of genetics.

In certain cases it would appear that we may have several allelomorphic pairs each of which is inherited independently of the others, and each of which is separately capable of forming the same character. When present in different numbers in different individuals, these units simply form quantitative differences. It may be objected that we do not know that two colors that appear the same physically are exactly the same chemically. That is true; but Nilsson-Ehle's case of several unit characters for presence of ligule in oats is certainly one where each of several Mendelian units forms exactly the same character. It may be that there is a kind of biological isomerism, in which, instead of molecules of the same formula having different physical properties, there are isomers capable of forming the same character, although, through difference in construction, they are not allelomorphic to each other. At least it is quite a probable supposition that through imperfections in the mechanism of heredity an individual possessing a certain character should give rise to different lines of descent so that in the F_n generation when individuals of these different lines are crossed, the character behaves as a di-hybrid instead of as a mono-hybrid. In other words, it is more probable that these units arise through variation in different individuals and are combined by hybridization, than that actually different structures for forming the same character arise in the same individual.

On the other hand, there is a possibility of an action just the opposite of this. Several of these quantitative units which produce the same character may become attached like a chemical radical and again behave as a single pair. Nilsson-Ehle gives one case which he does not attempt to explain, where the same cross gave a 4:1 ratio in one instance and 8.4:1 ratio in another instance. In his other work characters always behaved the same way; that is, either as one pair, two pairs, three pairs, etc. In my work, the yellow endosperm of maize has behaved differently in the same strain, but it is probably because the yellow parent is homozygous for one yellow and heterozygous for the other. They were known to be pure for one yellow, but it would take a long series of crosses to prove purity in two yellows.

Let us now consider what is the concrete result of the inter-action of several cumulative units affecting the same character. Where there is simple presence dominant to absence of a number n of such factors, in a cross where all are present in one parent and all absent in the other parent, there must be 4^n individuals to run an even chance of obtaining a single F_2 individual in which the character is absent. When four such units, $A_1A_2A_3A_4$ are crossed with $a_1a_2a_3a_4$, their absence, only one pure recessive is expected in 256 individuals. And 256 individuals is a larger number than is usually reported in genetic publications. When a smaller population is considered, it will appear to be a blend of the two parents with a fluctuating variability on each side of its mode. Of course if there is absolute dominance and each unit appears to affect the zygote in the same manner that they do when combined, the F_2 generation will appear like the dominant parent unless a very large number of progeny are under observation and pure recessives are obtained. This may be an explanation of the results obtained by Millardet; it is certainly as probable as the hypothesis of the non-formation of homozygotes. Ordinarily, however, there is not perfect dominance, and variation due to heterozygosis combined with fluctuating variation makes it almost impossible to classify the individuals except by breeding. The two yellows in the endosperm of maize is an example of how few characters are necessary to make classification difficult. First, there is a small amount of fluctuation in different ears due to varying light conditions owing to differences in thickness of the husk; second, all the classes having different gametic formulæ differ in the intensity of their yellow in the following order, $Y_1Y_1Y_2Y_2$, $Y_1y_1Y_2Y_2$ or $Y_1Y_1Y_2y_2$, Y_1Y_1, Y_2Y_2, Y_1y_1, Y_2y_2, y_1y_2. As dominance becomes less and less evident, the Mendelian classes vary more and more from the formula $(3+1)^n$, and approach the normal curve, with a regular gradation of individuals on each side of the mode. When there is no dominance and open fertilization, a state is reached in which the curve of variation simulates the fluctuation curve, with the difference that the gradations are heritable.

One other important feature of this class of genetic facts must be considered. If units $A_1A_2A_3a_4$ meet units $a_1a_2a_3A_4$, in the F_2 generation there will be one pure recessive, $a_1a_2a_3a_4$, in every 256 individuals. This explains an apparent paradox. Two individuals are crossed, both seemingly pure for presence of the same character, yet one individual out of 256 is a pure recessive. When we consider the rarity with which pure dominants or pure recessives (for all characters) are obtained when there are more than three factors, we can hardly avoid the suspicion that here is a perfectly logical way of accounting for many cases of so-called atavism. Furthermore, many apparently new characters may be formed by the gradual dropping of these cumulative factors without any additional hypothesis. For example, in *Nicotiana tabacum* varieties there is every gradation[6] of loss of leaf surface near the base of the sessile leaf, until in *N. tabacum fruticosa* the leaf is only one step removed from a petioled condition. If this step should occur the new plant would almost certainly be called a new species; yet it is only one degree further in a definite series of loss gradations that have already taken place. If it should be assumed that in other instances slight qualitative as well as quantitative changes take place as units are added, then it becomes very easy, theoretically, to account for quite different characters in the individual homozygous for presence of all dominant units, and in the individual in which they are all absent.

Unfortunately for these conceptions, although I feel it extremely probable that variations in *some* characters that seem to be continuous will prove to be combinations of segregating characters, it is exceedingly difficult to demonstrate the matter beyond a reasonable doubt. As an illustration of the difficulties involved in the analysis of pedigree cultures embracing such characters, I wish to discuss some data regarding the inheritance of the number of rows of kernels on the maize cob.

The maize ear may be regarded as a fusion of four or more spikes, each joint of the rachis bearing two spikelets. The rows are, therefore, distinctly paired, and no case is known where one of the pair has been aborted. This is a peculiar fact when we consider the great number of odd kinds of variations that occur in nature. The number of rows per cob has been considered to belong to continuous variations by DeVries, and a glance at the progeny from the seeds of a single selfed ear as shown in Table V seems to confirm this view.

There is considerable evidence, however, that this character is made up of a series of cumulative units, inde-

[6] It is not known at present how this character behaves in inheritance.

TABLE V.

PROGENY OF A SELFED EAR OF LEAMING MAIZE HAVING 20 ROWS

Classes of rows	12	14	16	18	20	22	24	26	28	30
No. of ears	1	0	5	4	53	35	19	5	2	1

pendent in their inheritance. There is no reason why it should not be considered to be of the same nature as various other size characters in which variation seems to be continuous, but in which relatively constant gradations may be isolated, each fluctuating around a particular mode. But this particular case possesses an advantage not held by most phenomena of its class, in that there is a definite discontinuous series of numbers by which each individual may be classified.

Previous to analyzing the data from pedigree cultures, however, it is necessary to take into consideration several facts. In the first place, what limits are to be placed on fluctuations?[7] From the variability of the progeny of single ears of dent varieties that have been inbred for several generations, it might be concluded that the deviations are very large. But this is not necessarily the case; these deviations may be due largely to gametic structure in spite of the inbreeding, since no conscious selection of homozygotes has been made. There is no such variation in eight-rowed varieties, which may be considered as the last subtraction form in which maize appears and therefore an extreme homozygous recessive. In a count of the population of an isolated maize field where Longfellow, an eight-rowed flint, had been grown for many years, 4 four-rowed, 993 eight-rowed, 2 ten-rowed and 1 twelve-rowed ears were found. Only seven aberrant ears out of a thousand had been produced, and some of these may have been due to vicinism.

On the other hand a large number of counts of the number of rows of both ears on stalks that bore two ears has shown that it is very rare that there is a change

[7] The word fluctuation is used to designate the somatic changes due to immediate environment, and which *are not inherited.*

greater than ± 2 rows. If conditions are more favorable at the time when the upper ear is laid down it will have two more rows than the second ear; if conditions are favorable all through the season, the ears generally have the same number of rows; while if conditions are unfavorable when the upper ear is laid down, the lower ear may have two more rows than the upper ear. Furthermore, seeds from the same ear have several times been grown on different soils and in different seasons, and in each case the frequency distribution has been the same. Hence it may be concluded that in the great majority of cases fluctuation is not greater than in ± 2 rows, although fluctuations of ± 4 rows have been found.

A second question worthy of consideration is: Do somatic variations due to varying conditions during development take place with equal frequency in individuals with a large number of rows and in individuals with a small number of rows? From the fact that several of my inbred strains that have been selected for three generations for a constant number of rows, increase directly in variability as the number of rows increases, the question should probably be answered in the negative. This answer is reasonable upon other grounds. The eight-rowed ear may vary in any one of four spikes, the sixteen-rowed ear may vary in any one of eight spikes; therefore the sixteen-rowed ear may vary twice as often as the eight-rowed ear. By the same reasoning, the sixteen-rowed ear may sometimes throw fluctuations twice as wide as the eight-rowed ear.

A third consideration is the possibility of increased fluctuation due to hybridization. Shull[8] and East[9] have shown that there is an increased stimulus to cell division when maize biotypes are crossed—a phenomenon apart from inheritance. There is no evidence, however, that

[8] Shull, G. H., "A Pure-line Method in Corn Breeding," *Rept. Amer. Breeders' Assn.*, 5, 51–59, 1909.

[9] East, E. M., "The Distinction between Development and Heredity in Inbreeding," AMER. NAT., 43, 173–181, 1909.

increased gametic variability results. Johannsen[10] has shown that there is no such increase in fluctuation when close-pollinated plants are crossed. I have crossed several distinct varieties of maize where the modal number of rows of each parent was twelve, and in every instance the F_1 progeny had the same mode and about the same variability.

Finally, a possibility of gametic coupling should be considered. Our common races of flint maize all have a low number of rows, usually eight but sometimes twelve; dent races have various modes running from twelve to twenty-four rows. When crosses between the two subspecies are made, the tendency is to separate in the same manner.

Attention is not called to these obscuring factors with the idea that they are universally applicable in the study of supposed continuous variation. But there are similar conditions always present that make analysis of these variations difficult, and the facts given here should serve to prevent premature decision that they do not show segregation in their inheritance.

Table VI shows the results from several crosses between maize races with different modal values for number of rows. Several interesting points are noticeable. The modal number is always divisible by four. This is also the case with some twenty-five other races that I have examined but which are not shown in the table. I suspect that through the presence of pure units zygotes having a multiple of four rows are formed, while heterozygous units cause the dropping of two rows. The eight-rowed races are pure for that character, the twelve-rowed races vary but little, but the races having a higher number of rows are exceedingly variable.

When twelve-rowed races are crossed with those having eight rows, the resulting F_1 generation always—or nearly

[10] Johannsen, W., "Does Hybridization Increase Fluctuating Variability?" *Rept. Third Inter. Con. on Genetics*, 98–113, London, Spottiswoode, 1907.

TABLE VI.

CROSSES BETWEEN MAIZE STRAINS WITH DIFFERENT NUMBERS OF ROWS

Parents. (Female Given First.)	Gen.	Row Classes.						
		8	10	12	14	16	18	20
Flint No. 5		100						
Flint No. 11		1	4	387	7	1		
Flint No. 24		100						
Flint No. 15		100						
Dent No. 6				6	31	51	18	4
Dent No. 8			3	54	36	12	2	
Sweet No. 53 [11]		1	5	25	4			
Sweet No. 54 [11]		25	2	1				
No. 5 × No. 53	F_1	1	7	13				
No. 5 × No. 6	F_1	11	18	27	3			
No. 11 × No. 5	F_1	2	4	18				
No. 11 × No. 53	F_1	2	5	17				
No. 24 × No. 53	F_1	57	8	3				
No. 15 × No. 8	F_1	1	14	26	3	1		
No. 15 × No. 8 (from 10-row ear)	F_2	14	15	28	9	1		
No. 15 × No. 8 (from 12-row ear)	F_2	4	13	25	6	3		
No. 8 × No. 54	F_1	1	6	14				
No. 8 × No. 54 (from 12-row ear)	F_2	11	25	38	2	1		

always—has the mode at twelve rows. In one case cited in Table VI, No. 24 × No. 53, nearly all the F_1 progeny were eight-rowed. It might appear from this, either that the low number of rows was in this case dominant, or that the female parent has more influence on the resulting progeny than the male parent. I prefer to believe, however, that the individual of No. 53 which furnished the pollen was due to produce eight-rowed progeny. Unfortunately no record was kept of the ear borne by this plant, but No. 53 sometimes does produce eight-rowed ears.

When a race with a mode higher than twelve is crossed with an eight-rowed race, the F_1 generation is always intermediate, although it tends to be nearer the high-rowed parent. Only one example is given in the table, but it is indicative of the class. These results are rather confusing, for there seems to be a tendency to dominance in the twelve-rowed form that is not found in the forms with a higher number of rows. I have seen cultures of other investigators where 12-row × 8-row resulted in a

[11] Approximately.

ten-rowed F_1 generation, so the complication need not worry us at present.

The results of the F_2 generation show a definite tendency toward segregation and reproduction of the parent types. I might add that in at least two cases I have planted extracted eight-rowed ears and have immediately obtained an eight-rowed race which showed only slight departures from the type. Selection from those ears having a high number of rows has also given races like the high-rowed parent without recrossing with it. It is regretted that commercial problems were on hand at the time and no exact data were recorded. It can be stated with confidence, however, that ears like each parent are obtained in the F_2 generation, from which with care *races* like each parent may be produced. *Segregation seems to be the best interpretation of the matter.*

These various items may seem disconnected and uninteresting, but they have been given to show the tangible basis for the following theoretical interpretation. No hard and fast conclusion is attempted, but I feel that this interpretation with possibly slight modifications will be found to aid the explanation of many cases where variation is apparently continuous.

Suppose a basal unit to be present in the gametes of all maize races, this unit to account for the production of eight rows. Let additional independent interchangeable units, each allelomorphic to its own absence, account for each additional four rows; and let the heterozygous condition of any unit represent only half of the homozygous condition, or two rows. Then the gametic condition of a homozygous twenty-rowed race would be $8 + AABBCC$, each letter actually representing two rows. When crossed with an eight-rowed race, the F_2 generation will show ears of from eight to twenty rows, each class being represented by the number of units in the coefficients in the binomial expansion where the exponent is twice the number of characters, or in this case $(a + b)^6$.

The result appears to be a blend between the characters

of the two parents with a normal frequency distribution of the deviants. Only one twenty-rowed individual occurs in 64 instead of the 27 expected by the interaction of three dominant factors in the usual Mendelian ratios. The remainder of the 27 will have different numbers of rows, and, by their gametic formulæ, different expectations in future breeding as follows:

1 $AABBCC = 20$ rows.
2 $AaBBCC = 18$ rows.
2 $AABbCC = 18$ rows.
2 $AABBCc = 18$ rows.
4 $AaBbCC = 16$ rows.
4 $AaBBCc = 16$ rows.
4 $AABbCc = 16$ rows.
8 $AaBbCc = 14$ rows.

There are four visibly different classes and eight gametically different classes. It must also be remembered that the probability that the original twenty-rowed ear in actual practise may have had more than three units in its gametes has not been considered. This point is illustrated clearly if we work out the complete ratio for the three characters, and note the number of gametically different classes which compose the modal class of fourteen

TABLE VII

THEORETICAL EXPECTATION IN F_2 WHEN A HOMOZYGOUS TWENTY-ROWED MAIZE EAR IS CROSSED WITH AN EIGHT-ROWED EAR

Classes	8	10	12	14	16	18	20
No. ears	1	6	15	20	15	6	1

rows in Table VII. It actually contains seven gametically different classes and not a single homozygote. If this conception of independent allelomorphic pairs affecting the same character proves true, it will sadly upset the biometric belief that the modal class is *the type* around which the variants converge, for there is actually less chance of these individuals breeding true than those from *any other* class.

The conception is simple and is capable theoretically of bringing in order many complicated facts, although the presence of fluctuating variation will be a great factor in preventing analysis of data. I have thought of only one fact that is difficult to bring into line. If 8*AA*, 8*BB* and 8*CC* all represent homozygous twelve-rowed ears—to continue the maize illustration—and none of these factors are allelomorphic to each other, sixteen-rowed ears should sometimes be obtained when crossing two twelve-rowed ears. I am not sure but that this would happen if we were to extract all the homozygous twelve-rowed strains after a cross between sixteen-row and eight-row, and after proving their purity cross them. In some cases the additional four-row units would probably be allelomorphic to each other and in other cases independent of each other. On the other hand, this is only an hypothesis, and while I have faith in its foundation facts, the details may need change.

Castle has raised the point that greater variation should be expected in the F_1 generation than in the P_1 generations when crossing widely deviating individuals showing variation apparently continuous. If the parents are strictly pure for a definite number of units, say for size, a greater variation should certainly be expected in the F_1 generation after crossing. But considering the difficulties that arise when even five independent units are considered, can it be said that anything has heretofore been known concerning the actual gametic status of parents which it is known do vary in the character in question and in which the variations are inherited, for the race can be changed by selection within it. It may be, too, that the correct criterion has not been used in size measurements, for, as others have suggested, solids vary as the cube root of their mass, whereas the sum of the weights of the body cells has usually been measured and compared directly with similar sums.

Attention should be called to one further point. Many characters in all probability are truly blending in their inheritance, but there is another interpretation which may apply in certain cases. I have repeatedly tried to cross Giant Missouri Cob Pipe maize (14 feet high) and Tom Thumb pop maize (2 feet high), but have always failed. They both cross readily with varieties intermediate in size, but are sterile between themselves. We may imagine that the gametes of each race, though varying in structure, are all so dissimilar that none of them can unite to form zygotes. Other races may be found where only part of the gametes of varying structure are so unlike that they will not develop after fusion. The zygotes that do develop will be from those more alike in construction. An apparent blend results, and although segregation may take place, no progeny as extreme as either of the parents will ever occur.

I may say in conclusion that the effect of the truth of this hypothesis would be to add another link to the increasing chain of evidence that the word mutation may properly be applied to any inherited variation, however small; and the word fluctuation should be restricted to those variations due to immediate environment which do not affect the germ cells, and which—it has been shown—are not inherited. In addition it gives a rational basis for the origin of *new* characters, which has hitherto been somewhat of a Mendelian stumbling-block; and also gives the term unit-character less of an irrevocably-fixed-entity conception, which is more in accord with other biological beliefs.

Part III

HAMMER AND TONGS

Editor's Comments on Papers 12 Through 19

12 NORTON and PUNNETT
Appendix I in Mimicry in Butterflies

13 FISHER
On the Dominance Ratio

14 WRIGHT
Coefficients of Inbreeding and Relationship

15 HALDANE
A Mathematical Theory of Natural and Artificial Selection. Part I

16 HALDANE
A Mathematical Theory of Natural and Artificial Selection. Part II: The Influence of Partial Self-fertilization, Inbreeding, Assortative Mating, and Selective Fertilisation on the Composition of Mendelian Populations, and on Natural Selection

17 HALDANE
A Mathematical Theory of Natural and Artificial Selection. Part III

18 CHETVERIKOV
On Certain Aspects of the Evolutionary Process from the Standpoint of Modern Genetics

19 CHETVERIKOV
On the Genetic Constitution of Wild Populations

To say that the conflicts had been resolved would be pointless, but the next two decades saw a gradual building up of both the theoretical and the experimental basis necessary to understand the elementary processes. A clearly identifiable step was the calculation of the number of generations required for certain gene frequency changes under various selection intensities. The table found in Appendix I of Punnett's book *Mimicry in Butterflies* was calculated by H. T. J. Norton, a Cambridge University mathematician. Norton's main contribution to genetics (1928) has been reprinted in *Demographic Genetics* (Vol. 3, "Benchmark Papers in Genetics"). Punnett was a major contributor to genetics and requires further treatment.

R. C. Punnett (1875–1967) studied medicine at Cambridge, changed to zoology, and established a reputation as a morphologist. He turned to genetics early in the century and replaced Bateson in the Chair of Biology in 1910. Punnett wrote an early text (1905) *Mendelism,* which contributed to the spread of the concepts. He helped found the *Journal of Genetics* and studied the genetics of domestic organisms, poultry, rabbit, duck, cattle, sweet pea, maize marigold, and man, and was the author of eight books and more than one hundred papers. Punnett believed that mimicry resulted from discontinuous variation and his ideas contributed considerable stimulus to the study of this subject.

Fisher's (1918) paper, "The correlation between relatives on the supposition of Mendelian inheritance," will appear in a forthcoming Benchmark volume, *Quantitative Genetics*; its importance to evolutionary genetics cannot be minimized. This is considered by many (Kempthorne, 1974) to be one of the benchmark papers of all scientific literature, stimulating future investigation and resolving and creating a basis for many questions. Its seminal nature has been so pervasive that Moran and Smith (1966) have provided an interesting commentary reprinting of the paper. Fisher considered the degree of dominance, epistacy, assortative mating, multiple alleles, multiple factors, linkage, and various interrelations as they influenced the correlation between relatives. Fisher used the partitioning of the components of variation earlier developed by Gauss, Pearson, Weinberg, and others. Later Fisher developed the significance tests for various degrees of freedom in the analysis of variance. Norton and Pearson (1976) reprinted the referees' comments on Fisher's 1918 paper. These suggest the difficulty in resolving a polarized issue. Wright (1977) suggests that the differences between the Mendelians and the statisticians were never so striking in the United States.

In 1922 Fisher examined the conditions under which the variance of the population may be maintained. He considered large populations with stable distribution and likened them to the velocities in the theory of gases. He considered selection, mutation, random effects, and the size of the population, and proposed that genotypic selection was balanced by occasional mutations. He noted that stability of the frequency ratio only exists when selection favors the heterozygote. His dominance ratio was defined in the 1918 paper as the ratio of the dominance deviations to the total genotypic variance. Paper 13 reprints Fisher's 1922 paper.

Ronald A. Fisher (1890-1962) trained in mathematics, statistical mechanics, and quantum theory at Cambridge. He worked on a farm in Canada, for an investment company, and as a public school teacher. In 1919 he became the statistician at Rothamsted Experimental Station, where he had the opportunity to apply statistical methods to biological problems. He succeeded Karl Pearson as Galton Professor at University College in 1933. In 1943 he succeeded Punnett as Arthur Balfour Professor of Genetics in Cambridge, where he remained until his retirement in 1957. He moved to CSIRO in Adelaide, Australia, where he remained active until his death. His genetical books included *The Genetical Theory of Natural Selection* and *The Theory of Inbreeding*. His major statistical studies include *Statistical Methods for Research Workers, The Design of Experiments,* and *Statistical Methods and Scientific Inference*. These, and his many papers have earned him the title of the founder of modern statistics (Mather, 1969).

S. Wright, a student of Castle, worked extensively on the genetics of coat color in mammals. In his work in the Animal Husbandry Division of the U.S. Department of Agriculture he became interested in measures of the degree of inbreeding relationships with various systems of mating. These he worked out in detail in a series of very significiant papers (Wright, 1921). In Paper 14, Wright gives a formula which provides a measure of the departure of the amount of homozygosis under random mating toward complete homozygosis with inbreeding. While the paper applies the analysis to pedigrees, the application of inbreeding to an understanding of the evolution and genetic structure of natural populations has been extensive in later literature (see Wright, 1969, for review).

Wright (1889–) studied at Lombard, Illinois, and Harvard, receiving the D.Sc. in 1915. He worked in the Department of Agriculture from 1915 to 1925. He was at the University of Chicago from 1926 to 1954, when he went to the University of Wisconsin where he is still active. His almost 200 papers have covered a variety of genetic topics including the interaction of genetic characters in guinea pigs, physiological genetics, and the theoretical and experimental analysis of inbreeding, migration, and selection. He has developed a general theory of evolution involving interaction systems rather than the evolution by mass selection of single gene mutations. The theory will be introduced later in this volume. The theory was first developed in the middle 1920s, but was not published until the early 1930s. A more up-to-date version of the theory is in the *Proceedings of the Royal Society* in 1965 and in Wright's recent books, two of which are still in press.

A real example of the hammer and tongs approach must be that of

J. B. S. Haldane, who, in a series of papers from 1924 to 1933, developed a Mathematical Theory of Natural and Artificial Selection. Part IV appears in the Benchmark volume *Demographic Genetics,* Part V appears in the Benchmark volume *Stochastic Models in Population Genetics* and and Part VI appears in the Benchmark volume *Genetics of Speciation.* Parts I, II, and III are presented here in their entirety as Papers 15, 16, and 17.

In the initial paper (1924, Paper 15) developing a mathematical theory of selection, Haldane considered first a single completely dominant allele with random breeding, constant selection intensity, and nonoverlapping generations. He then considered self-fertilization, within family selection, sex-limited and sex-linked characters. Haldane noted that selection is ineffective on rare recessive characters which may explain why there are so many deleterious autosomal recessive characters. Haldane suggested four circumstances under which selection of recessives may be more effective: self-fertilization, assortative fertilization, incomplete dominance, and heterozygote advantage. He noted that the isolation of small communities deserved study, a position arrived at independently by Wright.

In Part II Haldane (Paper 16) dealt with partial self-fertilization and partial inbreeding and in Part III (Paper 17) he considered partial dominance, multiple factors, linkage, and polyploidy. The remaining papers in this series are discussed in Part IV of this volume. Clearly Haldane felt that the essential elements of the mathematical theory required detailed analysis with clear assumptions before the more elegant possibilities of gene interaction could be discerned.

J. B. S. Haldane (1892–1964) was a man of diverse interests and enormous talents, contributing to mathematics, biochemistry, physiology, literature, and philosophy. He was educated at Oxford in mathematics and classics, served in the First World War, and returned to study at Oxford before becoming a reader in biochemistry at Cambridge. In 1932 he began an association with University College in London which included occupying the Weldon Chair of Biometry. In 1957 he went to India, where he remained active until his death. He served in the Spanish Civil War, and during World War II he studied the physiology of human respiration during military submarine disasters, and worked in civil defense. For some years he was a member of the Communist Party. M. J. D. White (1965), his biographer for *Genetics,* suggests that he was "the most erudite biologist of his generation, and perhaps of the century" (p. 6). Pirie (1966) provides a bibliography. In addition to the work summarized in this volume, Haldane contributed to genetic studies of the intensity of selection, and various aspects of genetic load, coadaptation, mutation, statistics, and enzymology, and to a wide spectrum of knowledge including the solar system, atomic warfare (1939), origin of life, popular science, and children's literature. His contribution to population and evolutionary genetics has been reviewed by Wright (1969). He was, perhaps, a last vestige of the renaissance man: we hope not, for the world can use more like him.

The genetic literature between 1900 and the early 1930s, as published in the various new journals started by geneticists (*Anals of Eugenics,* 1925– ; *Genetics,* 1916– ; *Genetica,* 1917– ; *Hereditas,* 1920– ; *Japanese Journal of Genetics,* 1925– ; *Journal of Genetics,* 1910– ; and *Journal of Heredity,* 1903–), provides extensive coverage of the expanding knowledge of Mendelism. Morgan (1932) discussed the advances of the period and referred to the significance of the discovery of linkage and the understanding of polyploidy, and the expansion of the basic knowledge of the nature of mutation. He noted that the genes of the laboratory were the same as those of nature and suggested a need for physiological studies, meiotic studies, mutation process studies, and for the applications to agriculture and horticulture. He was speaking to the International Genetics Congress, and noted his opinion that great progress could be made by not having congresses too often. Blakeslee (1936), speaking at the Botanical Congress, noted that the important developments had been the understanding of linkage and linearity or order of genes on chromosomes, the understanding of the structure and numbers of chromosomes, hybridization and polyploidy, and of the mutagenic effect of radiation. He called for a study of the species problem and suggested the eventual conscious control of evolution. Provine (1971) has reviewed some of these developments.

The beginnings of the applications of Mendelian principles to the evolution of natural populations are more difficult to trace during these years. They are mostly found in other journals (*American Naturalist, Nature, Science,* and various zoological and botanical journals). A few examples of this scattered literature deserve comment to describe the state of the art.

Sumner (1923, 1929, 1930) studied the genetics of the populations, races, and species of the genus *Peromyscus* and concluded that no form had arisen through a single act of mutation. He found that races and subraces differ by a considerable number of Mendelian factors and the differences have accumulated along geographic, climatic, and edaphic gradients. Schmidt (1917) studied the genetic basis of different races of fish by raising the offspring of the crosses in a variety of environments, and measuring, for example, the average number of vertebrae of parents and offspring. He concluded that the variability was composed of genetic (he called generative) and environmental (he called personal) factors and that most of the differences between racial characters were inherited. Goldschmidt (1934) investigated the genetic variability which accompanied the geographic variation in the moth *Lymantria*. He found most characters were determined by a number of genes; sometimes a single locus was interpreted to have greater effect than the others.

Haldane (1929) noted that much of the work of genes in nature was being carried out in Russia and therefore was not available to most workers. Apparently much of the work in Russia was stimulated by the efforts of S. S. Chetverikov (1926, 1927), who influenced the studies of Dubinin, Dobzhansky, N. Timofeff-Ressovsky, Astaurov, Romashov, and others. Chetverikov's 1926 paper was translated and published in the *Proceedings of the American Philosophical Society* in 1961 and is reprinted here (Paper 18). His 1927 paper to the *Fifth International Genetics Congress* summarizes his own studies with natural populations. It has been translated specifically for this volume (Paper 19). S. S. Chetverikov (1880–1959) studied at the University of Moscow and eventually joined the Institute of Experimental Biology, studying genetics and evolution. In 1929 he was politically removed from Moscow. He worked for a zoo, a junior college, and as a geneticist at the University of Gorky until 1948 when further political pressures removed him from all academic pursuits. Dobzhansky (1967) has provided a biography.

Chetverikov's summary of the state of evolutionary studies in the middle of the 1920s serves as an introduction to the development of a number of synthetic studies which later integrated various fields and solidified our understanding of the evolutionary process. These would certainly include *Genetics and the Origin of Species* by Th. Dobzhansky; *Systematics and the Origin of Species* by E. Mayr; *Evolution, the Modern Synthesis* by Julian Huxley; *Tempo and Mode in Evolution* by G. G. Simpson; and *Variation and Evolution in Plants* by G. L. Stebbins, Jr. However, before these books could be produced the theoretical constructs of evolutionary genetics would have to be expressed in integrated form in terms directly related to natural populations.

By the late 1920s the theory of natural selection of continuous characters resulting from the action of Mendelian units had replaced the positions of "either continuous or discontinuous" evolution. The mutation theory of de Vries had itself evolved to identify the ultimate source of variation rather than explain directly all of evolution. A synthetic statement of the interaction of Mendelism and quantitative genetics, and of mutation and selection was needed. This was provided by Fisher, Wright, and Haldane in the early 1930s.

12

Reprinted by permission from *Mimicry in Butterflies,* R. C. Punnett, ed., Cambridge University Press, 1915, pp. 154–156

MIMICRY IN BUTTERFLIES

APPENDIX I

H. T. J. Norton and R. C. Punnett

FOR the table on p. 155 I am indebted to the kindness of Mr H. T. J. Norton of Trinity College, Cambridge. It affords an easy means of estimating the change brought about through selection with regard to a given hereditary factor in a population of mixed nature mating at random. It must be supposed that the character depending upon the given factor shews complete dominance, so that there is no visible distinction between the homozygous and the heterozygous forms. The three sets of figures in the left-hand column indicate different positions of equilibrium in a population consisting of homozygous dominants, heterozygous dominants, and recessives. The remaining columns indicate the number of generations in which a population will pass from one position of equilibrium to another, under a given intensity of selection. The intensity of selection is indicated by the fractions $\frac{100}{50}$, $\frac{100}{75}$, etc. Thus $\frac{100}{75}$ means that where the chances of the favoured new variety of surviving to produce offspring are 100, those of the older variety against which selection is operating are as 75; there is a 25 % selection rate in favour of the new form.

The working of the table may perhaps be best explained by a couple of simple examples.

In a population in equilibrium consisting of homozygous dominants, heterozygous dominants and recessives the last named class comprises 2·8 % of the total: assuming that a 10 % selection rate now operates in its favour as opposed to the two classes of dominants—in how many generations will the recessive come to constitute one-quarter of the population? The answer is to be looked for in column B (since the favoured variety is recessive) under the fraction $\frac{100}{90}$. The recessive

Percentage of total population formed by old variety	Percentage of total population formed by the hybrids	Percentage of total population formed by the new variety	Number of generations taken to pass from one position to another as indicated in the percentages of different individuals in left-hand column							
			A. Where the new variety is dominant				B. Where the new variety is recessive			
			$\frac{100}{50}$	$\frac{100}{75}$	$\frac{100}{90}$	$\frac{100}{99}$	$\frac{100}{50}$	$\frac{100}{75}$	$\frac{100}{90}$	$\frac{100}{99}$
99·9	·09	·000								
98·0	1·96	·008	4	10	28	300	1920	5740	17,200	189,092
90·7	9·0	·03	2	5	15	165	85	250	744	8,160
69·0	27·7	2·8	2	4	14	153	18	51	149	1,615
44·4	44·4	11·1	2	4	12	121	5	13	36	389
25·	50·	25·	2	4	12	119	2	6	16	169
11·1	44·4	44·4	4	8	18	171	2	4	11	118
2·8	27·7	69·0	10	17	40	393	2	4	11	120
·03	9·0	90·7	36	68	166	1,632	2	6	14	152
·008	1·96	98·0	170	333	827	8,243	2	6	16	165
·000	·09	99·9	3840	7653	19,111	191,002	4	10	28	299

passes from 2·8 % to 11·1 % of the population in 36 generations,. and from 11·1 % to 25 % in a further 16 generations—*i.e.* under a 10 % selection rate in its favour the proportion of the recessive rises from 2·8 % to 25 % in 52 generations.

If the favoured variety is dominant it must be borne in mind that it can be either homozygous or heterozygous—that for these purposes it is represented in the left-hand column by the hybrids as well as by the homozygous dominant. In a population in equilibrium which contains about 2 % of a dominant form, the great bulk of these dominants will be heterozygous, and the relative proportion of recessives, heterozygous, and homozygous dominants is given in the second line of the left-hand column.

Let us suppose now that we want to know what will be the percentage of dominants after 1000 generations if they form 2 % of the population to start with, and if, during this period, they have been favoured with a 1 % selection advantage. After 165 generations the proportion of recessives is 90·7, so that the proportion of dominants has risen to over 9 %; after 153 further generations the percentage of dominants becomes $27{\cdot}7 + 2{\cdot}8 = 30{\cdot}5$; after 739 generations it is 88·8 %, and after 1122 generations it is $69{\cdot}0 + 27{\cdot}7 = 96{\cdot}7$. Hence the answer to our question will be between 89 % and 97 %, but nearer to the latter figure than the former.

Mr Norton has informed me that the figures in the table are accurate to within about 5 %.

13

Reprinted from *Roy. Soc. Edin. Proc.* **42**:321–341 (1922)

On the Dominance Ratio. By **R. A. Fisher**, M.A., Fellow of Gonville and Caius College. *Communicated* by Professor J. ARTHUR THOMSON.

(MS. received March 8, 1922. Read June 19, 1922.)

INTRODUCTION.

IN 1918, in a paper published in the *Transactions of the Royal Society of Edinburgh*, the author attempted an examination of the statistical effects in a mixed population of a large number of genetic factors, inheritance in which followed the Mendelian scheme. At that time, two misapprehensions were generally held with regard to this problem. In the first place, it was generally believed that the variety of the assumptions to be made about the individual factors—which allelomorph was dominant; to what extent did dominance occur; what were the relative magnitudes of the effects produced by the different factors; in what proportion did the allelomorphs occur in the general population; were the factors dimorphic or polymorphic; to what extent were they coupled,—besides the more general possibilities of preferential mating (homogamy), preferential survival (selection), and environmental effects, rendered it possible to reproduce any statistical resultant by a suitable specification of the population. It was, therefore, important to prove that when the factors are sufficiently numerous, the most general assumptions as to their individual peculiarities lead to the same statistical results. Although innumerable constants enter into the analysis, the constants necessary to specify the statistical aggregate are relatively few. The total variance of the population in any feature is made up of the elements of variance contributed by the individual factors, increased in a calculable proportion by the effects of homogamy in associating together allelomorphs of like effect. The degree of this association, together with a quantity which we termed the Dominance Ratio, enter into the calculation of the correlation coefficients between husband and wife, and between blood relations. Special causes, such as epistacy, may produce departures, which may in general be expected to be very small from the general simplicity of the results; the whole investigation may be compared to the analytical treatment of the Theory of Gases, in which it is possible to make the most varied assumptions as to the accidental circumstances, and even the essential nature of the individual molecules, and yet to develop the general laws as to the behaviour of gases, leaving but a few fundamental constants to be determined by experiment.

In the second place, it was widely believed that the results of biometrical investigation ran counter to the general acceptance of the Mendelian scheme of inheritance. This belief was largely due to the narrowly restricted assumptions as to the Mendelian factors, made by Pearson in his paper of 1903 (6). It was there assumed that the factors were all equally important, that the allelomorphs of each occurred in equal numbers, and that all the dominant genes had a like effect. The effect of homogamy was also left out of consideration, and it is to this that must be ascribed the much lower correlations given by calculation, compared to those actually obtained. When the more general system came to be investigated, it was found to show a surprisingly complete agreement with the experimental values, and to indicate with an accuracy which could not otherwise be attained, how great a proportion of the variance of these human measurements is to be ascribed to heritable factors.

At the time when the paper of 1918 was written, it was necessary, therefore, to show that the assumption of multiple, or cumulative, factors afforded a working hypothesis for the inheritance of such apparently continuous variates as human stature. This view is now far more widely accepted: Mendelian research has with increasing frequency encountered characters which are evidently affected by many separate factors. In some fortunate circumstances, as in *Drosophila*, it has been possible to isolate and identify the more important of these factors by experimental breeding on the Mendelian method; more frequently, however, and especially in the case of the economically valuable characters of animals and plants, no such analysis has been achieved. In these cases we can confidently fall back upon statistical methods, and recognise that if a complete analysis is unattainable it is also unnecessary to practical progress.

This fact is meeting with increasing recognition in the United States, and a considerable number of mathematical investigations have been published dealing with the statistical effects of various systems of mating (Wentworth and Remick, 1916; Jennings, 1916, 1917; Robbins, 1917, 1918). A number of the simpler results of my 1918 paper have since been confirmed by independent American investigators (Wright, 1921). The present note is designed to discuss the distribution of the frequency ratio of the allelomorphs of dimorphic factors, and the conditions under which the variance of the population may be maintained. A number of points of general interest are shown to flow from purely statistical premises

Recent work in genetics (East and Jones, 1920) leads unavoidably to the conclusion that inbreeding is not harmful in itself, but is liable to appear harmful only through the emergence of harmful recessive characters. This raises the question as to why recessive factors should tend to be harmful, or why harmful factors should tend to be recessive: unless this association exist we should expect to obtain great improvements by inbreeding ordinarily crossbred species, as often as great deterioration. The statistical reason for this association is clear from the distribution of the ratio of allelomorph frequency which occurs under genotypic selection, for, if we assume that adaptation is the result of selection, the majority of large mutations must be harmful, and these can only be incorporated in the common stock in the sheltered region where the rare recessives accumulate (fig. 4).* Similarly there are many well-attested cases of the crossbred (heterozygous) individual showing surprising vigour; but it is not obvious that there is any biological reason for the heterozygote to be more vigorous than the two homozygotes. From a consideration of the stability of the frequency ratios, however, it appears that there will only be stable equilibrium if the heterozygote is favoured by selection against both the homozygotes: naturally this will occur only in a minority of factors, but when it occurs such a factor will be conserved. In the opposite case it will certainly be eliminated.

Cases in which the heterozygote is favoured by selection in preference to both homozygous forms have an additional interest in that these cases, when the selection is intense, may form the basis upon which is built up a system of balanced lethal factors. Muller (1918) has shown that such systems will tend to be built up when selection strongly favours the heterozygote, and has explained how in the light of such systems the majority of the phenomena, including the "mutations," of *Œnothera*, find a genetic explanation.

The interesting speculation has recently been put forward that random survival is a more important factor in limiting the variability of species than preferential survival (Hagedoorn, 4). The ensuing investigation negatives this suggestion. The decay in the variance of a species breeding at random without selection, and without mutation, is almost inconceivably slow: a moderate supply of fresh mutations will be sufficient to maintain the variability. When selection is at work even to the most trifling extent, the new mutations must be much more numerous in order to

* On the Lamarckian theory of evolution, on the other hand, where most, or all, mutations are assumed to be beneficial, we should expect by inbreeding, which uncovers the accumulated mutations in this region, to make great and immediate progress.

maintain equilibrium. That such is the actual state of the case in mankind may be inferred from the fact that the frequency distribution of the numerical proportion of the allelomorphs, calculated on the assumption of selection maintained in equilibrium by occasional mutation, leads to the value of the Dominance Ratio which is actually observed. In all cases it is worth noting that the rate of mutation required varies as the variance of the species, but diminishes as the number of individuals is increased. Thus a numerous species, with the same frequency of mutation, will maintain a higher variability than will a less numerous species: in connection with this fact we cannot fail to remember the dictum of Charles Darwin, that "wide ranging, much diffused and common species vary most" (1, chap. ii).

1. Equilibrium under Selection.

Let the three phases of a dimorphic factor be born in any generation in the proportion

$$P : 2Q : R,$$

then the proportion of the two allelomorphic genes will be

$$P+Q : Q+R, \quad \text{or} \quad p : q;$$

if by selection those that become parents are in the proportion

$$aP : 2bQ : cR, \qquad \text{where} \qquad aP+2bQ+cR=1,$$

then the proportion born in the next generation will be

$$(aP+bQ)^2 : 2(aP+bQ)(bQ+cR) : (bQ+cR)^2;$$

equilibrium is thus only possible if $Q^2=PR$, *i.e.* $P=p^2$, $Q=pq$, $R=q^2$, and if $aP+bQ=p$, $bQ+cR=q$.

Hence it follows that, if

$$a=1+\alpha, \; b=1+\beta, \; c=1+\gamma,$$

$$\frac{\alpha}{p^2}=-\frac{\beta}{pq}=\frac{\gamma}{q^2}$$

specifies the condition of equilibrium.

If selection favours the homozygotes, no stable equilibrium will be possible, and selection will then tend to eliminate whichever gene is below its equilibrium proportion; such factors will therefore not commonly be found in nature: if, on the other hand, the selection favours the heterozygote, there is a condition of stable equilibrium, and the factor will continue in the stock. Such factors should therefore be commonly found, and may explain instances of heterozygote vigour, and to some extent the deleterious effects sometimes brought about by inbreeding.

If the selective action is sufficiently powerful, it may lead in these cases to the establishment of a balanced lethal system.

2. The Survival of Individual Genes.

If we consider the survival of an individual gene in such an organism as an annual plant, we may suppose that the chance of it appearing in the next generation in 0, 1, 2, 3 individuals to be

$$p_0, p_1, p_2, \ldots$$

where

$$p_0 + p_1 + p_2 + \ldots = 1.$$

If

$$f(x) = p_0 + p_1 x + p_2 x^2 + \ldots$$

then evidently if there were two such genes in the first generation, the chance of occurrence in r individuals, or more strictly, in r homologous loci, in the second generation, will be the coefficient of x^r in

$$(f(x))^2.$$

It follows that the chance of a single gene occurring in r homologous loci, in the third generation, will be coefficient of x^r in

$$f(f(x)).$$

The form of $f(x)$ will vary from species to species, and in the same species according to the stage of development on which we fix our attention. For simplicity we shall suppose that the successive generations are enumerated at the same stage of development. For the purpose of an evolutionary argument it is indifferent at what stage of development the enumeration is made: in general it will be most convenient to fix our attention on that stage at which the species is least numerous.

In certain important cases the form of $f(x)$ may be calculated. In a field of cross-fertilised grain each mature and ripened plant is the mother of a considerable number of grains, and the father, possibly, of an almost unlimited number. If the number of the species is nearly constant, the average number of its progeny which are destined to become mature is very nearly 2. Or since each gene of a homologous pair occurs in half the gametes, the average number of mature plants in the second generation in which it occurs is 1. Each ovule, therefore, or each pollen grain has individually a very small chance of surviving, and the proportions p_0, p_1, p_2, will be closely given by the Poisson series

$$e^{-1}\left(1, 1, \frac{1}{2!}, \frac{1}{3!}, \ldots\right)$$

In the more general case in which the number of the species is not stationary but increases in each generation in the ratio $m : 1$, m being near to unity, the series will be

$$e^{-m}\left(l, m, \frac{m^2}{2!}, \frac{m^3}{3!}, \ldots\right)$$

and $f(x) = e^{m(x-1)}$. The chance of extinction of a single gene in one generation is e^{-m}, where m is near to unity. In other species in which an individual may survive for many breeding seasons, or in which the generation is of indeterminate length, the form of the function $f(x)$ will be modified: it is sufficiently clear, however, that if we consider that stage in an animal's or plant's life-history at which reproduction is about to commence, the form of the function will not be very different, and the chance of extinction of a particular gene, thus far established in the species, will be

$$e^{-l},$$

where l is a small number not greatly different from unity.* The arbitrary element thus introduced into the question of the survival of a mutant gene is due to the fact that in the first place its survival depends on that of the individual in which it occurs, and this chance is variable from species to species; once, however, it has reached the point of existing in an adult individual capable of leaving many offspring, the conditions of its survival are closely similar in all cases. While it is rare, its survival will be at the mercy of chance, even if it is well fitted to survive. Using the above expression,

$$f(x) = e^{x-1},$$

it may be seen that only about 2 per cent. will survive 100 generations, while those that do will on the average be represented in some 50 individuals. Only when the number of individuals affected becomes large will the effect of selection predominate over that of random survival, though even then only a very small minority of the population may be affected.

3. Factors not acted on by Selection.

If p be the proportion of any gene, and q of its allelomorph in a dimorphic factor, then in n individuals of any generation we have $2np$ genes scattered at random. Let

$$\cos\theta = 1 - 2p$$

where θ lies between 0 and π.

* An upper limit can be set to l by the mere fact of segregation, for in the case of the most uniform possible reproduction, when each individual bears 2 offspring the chance of extinction of any gene is $\frac{1}{4}$, so that l cannot exceed 1·4.

Further, if a second generation of n individuals be now formed at random, the standard departure of p from its previous value will be

$$\sigma_p = \sqrt{\frac{pq}{2n}},$$

hence,

$$\sigma_\theta = \sqrt{\frac{pq}{2n}}\frac{d\theta}{dp} = \frac{1}{\sqrt{2n}}.$$

The fact that this is independent of θ makes it easy to calculate the changes in the distribution of θ, in the absence of selection, for let $y(\theta)\,d\theta$ represent the distribution of θ in any one generation, the distribution in the next will be given by

$$y + \Delta y = \int_0^\pi \frac{1}{\sqrt{2\pi\sigma}} e^{-\frac{\delta\theta^2}{2\sigma^2}}\left(y + y'\delta\theta + \frac{\delta\theta^2}{2\,!}y'' + \ldots\right)$$

$$= y + \frac{\sigma^2}{2}y'' + \ldots$$

Now σ^2 is very small, being $\frac{1}{2n}$, so that measuring time in generations, we have

$$\frac{\partial y}{\partial \mathrm{T}} = \frac{1}{4n}\frac{\partial^2 y}{\partial \theta^2}.$$

Since we have drawn no distinction between the gene and its allelomorph, we are only concerned with symmetrical solutions: the stationary case is

$$y = \frac{\mathrm{A}}{\pi},$$

where A is the number of factors present.

Besides this, we have when y is increasing

$$y = \mathrm{A}_0 e^{k\mathrm{T}} \frac{p}{2\sinh \frac{1}{2}p\pi} \cdot \cosh p\left(\theta - \frac{\pi}{2}\right),$$

and when y is decreasing

$$y = \mathrm{A}_0 e^{-k\mathrm{T}} \frac{p}{2\sin \frac{1}{2}p\pi} \cdot \cos p\left(\theta - \frac{\pi}{2}\right),$$

for which

$$k = \frac{p^2}{4n}.$$

4. Terminal Conditions.

If we represent by e^{-l} the chance that a particular gene borne by a single individual will not be represented in the next generation, the chance of extinction for a factor of which b genes are in existence will be

$$e^{-bl}.$$

When θ is near to 0, p which is always equal to $\sin^2\frac{\theta}{2}$, will be very nearly equal to $\frac{1}{4}\theta^2$. Let

$$t = \sin\tfrac{1}{2}\theta,$$

then the number of genes in existence is $2nt^2$, and the chance of their extinction in one generation is e^{-2nlt^2}.

This chance is negligible save when t is very small, and may be equated to $\frac{1}{2}\theta$; hence the number of genes exterminated in any one generation

$$2\int_0 ye^{-2nlt^2}d\theta$$

$$=4\int_0 ye^{-2nlt^2}dt.$$

In the stationary case $y=\frac{A}{\pi}$, and the number of genes exterminated will be

$$\frac{A}{\pi}\cdot\frac{2\sqrt{2\pi}}{\sqrt{4ln}}=A\sqrt{\frac{2}{\pi ln}},$$

if new mutations occur at a rate $n\mu$, then this equilibrium will be possible if

$$A=\sqrt{\frac{\pi l}{2}}n^{\frac{3}{2}}\mu.$$

For species in this stationary state the variance will vary (1) as the rate of mutation, (2) as the number of the population raised to the power of $\frac{3}{2}$, (3) as $\sqrt{l}$, a quantity which will seldom differ much from unity. Using the variate $z=\log_e\frac{p}{q}$, the distribution for this case is shown in fig. 1.

5. The Hagedoorn Effect.

In the absence of mutation, extinction will still go on, and the number of factors must diminish, hence we may put for this case

$$y=A_0e^{-kT}\cdot\frac{p'}{2\sin\frac{1}{2}p\pi}\cdot\cos p\left(\theta-\frac{\pi}{2}\right).$$

If θ is small,

$$\cos p\left(\theta-\frac{\pi}{2}\right)=\cos\tfrac{1}{2}p\pi+p\theta\sin\tfrac{1}{2}p\pi-\tfrac{1}{2}p^2\theta^2\cos\tfrac{1}{2}p\pi\ \ldots$$

$$=\cos\tfrac{1}{2}p\pi+2p\sin\tfrac{1}{2}p\pi\,.\,t-2p^2\cos\tfrac{1}{2}p\pi\,.\,t^2\ \ldots,$$

so that the rate of extinction is

$$A_0e^{-kT}\frac{p}{2\sin\frac{1}{2}p\pi}\cdot\sqrt{\frac{2\pi}{ln}}\left\{\cos\tfrac{1}{2}p\pi+2p\sin\tfrac{1}{2}p\pi\cdot\sqrt{\frac{2}{4\,ln}}\right\}$$

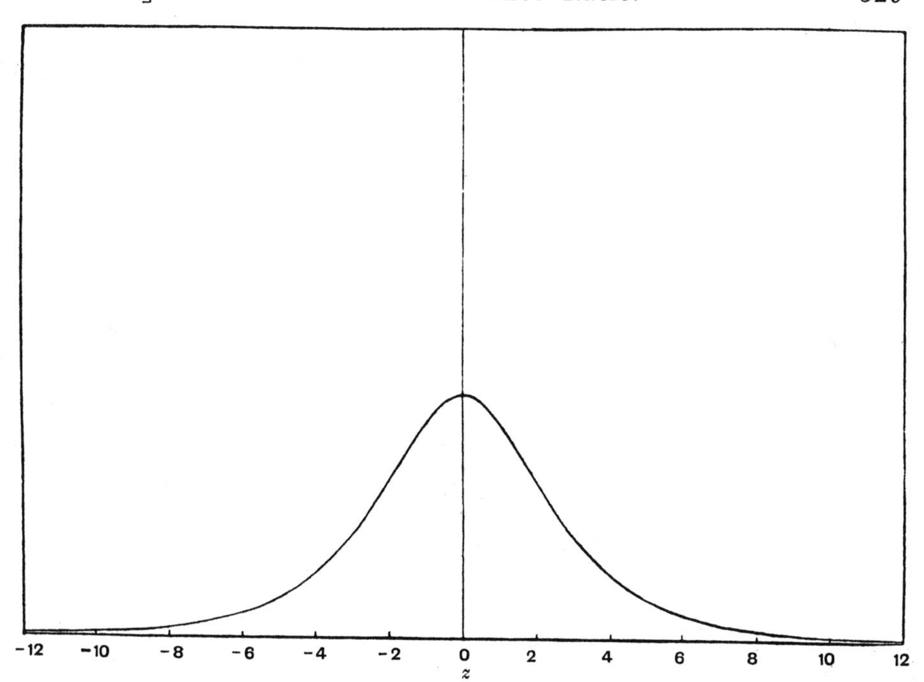

Distribution of the logarithmic frequency ratio $\left(z = \log \frac{p}{q}\right)$ of the allelomorphs of a dimorphic factor.

FIG. 1.—$df = \frac{1}{2\pi} \operatorname{sech} \frac{1}{2}z dz$;

represents the distribution when, in the absence of selection, fortuitous extinction is counterbalanced by mutation. Dominance Ratio = ·2308.

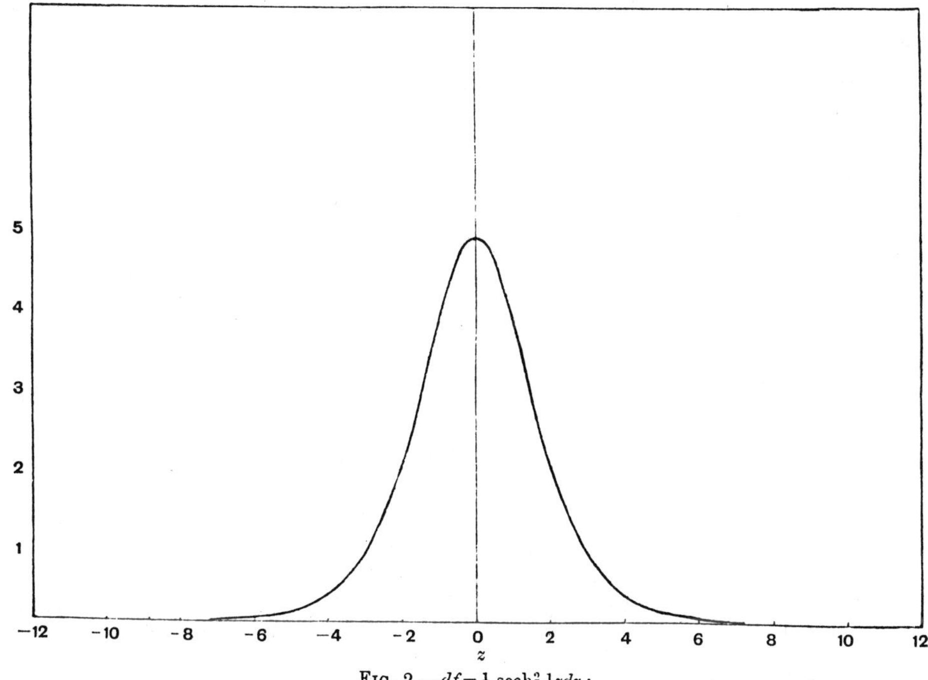

FIG. 2.—$df = \frac{1}{4} \operatorname{sech}^2 \frac{1}{2}z dz$;

represents the distribution when, in the absence of selection and mutation, the variance is steadily decaying owing to fortuitous extinction of genes. Dominance Ratio = ·2500. This is the condition emphasised by Hagedoorn.

the third term being evidently negligible compared to the first. For equilibrium, therefore,

$$k = p\sqrt{\frac{2\pi}{ln}}\left\{\tfrac{1}{2}\cot\tfrac{1}{2}p\pi + \frac{p}{2\pi ln}\right\}.$$

Remembering that $k = \frac{p^2}{4n}$, we have

$$\frac{p^2}{n}\left(\frac{1}{4} - \frac{1}{l}\right) = \sqrt{\frac{2\pi}{ln}}\frac{p}{2}\cot\tfrac{1}{2}p\pi.$$

Hence $\cot\frac{1}{2}p\pi$ is of the order $\frac{1}{\sqrt{n}}$ and is very small, so that p is near to 1. Then

$$k = \frac{1}{4n}.$$

This is a very slow rate of diminution, a population of n individuals breeding at random would require $4n$ generations to reduce its variance in the ratio 1 to e, or 2·8 n generations to halve it. As few specific groups contain less than 10,000 individuals between whom interbreeding takes place, the period required for the action of the Hagedoorn effect, in the entire absence of mutation, is immense. It will be noticed that since l is always less than 1·4 in species stationary in number, the solution above makes p slightly greater than l, which strictly would indicate negative frequencies at the extremes: the value of k is, however, connected with the curvature in the central portion of the curve, and the small distortion at the extremes, where the assumptions, upon which our differential equation is based, break down, will not affect its value. (Fig. 2 shows the distribution of $z = \log\frac{p}{q}$.)

The number by which the number of factors current is reduced in each generation s $\frac{A}{4n}$, and since this number depends on the general form of the distribution curve, it will not be diminished by a number of mutations of the same order. The effect of such very rare mutations would merely be to adjust the terminal of the curve until the rate of extinction is increased sufficiently to counterbalance the additional mutations. It is probable, however, that μ is always far greater than is necessary to make this state of affairs impossible, save in the case of a small colony recently isolated from a very variable species. In this case, with n small and A large, μ might for a time be of the order An^{-2}, rather than of the order $An^{-\frac{3}{2}}$, or An^{-1}.

In the case of a population with A factors, with a supply of fresh mutations sufficient only to be in equilibrium with a smaller number B factors, we may put

$$B\sqrt{\frac{2}{\pi l n}}=\frac{Ap}{2\sin\frac{1}{2}p\pi}\cdot\cos\tfrac{1}{2}p\pi\cdot\sqrt{\frac{2\pi}{ln}},$$

or,

$$\frac{B}{A}=\tfrac{1}{2}p\pi\cot\tfrac{1}{2}p\pi,$$

so that if

$$\frac{a}{\tan a}=\frac{B}{A},\quad 0<a<\frac{\pi}{2},$$

$$p=\frac{2a}{\pi},$$

and

$$k=\frac{a^2}{\pi n}.$$

Similarly, if B>A, the rate of increase in variance may be calculated from the equations

$$\frac{a}{\tanh a}=\frac{B}{A},$$

$$k=\frac{a^2}{\pi n}.$$

The rate of decrease, therefore, cannot, in the absence of selection, exceed the value indicated by $k=\frac{1}{4n}$; no such limit can be assigned to the rate of increase.

6. Uniform Genetic Selection.

In section 1 we have seen that the effects of selection on any Mendelian factor may be expressed by the triple ratio $a:b:c$ representing the relative fitness of the three phases. Only when b exceeds both a and c is there a condition of stable equilibrium; when b is less than both a and c there is a condition of unstable equilibrium; and such factors will tend rapidly to disappear from the stock. Generally, however, we may expect that either b will be intermediate, or equal to a, the value for the dominant homozygote. Two hypothetical cases may, therefore, be considered: (1), in which b is the geometric mean of a and c, and the selection merely affects the proportion of the allelomorphic genes; we may call this uniform genetic selection; and (2), in which b is equal to a, which we may call uniform genotypic selection.

In uniform genetic selection the genetic ratio will be altered in a constant ratio r in each generation, so that after n generations of selection we have

$$\frac{p}{q}=r^n\frac{p_0}{q_0},$$

evidently $r=\frac{a}{b}=\frac{b}{c}$ of section 1.

We may suppose that usually r is near to unity, and $\log r$, which may be positive or negative, may be considered to be of the order of 1 per cent. Let $\log r = a$, then for different factors a will have different values, indifferently positive and negative, since we have no reason to suppose that the selection favours either dominant or recessive characters. The mean square value of a for different factors we shall write σ_a^2.

For any factor

$$\frac{d}{dT}\log\frac{p}{q} = a;$$

therefore

$$\frac{dp}{dT} = pqa,$$

$$\frac{d\theta}{dT} = a\sqrt{pq}.$$

The factors which in one generation are at θ, will in the next be scattered owing to two causes: (1) random survival causing variance, $\frac{1}{2n}$; (2) selection causing variance, $pq\,\sigma_a^2 (= \frac{1}{4}\sin^2\theta \,.\, \sigma_a^2)$. The total variance at any point will be

$$\frac{1}{2n} + \tfrac{1}{4}\sigma_a^2 \sin^2\theta;$$

and so long as σ_a^2 is small as we have supposed, the equilibrium distribution will be

$$y \propto \frac{1}{\sqrt{\sin^2\theta + \dfrac{2}{n\sigma_a^2}}},$$

or nearly

$$y = \frac{A}{2\log(\sigma_a\sqrt{8n})} \cdot \frac{1}{\sqrt{\sin^2\theta + \dfrac{2}{n\sigma_a^2}}}.$$

n being large compared with $\frac{1}{\sigma_a^2}$, the effects of selection are, for the more important factors, much more influential than those of random survival. At the extremes, however, for very unequally divided factors the latter is the more important cause of variation. (The distribution of $z = \log\frac{p}{q}$ is shown in fig. 3.)

The amount of mutation needed to maintain the variability with this amount of selection may be calculated from the terminal ordinate

$$\frac{A\sigma_a\sqrt{\dfrac{n}{2}}}{2\log(\sigma_a\sqrt{8n})},$$

whence

$$n\mu = \sqrt{\frac{2\pi}{ln}} \cdot \frac{A\sigma_a\sqrt{\dfrac{n}{2}}}{2\log(\sigma_a\sqrt{8n})} = \frac{A\sigma_a\sqrt{\dfrac{\pi}{l}}}{2\log(\sigma_a\sqrt{8n})}.$$

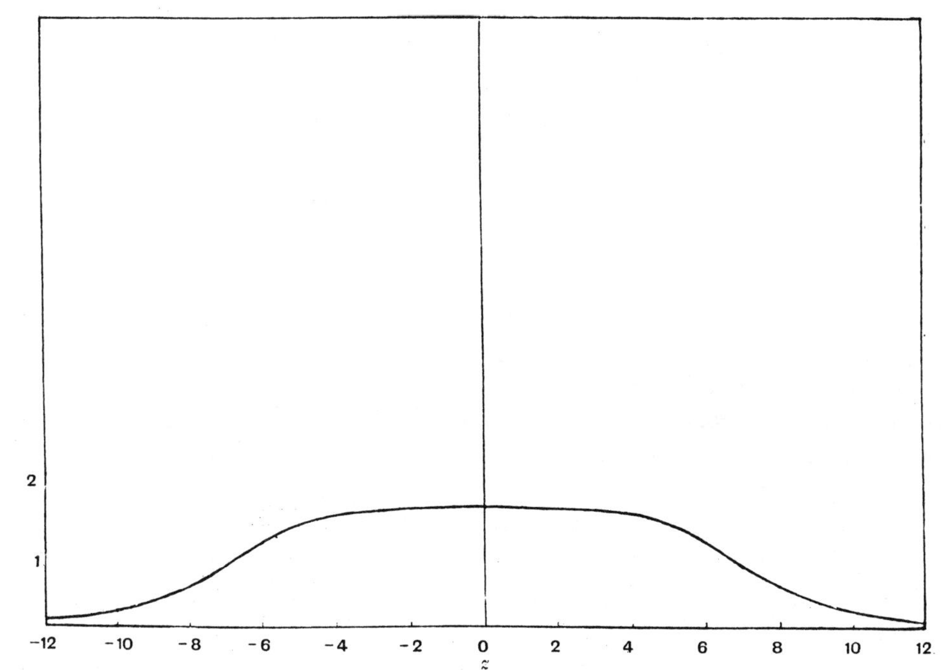

Distribution of the logarithmic frequency ratio.

FIG. 3.—$df \propto \dfrac{dz}{\sqrt{1+k^2 \cosh^2 \frac{1}{2}z}}$; $k = \cdot 1$; genetic selection counterbalanced by mutation. Dominance Ratio, ·2000.

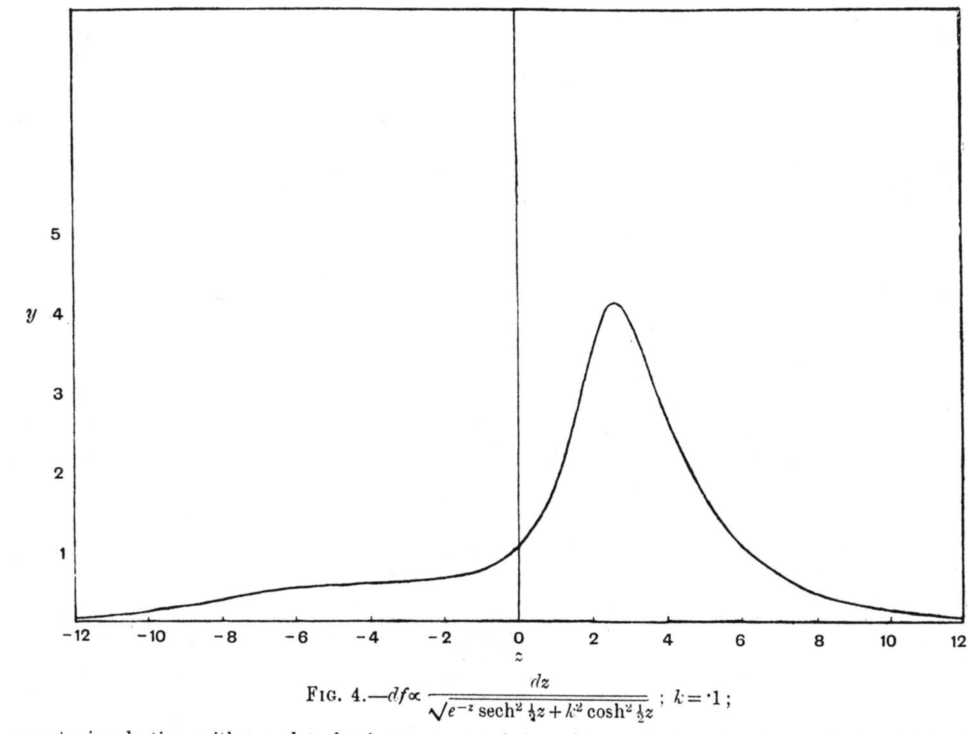

FIG. 4.—$df \propto \dfrac{dz}{\sqrt{e^{-z} \operatorname{sech}^2 \frac{1}{2}z + k^2 \cosh^2 \frac{1}{2}z}}$; $k = \cdot 1$; genotypic selection, with complete dominance, counterbalanced by mutation. Dominance Ratio, ·3333. This is the probable condition of natural species, including man. Note the accumulation of rare recessives.

Since the logarithm does not increase very rapidly, we may say approximately that A is proportional to $\frac{n\mu}{\sigma_a}$.

It will be seen that to maintain the same amount of variability, as in the case of equilibrium in the absence of selection (section 4), the rate of mutation must be increased by a factor of the order $\sigma_a \sqrt{n}$. Even in the low estimate we have made of the intensity of selection on the majority of factors, this quantity will usually be considerable. The existence of even the slightest selection is in large populations of more influence in keeping variability in check than random survival.

A further effect of selection is to remove preferentially those factors for which a is high, and to leave a predominating number in which a is low. In any factor a may be low for one of two reasons: (1) the effect of the factor on development may be very slight, or (2) the factor may effect changes of little adaptive importance. It is therefore to be expected that the large and easily recognised factors in natural organisms will be of little adaptive importance, and that the factors affecting important adaptations will be individually of very slight effect. We should thus expect that variation in organs of adaptive importance should be due to numerous factors, which individually are difficult to detect.

Owing to this preferential removal of important factors the above solution only truly represents an equilibrium of the variability of the species under absolutely uniform conditions of selection when the new mutations which arise have the same frequency distribution of relative importance as those removed by selection. It must be remembered, however, that the change of variability even by selection is a very slow process, and that gradual changes in the physical and biological environment of a species will alter the values of a for each factor, so tending to neutralise the tendency of selection to lower the value of σ_a. Nevertheless, a will be on the whole numerically smaller for factors in the current stock than it is for fresh mutations.

7. Uniform Genotypic Selection.

If the heterozygote is selected to the same extent as the dominant, or $b=a$, it is easy to see by writing down the first generation, that a genetic ratio $p:q$, becomes in one generation by selection $\frac{p}{q}\frac{a}{ap+cq}$; or, writing $1+\beta$ for $\frac{a}{c}$,

$$\frac{p}{q}\frac{1+\beta}{1+p\beta};$$

or, when β is small,

$$\frac{p}{q}(1+q\beta).$$

Such selection is therefore equivalent to a genetic selection

$$a = q\beta.$$

Now

$$\frac{d\theta}{dT} = a\sqrt{pq} = \beta q\sqrt{pq},$$

and for the variance caused by selection, instead of $pq\sigma_a^2$, as in Section 6, we now write $pq^3\sigma_\beta^2$: we have then for the total variance produced in one generation in the value of θ,

$$\frac{1}{2n}+\frac{1}{16}\sin^2\theta(1+\cos\theta)^2\sigma_\beta^2$$

$$=\frac{1}{2n}+\sin^2\tfrac{1}{2}\theta\cos^6\tfrac{1}{2}\theta\,.\,\sigma_\beta^2,$$

and the equilibrium distribution will be

$$y \propto \frac{1}{\sqrt{\sin^2\frac{1}{2}\theta\cos^6\frac{1}{2}\theta+\dfrac{1}{2n\sigma_\beta^2}}}.$$

It is important to notice that this distribution, unlike those hitherto considered, is unsymmetrical, factors of which the dominant phase is in excess are in the majority. This has an important influence on the value of the dominance ratio.

If $2n\sigma_\beta^2$ is large, we can write with sufficient accuracy *

$$y = \frac{A}{1\cdot4022(2n\sigma_\beta^2)^{\frac{1}{4}}+\frac{2}{3}\log(8n\sigma_\beta^2)-\frac{2}{3}}\cdot\frac{1}{\sqrt{\sin^2\frac{1}{2}\theta\cos^6\frac{1}{2}\theta+\dfrac{1}{2n\sigma_\beta^2}}}.$$

The terminal ordinate therefore varies nearly as $(2n\sigma_\beta^2)^{\frac{1}{4}}$, and for large populations in equilibrium, μ varies as $n^{-\frac{3}{4}}$ and as $\sigma_\beta^{\frac{1}{2}}$.

Genotypic selection resembles genetic selection in diminishing the amount of variability which a given frequency of mutation can maintain, or *per contra*, increasing the amount of mutation needed to maintain a given amount of variability; it differs, however, in being comparatively inactive in respect of factors in which the dominant allelomorph is in excess, and consequently in allowing a far greater number of factors to exist in this region (see fig. 4).

* I am indebted to Mr E. Gallop, Gonville and Caius College, Cambridge, for the value of the definite integral. Mr Gallop has shown that the three terms given are the heads of three series in descending powers of $n\sigma_\beta^2$, in which the integral may be expanded.

Now when dominance is complete, the dominance ratio from a group of factors having the same ratio $\frac{p}{q}$ is

$$\frac{1}{1+2\frac{q}{p}},$$

for in the notation of our previous paper

$$\delta^2 = 4p^2q^2a^2,$$

and

$$a^2 = 4p^2q^2a^2\left(1+2\frac{q}{p}\right),$$

where a is half the difference between the two homozygous forms (3, p. 404).

The dominance ratio is therefore raised by an excess of factors in which the dominant gene is the more numerous, such as occurs under genotypic selection.

8. The Dominance Ratio.

The distribution found for the ratio $\frac{p}{q}$ or for the value of θ, which indicates the same quantity, in sections 3 to 7, enable us to calculate the value attained by the dominance ratio under each of the suppositions there considered.

1. In the Hagedoorn condition, where the variance is steadily decaying by random survival, in the absence of mutations or selection,

$$df = \tfrac{1}{2}A \sin\theta d\theta,$$

writing $\phi = \frac{1}{2}\theta$, then $p = \sin^2\phi$, $q = \cos^2\phi$, whence

$$\epsilon^2 = S(\delta^2) = 8A\overline{a^2}\int_0^{\frac{1}{2}\pi} \sin^5\phi \cos^5\phi d\phi,$$

$$\sigma^2 = S(a^2) = 8A\overline{a^2}\int_0^{\frac{1}{2}\pi} (\sin^5\phi \cos^5\phi + 2\sin^3 \cos^7\phi) d\phi,$$

and

$$\frac{\epsilon^2}{\sigma^2} = \frac{1}{1+2\cdot\frac{3}{2}} = \cdot 2500.$$

2. When in the absence of selection, sufficient mutations take place to counteract the effect of random survival

$$df = \frac{2A}{\pi} d\phi,$$

and we have to consider the ratio of the integrals

$$\int_0^{\frac{1}{2}\pi} \sin^4\phi \cos^4\phi d\phi, \quad \int_0^{\frac{1}{2}\pi} \sin^2\phi \cos^6\phi d\phi,$$

which are in the ratio 3 : 5.

The dominance ratio is therefore

$$\frac{3}{3+2(5)} = \cdot 2308 ;$$

the greater variation in the ratio $\frac{p}{q}$ showing itself in a lower dominance ratio.

3. In the third symmetrical case, when genetic selection is at work, the variation of $\frac{p}{q}$ is even greater (fig. 3); since both δ^2 and a^2 contain the factor p^2q^2, the factors in which p or q is very small, make no appreciable contribution to these quantities, consequently we only consider the central portion of the distribution, where

$$df \propto \frac{d\phi}{\sin\phi\cos\phi},$$

the intensity of selection appearing only as a constant factor, and therefore influencing the range of variation of the species, but not its dominance ratio. Here we have the integrals

$$\int_0^{\frac{1}{2}\pi} \sin^3\phi\cos^3\phi d\phi \quad \text{and} \quad \int_0^{\frac{1}{2}\pi} \sin\phi\cos^5\phi d\phi,$$

leading to a dominance ratio

$$\frac{1}{1+4} = \cdot 2000.$$

4. In the case of genotypic selection, which case most nearly reproduces natural conditions, the distribution in the centre of the range is

$$df \propto \frac{d\phi}{\sin\phi\cos^3\phi},$$

consequently the two integrals with which we are concerned

$$\int_0^{\frac{1}{2}\pi} \sin^3\phi\cos\phi d\phi, \quad \int_0^{\frac{1}{2}\pi} \sin\phi\cos^3\phi d\phi$$

are now equal, and the dominance ratio is raised to $\frac{1}{3}$.

In considering the interpretation of the dominance ratio, in our previous inquiry, we found that for symmetrical distribution the value $\frac{1}{3}$ occurred as a limiting value when the standard deviation of $z\left(=\log\frac{p}{q}\right)$ was made zero. Since the dominance ratio calculated from observed human correlations averaged ·32, with a standard error about ·03, we were led to consider that either the allelomorphs concerned occurred usually in nearly equal numbers, a supposition for which we saw no rational explanation, or that the value of the dominance ratio had been raised by the prevalence of epistacy (non-linear interaction of factors), a suggestion for which no direct evidence could be adduced.

In the light of the above discussion in which we have deduced the distribution of allelomorphic ratios from the conditions of equilibrium with selective influences, from which condition it is probable that natural species do not widely depart, we find that the value $\frac{1}{3}$ for the dominance ratio is produced by the asymmetry of the distribution, and in such a manner as to be independent of the activity of the selective agencies, provided that this exceeds a certain very low level. When differential survival to the extent of only about 1 per cent. in a generation affects the different Mendelian factors, in a population of only a million, and far more for more powerful selection, or a larger population, the dominance ratio will be very close to its characteristic value of $\frac{1}{3}$.

The importance of the fact that this ratio is independent of the intensity of selection, lies not only in the fact that the intensity of selection is usually incapable of numerical estimation, but in the fact that factors having effects of different magnitudes on the soma, which are therefore exposed to selection of varying intensity, and contribute very different quota to the variance, are all affected in the same manner; those factors which by their insignificance might be exposed to selective influences which are not large compared to the effects of random survival will be precisely those which have little weight in computing the dominance ratio.

9. Assortative Mating.

With assortative mating it has been shown (3, p. 414) that the deviations from the mean of the three phases of any factor have, owing to association with similar factors, mean genotypic values given by the formula

$$I = i + \frac{A}{1-A} \cdot \frac{iP - kR}{p},$$

$$J = j - \frac{A}{1-A} \cdot \frac{p-q}{2pq}(iP - kR),$$

$$K = k - \frac{A}{1-A} \cdot \frac{iP - kR}{q},$$

when i, j, k are the deviations in the absence of association, A measures the degree of association produced by assortative mating: p, q are the gene frequencies, and P, R the corresponding phase frequencies for the homozygous phases.

Writing $j=i$ to represent complete dominance, and $P=p^2$, $R=q^2$, since

$$(p^2+2pq)i+q^2k=0,$$

$$\frac{i}{q^2}=-\frac{k}{p(p+2q)}=\frac{i-k}{1}=\frac{p^2i-q^2k}{2pq^2};$$

and since $i-k=2a$, we have

$$I=i+\frac{A}{1-A}\cdot 4aq^2,$$

$$J=i-\frac{A}{1-A}\cdot 2aq(p-q),$$

$$K=k-\frac{A}{1-A}\cdot 4apq;$$

or

$$I-J=2a\cdot\frac{A}{1-A}\cdot q,$$

$$J-K=2a\left(1+\frac{A}{1-A}q\right).$$

If now the survival factors of the three phases are a, b, c, the effect of one generation's selection is given by

$$\frac{p_1}{q_1}=\frac{p_0}{q_0}\frac{ap+bq}{bp+cq}=\frac{p_0}{q_0}(1+p\overline{a-b}+q\overline{b-c}),$$

since a, b, and c are near to 1;
hence

$$\alpha=p(a-b)+q(b-c).$$

Now as $I-J$, $J-K$, the mean differences in any trait due to a single factor, are small compared with the whole variation within the population, we must take $a-b$, $b-c$ proportional to $I-J$ and $J-K$. In other words,

$$a-b=(I-J)\gamma,$$
$$b-c=(J-K)\gamma,$$

where γ measures the intensity of selection per unit change in the trait.

Hence

$$\alpha=\gamma(p\overline{I-J}+q\overline{J-K})$$
$$=\gamma\cdot\frac{2a}{1-A}\cdot q.$$

The general case of uniform genotypic selection when the mean values of the phases are modified by homogamy, therefore, reduces to the case already considered in which homogamy is absent. The total effect of homogamy is to increase the effect of selection by the factor $\frac{1}{1-A}$. The distribution of frequency ratios is unaltered, for although by introducing a difference between I and J the selective effect is made more intense when

p is large, which would tend to make the distribution more symmetrical, this effect is exactly balanced by the increased effect of selection when p is small. The dominance ratio is therefore unaltered by the direct effect of assortative mating.

Summary.

The frequency ratio of the allelomorphs of a Mendelian factor is only stable if selection favours the heterozygote: such factors, though occurring rarely, will accumulate in the stock, while those of opposite tendency will be eliminated.

The survival of a mutant gene although established in a mature and potent individual is to a very large extent a matter of chance; only when a large number of individuals have become affected does selection, dependent on its contribution to the fitness of the organism, become of importance. This is so even for dominant mutants; for recessive mutants selection remains very small so long as the mutant form is an inconsiderable fraction of the interbreeding group.

The distribution of the frequency ratio for different factors may be calculated from the condition that this distribution is stable, as is that of velocities in the Theory of Gases: in the absence of selection the distribution of $\log \frac{p}{q}$ is given in fig. 1. Fig. 2 represents the case of steady decay in variance by the action of random survival (the Hagedoorn effect).

Fig. 3 shows the distribution in the somewhat artificial case of uniform genetic selection: this would be the distribution to be expected in the absence of dominance. Fig. 4 shows the asymmetrical distribution due to uniform genotypic selection with or without homogamy.

Under genotypic selection the dominance ratio for complete dominance comes to be exactly $\frac{1}{3}$, in close agreement with the value obtained from human measurements.

The rate of mutation necessary to maintain the variance of the species may be calculated from these distributions. Very infrequent mutation will serve to counterbalance the effect of random survival; for equilibrium with selective action a much higher level is needed, though still mutation may be individually rare, especially in large populations.

It would seem that the supposition of genotypic selection balanced by occasional mutations fitted the facts deduced from the correlations of relatives in mankind.

REFERENCES.

(1) Darwin, Charles, *The Origin of Species.*

(2) East, E. A., and Jones, D. F., *Inbreeding and Outbreeding* (1920).

(3) Fisher, R. A., "The Correlation between Relatives on the Supposition of Mendelian Inheritance," *Trans. Royal Soc. Edin.*, 1918, lii, pp. 399–433.

(4) Hagedoorn, A. L. and A. C., *The Relative Value of the Processes causing Evolution* (The Hague: Martinus Nijhoff), p. 294 (1921).

(5) Jennings, H. S., "Numerical Results of Diverse Systems of Breeding," *Genetics*, i, 53–89 (1916).

Jennings, H. S., "The Numerical Results of Diverse Systems of Breeding with respect to two pairs of characters, linked or independent; with special relation to the effects of linkage," *Genetics*, ii, 97–154 (1917).

(6) Pearson, K., "On a Generalised Theory of Alternative Inheritance, with special reference to Mendel's Laws," *Phil. Trans.*, cciii, A, pp. 53–87 (1903).

(7) Muller, H. J., "Genetic Variability: twin hybrids and constant hybrids in a case of balanced lethal factors," *Genetics*, iii, 422–499 (1918).

(8) Robbins, R. B., "Some Applications of Mathematics to Breeding Problems," *Genetics*, ii, 489–504 (1917).

Robbins, R. B., "Applications of Mathematics to Breeding Problems, II," *Genetics*, iii, 73–92 (1918).

Robbins, R. B., "Some Applications of Mathematics to Breeding Problems, III," *Genetics*, iii, 375–306 (1918).

(9) Wentworth, E. N., and Remick, B. L., "Some Breeding Properties of the Generalised Mendelian Population," *Genetics*, i, 608–616 (1916).

(10) Wright, S., "Systems of Mating," *Genetics*, vi, 111–178 (1921).

14

Reprinted from *Amer. Naturalist* **56**:330–338 (1922)

COEFFICIENTS OF INBREEDING AND RELATIONSHIP

DR. SEWALL WRIGHT

BUREAU OF ANIMAL INDUSTRY, UNITED STATES DEPARTMENT OF AGRICULTURE

IN the breeding of domestic animals consanguineous matings are frequently made. Occasionally matings are made between very close relatives—sire and daughter, brother and sister, etc.—but as a rule such close inbreeding is avoided and there is instead an attempt to concentrate the blood of some noteworthy individual by what is known as line breeding. No regular system of mating such as might be followed with laboratory animals is practicable as a rule.

The importance of having a coefficient by means of which the degree of inbreeding may be expressed has been brought out by Pearl[1] in a number of papers published between 1913 and 1917. His coefficient is based on the smaller number of ancestors in each generation back of an inbred individual, as compared with the maximum possible number. A separate coefficient is obtained for each generation by the formula

$$Z_n = 100\left(1 - \frac{q_{n+1}}{p_{n+1}}\right) = 100\left(1 - \frac{q_{n+1}}{2^{n+1}}\right)$$

where $q_{n+1}/2^{n+1}$ is the ratio of actual to maximum possible ancestors in the $n+1$st generation. By finding the ratio of a summation of these coefficients to a similar summation for the maximum possible inbreeding in higher animals, *viz.*, brother-sister mating, he obtains a single coefficient for the whole pedigree.

This coefficient has the defect, as Pearl himself pointed

[1] AMERICAN NATURALIST, 1917, 51: 545–559; 51: 636–639.

out, that it may come out the same for systems of breeding which we know are radically different as far as the effects of inbreeding are concerned. For example, in the continuous mating of double first cousins, an individual has two parents, four grandparents, four great grandparents and four in every generation, back to the beginning of the system. Exactly the same is true of an individual produced by crossing different lines, in each of which brother-sister mating has been followed. Yet in the first the individual will be homozygous in all factors if the system has been in progress sufficiently long; in the second he will be heterozygous in a maximum number of respects.

In order to overcome this objection Pearl has devised a partial inbreeding index which is intended to express the percentage of the inbreeding which is due to relationship between the sire and dam, inbreeding being measured as above described. A coefficient of relationship is used in this connection. These coefficients have been discussed by Ellinger[2] who suggests certain alterations and extensions by means of which the total inbreeding coefficient, a total relationship coefficient and a total relationship-inbreeding index for a given pedigree can be compared on the same scale.

An inbreeding coefficient to be of most value should measure as directly as possible the effects to be expected on the average from the system of mating in the given pedigree.

There are two classes of effects which are ascribed to inbreeding: First, a decline in all elements of vigor, as weight, fertility, vitality, etc., and second, an increase in uniformity within the inbred stock, correlated with which is an increase in prepotency in outside crosses. Both of these kinds of effects have ample experimental support as average (not necessarily unavoidable) consequences of inbreeding. The best explanation of the decrease in vigor is dependent on the view that Mendelian

[2] AMERICAN NATURALIST, 1920, 54: 540–545.

factors unfavorable to vigor in any respect are more frequently recessive than dominant, a situation which is the logical consequence of the two propositions that mutations are more likely to injure than improve the complex adjustments within an organism and that injurious dominant mutations will be relatively promptly weeded out, leaving the recessive ones to accumulate, especially if they happen to be linked with favorable dominant factors. On this view it may readily be shown that the decrease in vigor on starting inbreeding in a previously random-bred stock should be directly proportional to the increase in the percentage of homozygosis. Numerous experiments with plants and lower animals are in harmony with this view. Extensive experiments with guinea-pigs conducted by the Bureau of Animal Industry are in close quantitative agreement. As for the other effects of inbreeding, fixation of characters and increased prepotency, these are of course in direct proportion to the percentage of homozygosis. Thus, if we can calculate the percentage of homozygosis which would follow on the average from a given system of mating, we can at once form the most natural coefficient of inbreeding. The writer[3] has recently pointed out a method of calculating this percentage of homozygosis which is applicable to the irregular systems of mating found in actual pedigrees as well as to regular systems. This method, it may be said, gives results widely different from Pearl's coefficient, in many cases even as regards the relative degree of inbreeding of two animals.

Taking the typical case in which there are an equal number of dominant and recessive genes (A and a) in the population, the random-bred stock will be composed of 25 per cent. AA, 50 per cent. Aa and 25 per cent. aa. Close inbreeding will tend to convert the proportions to 50 per cent. AA, 50 per cent. aa, a change from 50 per cent. homozygosis to 100 per cent. homozygosis. For a natural coefficient of inbreeding, we want a scale which

[3] *Genetics*, 1921, 6: 111–178.

runs from 0 to 1, while the percentage of homozygosis is running from 50 per cent. to 100 per cent. The formula $2h-1$, where h is the proportion of complete homozygosis, gives the required value. This can also be written $1-2p$ where p is the proportion of heterozygosis. In the above-mentioned paper it was shown that the coefficient of correlation between uniting egg and sperm is expressed by this same formula, $f=1-2p$. We can thus obtain the coefficient of inbreeding f_b for a given individual B, by the use of the methods there outlined.

The symbol r_{bc}, for the coefficient of the correlation between B and C, may be used as a coefficient of relationship. It has the value 0 in the case of two random individuals, .50 for brothers in a random stock and approaches 1.00 for individuals belonging to a closely inbred subline of the general population.

In the general case in which dominants and recessives are not equally numerous, the composition of the random-bred stock is of the form x^2 AA, $2xy$ Aa, y^2 aa. The percentage of homozygosis is here greater than 50 per cent. The rate of increase, however, under a given system of mating, is always exactly proportional to that in the case of equality. The coefficient is thus of general application.

If an individual is inbred, his sire and dam are connected in the pedigree by lines of descent from a common ancestor or ancestors. The coefficient of inbreeding is obtained by a summation of coefficients for every line by which the parents are connected, each line tracing back from the sire to a common ancestor and thence forward to the dam, and passing through no individual more than once. The same ancestor may of course be involved in more than one line.

The path coefficient, for the path, sire (S) to offspring (O), is given by the formula $p_{o.s}=\frac{1}{2}\sqrt{(1+f_s)/(1+f_o)}$, where f_s and f_o are the coefficients of inbreeding for sire and offspring, respectively. The coefficient for the path, dam to offspring, is similar.

In the case of sire's sire (G) and individual, we have $p_{o.g} = p_{o.s}\, p_{s.g} = \frac{1}{4}\sqrt{(1+f_g)/(1+f_o)}$, and for any ancestor ($A$) we have for the coefficient pertaining to a given line of descent $p_{o.a} = (\frac{1}{2})^n\sqrt{(1+f_a)/(1+f_o)}$, where n is the number of generations between them in this line.

The correlation between two individuals (r_{bo}) is obtained by a summation of the coefficients for all connecting paths.

Thus

$$r_{bc} = \Sigma p_{ba} p_{ca}$$
$$= \Sigma \left(\frac{1}{2}\right)^{n+n'} \frac{1+f_a}{\sqrt{(1+f_b)(1+f_c)}},$$

where n and n' are the number of generations in the paths from A to B and from A to C, respectively.

The formula for the correlation between uniting gametes, which is also the required coefficient of inbreeding, is

$$f_o = \tfrac{1}{2} r_{sd}\sqrt{(1+f_s)(1+f_d)},$$

where r_{sd} is the correlation between sire and dam and f_s and f_d are coefficients of inbreeding of sire and dam. Substituting the value of r_{sd} we obtain

$$f_o = \Sigma\,(\tfrac{1}{2})^{n+n'+1}(1+f_a).$$

If the ancestor (A) is not inbred, the component for the given path is simply $(\frac{1}{2})^{n+n'+1}$ where n and n' are the number of generations from sire and dam respectively to the ancestor in question. If the common ancestor is inbred himself, his coefficient of inbreeding (f_a) must be worked out from his pedigree.

This formula gives the departure from the amount of homozygosis under random mating toward complete homozygosis. The percentage of homozygosis (assuming 50 per cent. under random mating) is $\frac{1}{2}(1+f_o) \times 100$.

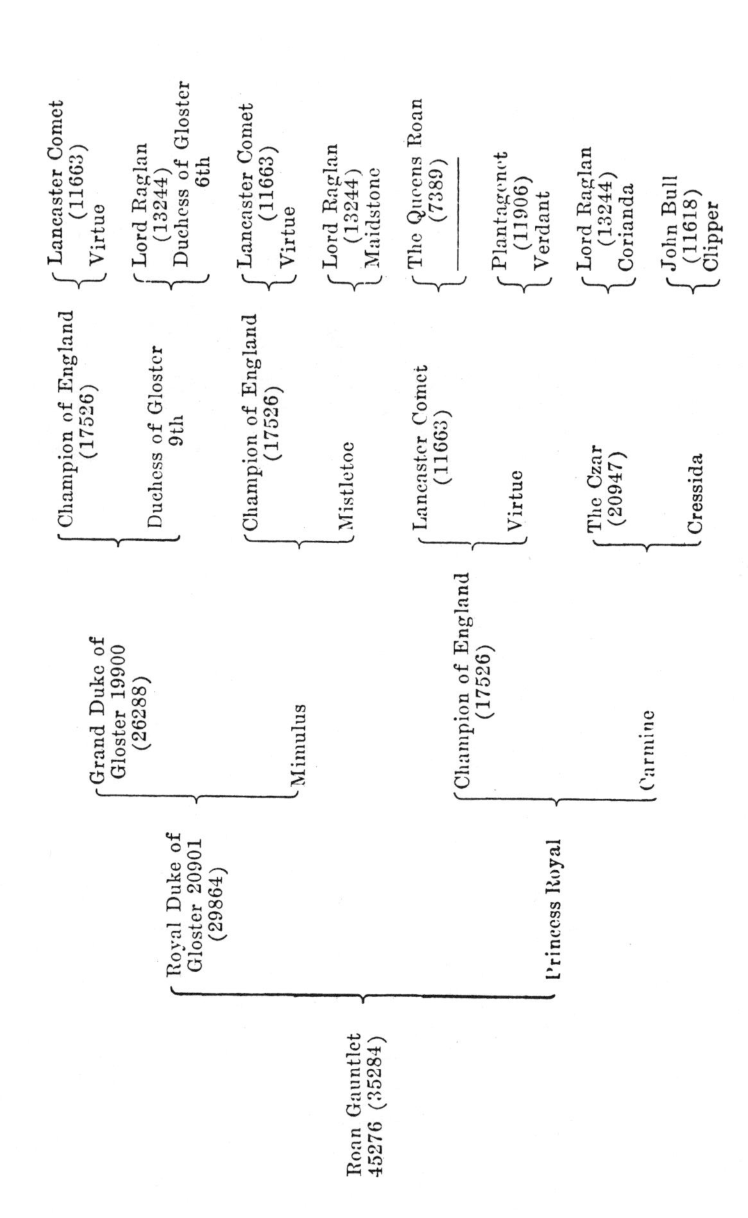
Roan Gauntlet 45276 (35284)
Royal Duke of Gloster 20901 (29864)
Princess Royal
Grand Duke of Gloster 19900 (26288)
Mimulus
Champion of England (17526)
Carmine
Champion of England (17526)
Duchess of Gloster 9th
Champion of England (17526)
Mistletoe
Lancaster Comet (11663)
Virtue
The Czar (20947)
Cressida
Lancaster Comet (11663)
Virtue
Lord Raglan (13244)
Duchess of Gloster 6th
Lancaster Comet (11663)
Virtue
Lord Raglan (13244)
Maidstone
The Queens Roan (7389)
Plantagenet (11906)
Verdant
Lord Raglan (13244)
Corianda
John Bull (11618)
Clipper

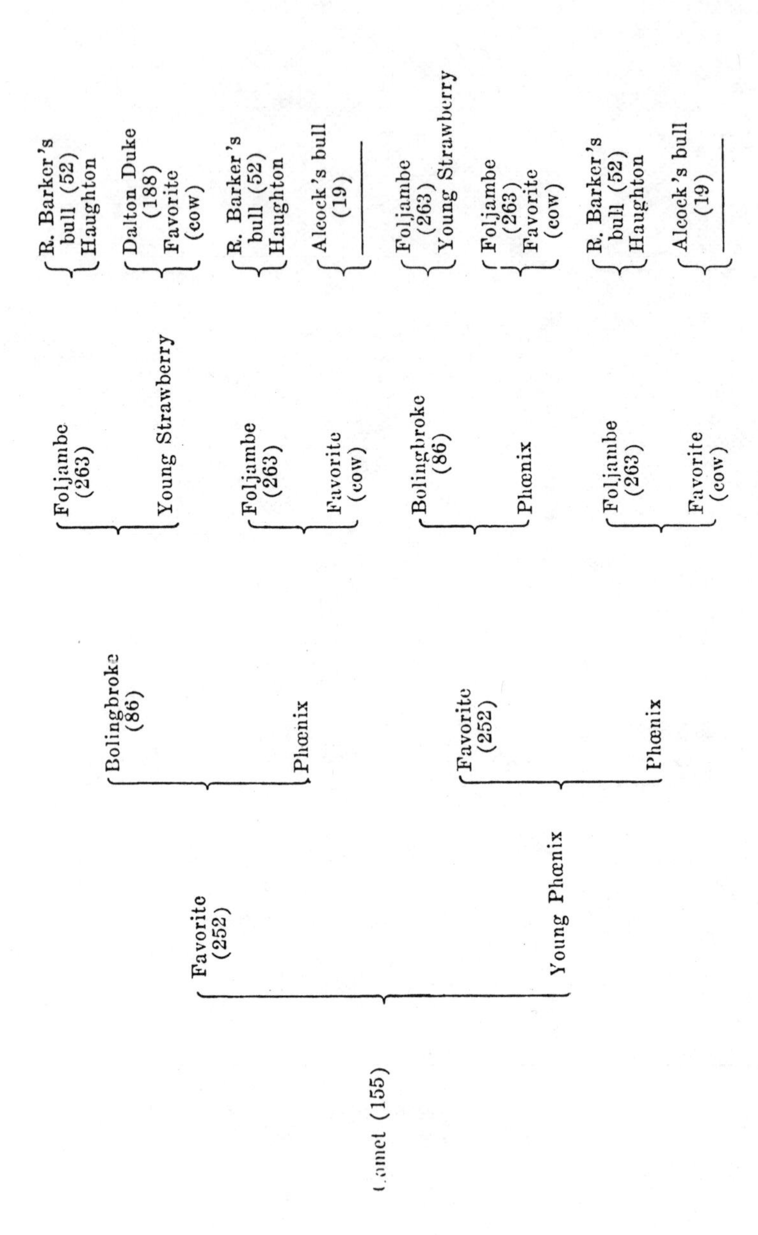

(155)
Favorite (252)
Young Phœnix
Bolingbroke (86)
Phœnix
Favorite (252)
Phœnix
Foljambe (263)
Young Strawberry
Foljambe (263)
Favorite (cow)
Bolingbroke (86)
Phœnix
Foljambe (263)
Favorite (cow)
R. Barker's bull (52)
Haughton
Dalton Duke (188)
Favorite (cow)
R. Barker's bull (52)
Haughton
Alcock's bull (19)
Foljambe (263)
Young Strawberry
Foljambe (263)
Favorite (cow)
R. Barker's bull (52)
Haughton
Alcock's bull (19)

By this means the inbreeding in an actual pedigree, however irregular the system of mating, can be compared accurately with that under any regular system of mating.

As an illustration, take the pedigree of Roan Gauntlet, a famous Shorthorn sire, bred by Amos Cruickshank. This bull traces back in every line to a mating of Champion of England with a daughter or granddaughter of Lord Raglan. For the present purpose we will assume that these bulls were not at all inbred themselves and not related to each other. Since the sire traces twice to Champion of England and twice to Lord Raglan and the dam once to each bull, there are in all four lines by which the sire and dam are connected.

Individual	Common Ancestors of Sire and Dam	f_a	n	n'	$(\frac{1}{2})^{n+n'+1} \times (1+f_a)$
Roan Gauntlet 45,276 (35,284)	Champion of England (17,526)...........	0	2	1	.062500
			2		.062500
	Lord Raglan (13,244)..	0	3	3	.007812
			3		.007813
					.140625

The coefficient of inbreeding comes out 14.1 per cent., a rather low figure when compared to such systems as brother-sister mating (one generation 25 per cent., two generations 37.5 per cent., three generations 50 per cent., ten generations 88.6 per cent.) or parent-offspring mating, (one generation 25 per cent., two generations 37.5 per cent., three generations 43.8 per cent., approaching 50 per cent. as a limit).

As an example of closer inbreeding, take the pedigree of Charles Collings' bull, Comet. The sire was the bull Favorite and the dam was from a mating of Favorite with his own dam. As Favorite was himself inbred to some extent, it is necessary to calculate first his own coefficient of inbreeding.

Individual	Common Ancestors of Sire and Dam	f_a	n	n'	$(\frac{1}{2})^{n+n'+1} \times (1+f_a)$
Favorite (252)	Foljambe (263)........	0	1	1	.1250
	Favorite (cow)........	0	2	1	.0625
					.1875
Comet (115)	Favorite (252)........	.1875	0	1	.2969
	Phœnix..............	0	1	1	.1250
	Foljambe............	0	2	2	.0312
	Favorite (cow)........	0	3	2	.0156
					.4687

In the case of Comet, Foljambe and Favorite (cow) each appears twice in the pedigree of the sire and three times in the pedigree of the dam. However, only those pedigree paths which connect sire and dam and which do not pass through the same animal twice are counted. The listing of Favorite (252) and Phœnix as common ancestors eliminates all but one path in each case as regards Foljambe and Favorite cow. The remaining paths are those due to the common descent of Bolingbroke, the sire's sire and Phœnix as the dam's dam from the above two animals.

By tracing the pedigrees back to the beginning of the herd book, the coefficients of inbreeding are slightly increased. This meant going back to the seventh generation for one common ancestor of the sire and dam of Favorite. The coefficient in the case of Favorite becomes .192 instead of .188 and that of Comet .471 instead of .469. Remote common ancestors in general have little effect on the coefficient. It will be noticed that Comet has a degree of inbreeding almost equal to three generations of brother-sister mating or an indefinite amount of sire-daughter mating where the sire is not himself inbred.

15

Reprinted from *Cambridge Phil. Soc. Trans.* 23:19–41 (1924)

A MATHEMATICAL THEORY OF NATURAL AND ARTIFICIAL SELECTION. PART I

J. B. S. Haldane
Trinity College, Cambridge

A SATISFACTORY theory of natural selection must be quantitative. In order to establish the view that natural selection is capable of accounting for the known facts of evolution we must show not only that it can cause a species to change, but that it can cause it to change at a rate which will account for present and past transmutations. In any given case we must specify:

(1) The mode of inheritance of the character considered,
(2) The system of breeding in the group of organisms studied,
(3) The intensity of selection,
(4) Its incidence (e.g. on both sexes or only one), and
(5) The rate at which the proportion of organisms showing the character increases or diminishes.

It should then be possible to obtain an equation connecting (3) and (5).

The principal work on the subject so far is that of Pearson (1), Warren (2), and Norton. Pearson's work was based on a pre-Mendelian theory of variation and heredity, which is certainly inapplicable to many, and perhaps to all characters. Warren has only considered selection of an extremely stringent character, whilst Norton's work is as yet only available in the table quoted by Punnett (3).

In this paper we shall only deal with the simplest possible cases. The character dealt with will be the effect of a single completely dominant Mendelian factor or its absence. The system of breeding considered will be random mating on the one hand or self-fertilization, budding, etc. on the other. Moreover we shall confine ourselves to organisms such as annual plants, and many insects and fish in which different generations do not interbreed. Even so it will be found that in most cases we can only obtain rigorous solutions when selection is very rapid or very slow. At intermediate rates we should require to use functions of a hitherto unexplored type. Indeed the mathematical problems raised in the more complicated cases to be dealt with in subsequent papers seem to be as formidable as any in mathematical physics. The approximate solutions given in this paper are however of as great an order of accuracy as that of the data hitherto available.

It is not of course intended to suggest that all heredity is Mendelian, or all evolution by natural selection. On the other hand we know that besides non-Mendelian differences between species (e.g. in chromosome number) there are often Mendelian factor-differences. The former are important because they often lead to total or partial sterility in crosses, but their somatic expression is commonly less striking than that of a single factor-difference. Their behaviour in crosses is far from clear, but where crossing does not occur evolution takes place according to equations (1·0)—(1·2).

SPECIFICATION OF THE INTENSITY OF SELECTION.

If a generation of zygotes immediately after fertilization consists of two phenotypes A and B in the ratio $pA:1B$, and the proportion which form fertile unions is $pA:(1-k)B$, we shall describe k as the coefficient of selection. Thus if $k = \cdot 01$, a population of equal numbers of A and B would survive to form fertile unions in the proportion $100A:99B$, the A's thus having a slight advantage. k may be positive or negative. When it is small selection is slow. When $k=1$ no B's reproduce, when $k=-\infty$ no A's reproduce. It will be convenient to refer to these two cases as "complete selection." They occur in artificial selection if the character is well marked.

If the character concerned affects fertility, or kills off during the breeding period, we can use just the same notation. In this case each B on the average leaves as many offspring as $(1-k)$ A's, e.g. if $k = \cdot 01$ then 100 B's leave as many as 99 A's. The effect is clearly just the same as if one of the B's had died before breeding. It will be observed that no assumption is made as to the total number of the population. If this is limited by the environment, natural selection may cause it to increase or diminish. It will for example tend to increase if selection renders the organism smaller or fitter to cope with its environment in general. If on the other hand selection increases its size, or merely arms it in the struggle with other members of its species for food or mates, the population will tend to diminish or even to disappear.

Warren (2) considered the case where the total population is fixed. He supposes that the parents produce l times their number of offspring, and that type A is p times as numerous as type B, but $\frac{1}{m}$ as likely to die. In this case it can be shown that

$$k = \frac{(l-1)(m-1)(p+1)}{lm-l+p+1}.$$

Hence the advantage of one type over the other as measured by k is not independent of the composition of the population unless $m-1$ is very small, when $k=(l-1)(m-1)$ approximately. Hence when selection is slow—the most interesting case—the two schemes of selection lead to similar results. On the other hand the mathematical treatment of selection on our scheme is decidedly simpler.

FAMILIAL SELECTION.

The above notation may easily be applied to the cases, such as Darwinian sexual selection, where one sex only is selected. There is however another type of selection which so far as I know has not been considered in any detail by former authors, but which must have been of considerable importance in evolution. So far we have assumed that the field of struggle for existence is the species as a whole, or at least those members of it living within a given area. But we have also to consider those cases where the struggle occurs between members of the same family. Such cases occur in many mammals, seed-plants, and nematodes, to mention no other groups. Here the size of the family is strictly limited by the food or space available for it, and more embryos are produced than can survive to enter into the struggle with members of other families. Thus in the mouse Ibsen and Steigleder (4) have shown that some embryos of any litter perish in utero. Their deaths are certainly sometimes selective. In litters from the mating yellow × yellow one-quarter of the embryos die in the blastula stage, yet as Durham (5) has shown, such litters are no smaller than the normal, because the death of the YY embryos allows others to survive which would normally have perished.

The above is a case of complete selection. Where the less viable type of embryo, instead of perishing inevitably, is merely at a slight disadvantage, it is clear that selection will only be effective, or at any rate will be much more effective, in the mixed litters. Thus let us consider 3 litters of 20 embryos each, the first consisting wholly of the stronger type, the second containing 10 strong and 10 weak, the third wholly of the weaker type. Suppose that in each case there is only enough food or space for 10 embryos, and that the strong type has an advantage over the weak such that, out of equal numbers, 50 °/₀ more of the strong will survive, i.e. $k = \frac{1}{3}$. Then the survivors will be 10 strong from the first litter, 6 strong and 4 weak from the second, and 10 weak from the third, or 16 strong and 14 weak. If the competition had been free, as with pelagic larvae, the numbers would have been 18 strong and 12 weak. Clearly with familial selection the same advantage acts more slowly than with normal selection, since it is only effective in mixed families.

The "family" within which selection acts may have both parents in common, as in most mammals, or many different male parents, as in those plants whose pollen, but not seeds, is spread by the wind. In this case the seeds from any one plant will fall into the same area, and unless the plants are very closely packed, will compete with one another in the main. In rare cases familial sexual selection may occur. Thus in *Dinophilus* the rudimentary males fertilize their sisters before leaving the cocoon. Clearly so long as every female gets fertilized before hatching selection can only occur in the male sex between brothers, and must tend to make the males copulate at as early a date as possible.

The survival of many of the embryonic characters of viviparous animals and seed-plants must have been due to familial selection.

Selection in the absence of amphimixis.

The simplest form of selection is uncomplicated either by amphimixis or dominance. It occurs in the following cases:

(1) Organisms which do not reproduce sexually, or are self-fertilizing.

(2) Species which do not cross, but compete for the same means of support.

(3) Organisms in which mating is always between brother and sister.

(4) Organisms like *Bryophyta* which are haploid during part of the life cycle, provided that selection of the character considered only occurs during the haploid phase.

(5) Heterogamous organisms in which the factor determining the character selected occurs in the gametes of one sex only. For example Renner (6) has shown that *Oenothera muricata* transmits certain characters by the pollen only, others by the ovules only. Schmidt (7) has found a character in *Lebistes* transmitted by males to males only, and Goldschmidt (8) has postulated sex-factors in *Lymantria* transmitted only by females to females. As far as the characters in question are concerned there is no amphimixis, and these organisms behave as if they were asexual. Other species of *Oenothera* which are permanently heterozygous for other reasons would probably be selected in much the same way.

Let the nth generation consist of types A and B in the ratio $u_n A : 1B$, and let the coefficient of selection be k, i.e. $(1-k)$ B's survive for every A. Then the survivors of the nth generation, and hence the first numbers of the $(n+1)$th, will be $u_n A : (1-k) B$.

$$\therefore \quad u_{n+1} = \frac{u_n}{1-k}, \qquad \text{(1·0)}$$

and if u_0 be the original ratio $u_n = (1-k)^{-n} u_0$.

Now if we write y_n for the proportion of B's in the total population of the nth generation,

$$y_n = \frac{1}{1+u_n} = \frac{1}{1+(1-k)^{-n}u_0} = \frac{y_0}{y_0+(1-k)^{-n}(1-y_0)},$$

or if we start with equal numbers of A and B, $y_0 = \frac{1}{2}$, and

$$y_n = \frac{1}{1+(1-k)^{-n}}. \qquad \text{.........(1·1)}$$

If k is very small, i.e. selection slow, then approximately

$$\left.\begin{aligned} y_n &= \frac{1}{1+e^{kn}} \\ \text{or} \qquad kn &= \log_e\left(\frac{1-y_n}{y_n}\right) \end{aligned}\right\}. \qquad \text{.........(1·2)}$$

Hence the proportion of B's falls slowly at first, then rapidly for a short time, then slowly again, the rate being greatest when $y = \frac{1}{2}$. Before $y = \frac{1}{2}$, n is of course taken as negative. So long as k is small the time taken for any given change in the proportions varies inversely as k. The curve representing graphically the change of the population is symmetrical about its middle point, and is shown in Fig. 1 for the case where $k = \cdot 001$, i.e. 999 B's survive for every 1000 A's. 9,184 generations are needed for the proportion of A's to increase from 1 °/₀ to 99 °/₀. Equation (1·2) gives an error of only 4 in this number.

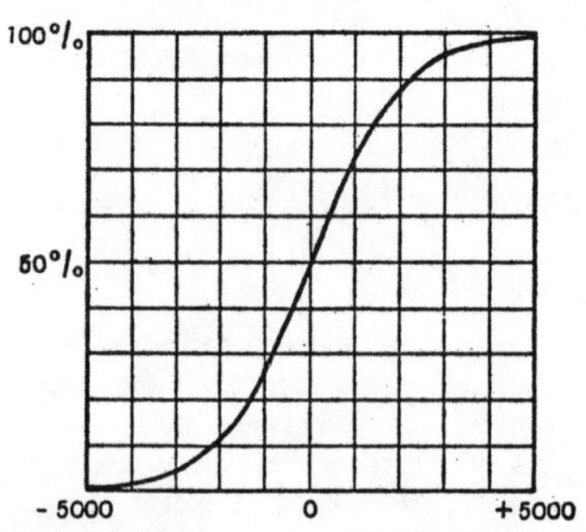

Fig. 1. Effect of selection on a non-amphimictic character. $k = \cdot 001$.
Abscissa = generations.
Ordinate = percentage of population with the favoured character.

As will be shown below, selection proceeds more slowly with all other systems of inheritance. In this case the speed must compensate to some extent for the failure to combine advantageous factors by amphimixis. Where occasional amphimixis occurs, as for example in wheat, conditions are very favourable for the evolution of advantageous combinations of variations.

SELECTION OF A SIMPLE MENDELIAN CHARACTER.

Consider the case of a population which consists of zygotes containing two, one, or no "doses" of a completely dominant Mendelian factor A, mating is at random, and selection acts to an equal degree in both sexes upon the character produced by the factor. Pearson (9) and Hardy (10) have shown that in a population mating at random the square of the number of heterozygotes is equal to four times the product of the numbers of the two homozygous classes. Let $u_n A : 1a$ be the proportion of the two types of gametes produced by the $(n-1)$th generation. Then in the nth generation the initial proportions of the three classes of zygotes are:

$$u_n^2 AA : 2u_n Aa : 1aa.$$

The proportion of recessives to the whole population is:

$$y_n = (1+u_n)^{-2}. \quad \text{.......................................(2·0)}$$

Now only $(1-k)$ of the recessives survive to breed, so that the survivors are in the proportions:

$$u_n^2 AA : 2u_n Aa : (1-k)\, aa.$$

The numbers of the next generation can be most easily calculated from the new gametic ratio u_{n+1}. This is immediately obvious in the case of aquatic organisms which shed their gametes into the water. If each zygote produces N gametes which conjugate, the numbers are clearly:

$$(Nu_n^2 + Nu_n)\, A, \text{ and } (Nu_n + N\overline{1-k})\, a.$$

So the ratio

$$u_{n+1} = \frac{u_n(1+u_n)}{1+u_n-k}. \quad \text{.......................................(2·1)}$$

It can easily be shown that this result follows from random mating, for matings will occur in the following proportions:

$AA \times AA$		$u_n^2 \times u_n^2$	or u_n^4,
$AA \times Aa$	and reciprocally,	$2 \times u_n^2 \times 2u_n$	„ $4u_n^3$,
$AA \times aa$	„ „	$2 \times u_n^2 \times (1-k)$	„ $2(1-k)\, u_n^2$,
$Aa \times Aa$		$2u_n \times 2u_n$	„ $4u_n^2$,
$Aa \times aa$	and reciprocally,	$2 \times 2u_n \times (1-k)$	„ $4(1-k)\, u_n$,
$aa \times aa$		$(1-k) \times (1-k)$	„ $(1-k)^2$.

Hence zygotes are formed in the following proportions:

AA	$u_n^4 + 2u_n^3 + u_n^2$	or $u_n^2(1+u_n)^2$,
Aa	$2u_n^3 + 2(1-k)\, u_n^2 + 2u_n^2 + 2(1-k)\, u_n$	„ $2u_n(1+u_n)(1+u_n-k)$,
aa	$u_n^2 + 2(1-k)\, u_n + (1-k)^2$	„ $(1+u_n-k)^2$.

These ratios may be written:

$$\left[\frac{u_n(1+u_n)}{1+u_n-k}\right]^2 AA : \frac{2u_n(1+u_n)}{1+u_n-k} Aa : 1aa,$$

or

$$u_{n+1}^2 AA : 2u_{n+1} Aa : 1aa,$$

where

$$u_{n+1} = \frac{u_n(1+u_n)}{1+u_n-k}, \quad \text{.......................................(2·1)}$$

as above. It is however simpler to obtain u_{n+1} directly from the ratio of A to a among the gametes of the population as a whole, and this will be done in our future calculations.

Now if we know the original proportion of recessives y_0, we start with a population:

$$u_0^2 AA : 2u_0 Aa : 1aa,$$

where

$$u_0 = y_0^{-\frac{1}{2}} - 1,$$

and we can at once calculate

$$u_1 = \frac{u_0(1+u_0)}{1+u_0-k},$$

and thence u_2 and so on, obtaining y_1, y_2, etc. from equation (2·0). Thus if we start with 25 °/。 of recessives, and $k = ·5$, i.e. the recessives are only half as viable as the dominants, then $u_0 = 1$, and

$$u_1 = \frac{1(1+1)}{1+1-\frac{1}{2}} = \tfrac{4}{3},$$

$$y_1 = (1+\tfrac{4}{3})^{-2} = \tfrac{9}{49} = ·184, \text{ or } 18·4\ °/_。.$$

Similarly $y_2 = 13{\cdot}75$ %, $y_3 = 10{\cdot}9$ %, and so on. Starting from the same population, but with $k = -1$, so that the recessives are twice as viable as the dominants, we have $y_1 = 36$ %, $y_2 = 49{\cdot}8$ %, $y_3 = 64{\cdot}6$ %, $y_4 = 77{\cdot}5$ %, $y_5 = 87{\cdot}0$ %, and so on. If k is small this method becomes very tedious, but we can find a fairly accurate formula connecting y_n with n.

The case of complete selection is simple. If all the dominants are killed off or prevented from breeding we shall see the last of them in one generation, and $y_n = 1$. Punnett (11) and Hardy have solved the case where the recessives all die. Here $k = 1$, and

$$u_{n+1} = \frac{u_n(1+u_n)}{1+u_n-1} = 1 + u_n.$$

$$\therefore \quad u_n = n + u_0;$$

$$\therefore \quad y_n = (n+1+u_0)^{-2}$$
$$= (n + y_0^{-\frac{1}{2}})^{-2}$$
$$= y_0(1 + n y_0^{\frac{1}{2}})^{-2}. \qquad \text{.......(2·2)}$$

Thus if we start with a population containing $\frac{1}{4}$ recessives the second generation will contain $\frac{1}{9}$, the third $\frac{1}{16}$, the nth $\frac{1}{(n+1)^2}$. Thus 999 generations will be needed to reduce the proportion to one in a million, and we need not wonder that recessive sports still occur in most of our domestic breeds of animals.

When selection is not very intense, we can proceed as follows:

$$u_{n+1} = \frac{u_n(1+u_n)}{1+u_n-k}; \qquad \text{.......(2·1)}$$

$$\therefore \quad \Delta u_n \equiv u_{n+1} - u_n = \frac{k u_n}{1+u_n-k}.$$

When k is small we can neglect it in comparison with unity, and suppose that u_n increases continuously and not by steps, i.e. take $\Delta u_n = \frac{du_n}{dn}$.

$$\therefore \quad \frac{du_n}{dn} = \frac{k u_n}{1+u_n} \text{ approximately;}$$

$$\therefore \quad kn = \int_{u_0}^{u_n} \frac{1+u}{u}\,du$$
$$= u_n - u_0 + \log_e\left(\frac{u_n}{u_0}\right). \qquad \text{.......(2·3)}$$

If we start from or work towards a standard population containing 25 % of recessives, and hence $u_0 = 1$, we have

$$kn = u_n + \log_e u_n - 1. \qquad \text{.......(2·4)}$$

This equation is accurate enough for any practical problem when $|k|$ is small, and as long as k lies between $\pm 0{\cdot}1$, i.e. neither phenotype has an advantage of more than 10 %, it may be safely used. When $|k|$ is large the equation

$$kn = u_n + (1-k)\log_e u_n - 1 \qquad \text{.......(2·5)}$$

is fairly accurate for positive values of n. Thus when $k = \frac{1}{2}$, the error is always under 4 %. For large values of $|k|$ and negative values of n the equation

$$kn = u_n + \left(1 - \frac{k}{2}\right)\log_e u_n - 1 \qquad \text{.......(2·6)}$$

gives results with a very small error. But for every case so far observed equation (2·4) gives results within the limits of observational error.

In the above equations we have only to make k negative to calculate the effects of a selection which favours recessives at the expense of dominants. For the same small intensity of selection the same time is clearly needed to produce a given change in the percentage of recessives whether dominants or recessives are favoured. Fig. 2 shows graphically the rate of increase of dominants and recessives respectively when $k = \pm \cdot 001$, i.e. the favoured type has an advantage of one in a thousand, as in Fig. 1. In each case 16,582 generations are required to increase the proportion of the favoured type from 1 °/。 to 99 °/。, but dominants increase more rapidly than recessives when they are few, more slowly when they are numerous. The change occurs most rapidly when y_n, the proportion of recessives, is 56·25 °/。. When selection is ten times as intense, the population will clearly change ten times as fast, and so on.

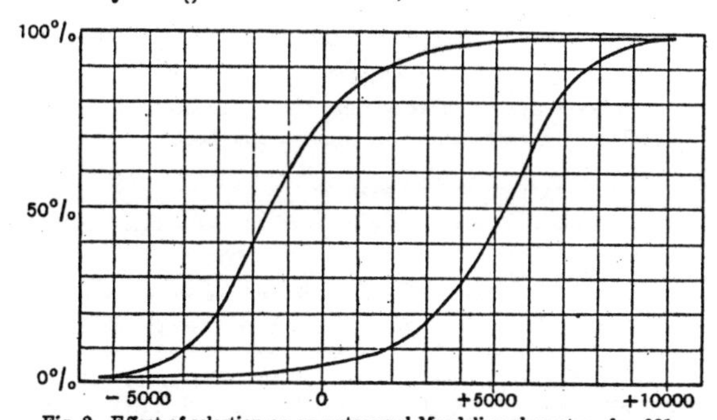

Fig. 2. Effect of selection on an autosomal Mendelian character. $k = \cdot 001$.
Upper curve, dominants favoured; lower curve, recessives favoured.
Abscissa = generations. Ordinate = percentage of population with the favoured character.

TABLE I.

Effect of slow selection on an autosomal Mendelian character.

kn (number of generations × k)	− 1000	− 100	− 50	− 20	− 15	− 10
°/。 of recessives when dominants are favoured	...	...	...	...	99·9998	99·975
" " " recessives " "	·0001	·0105	·0427	·2773	·4215	1·036

kn	− 9	− 8	− 7	− 6	− 5	− 4·5	− 4	− 3·5	− 3	− 2·5	− 2	− 1·5
dominants favoured	99·933	99·82	99·50	98·68	96·50	94·38	91·14	86·36	79·71	71·24	61·53	50·68
recessives favoured	1·254	1·545	1·940	2·497	3·308	...	4·537	...	6·528	...	9·718	12·11

kn	− 1	− 0·5	0	0·5	1	1·5	2	2·5	3	3·5	4	4·5
dominants favoured	40·98	32·05	25·0	19·53	15·30	12·11	9·718	...	6·528	...	4·537	...
recessives favoured	15·30	19·53	25·0	32·05	40·98	50·68	61·53	71·24	79·71	86·36	91·14	94·38

kn	5	6	7	8	9	10	15	20	50	100	1000
dominants favoured	3·308	2·497	1·940	1·545	1·254	1·036	·4215	·2773	·0427	·0105	·0001
recessives favoured	96·50	98·68	99·50	99·82	99·933	99·975	99·9998	...	...	...	...

In Table I the values of y_n calculated from equations (2·4) and (2·0) are given in terms of kn. In Table II kn is given in terms of y_n. The number of generations (forwards or backwards) is reckoned from a standard population containing 75 °/。 of dominants and 25 °/。 of recessives. A few examples will make the use of these tables clear.

1. Detlefsen (12) has shown that in a mixed population of mice about 95·9 without the factor G, causing light bellies and yellow-tipped hair, survive for every 100 with it. Hence $k = \cdot 041$. It is required to find how many recessives will be left after 100 generations, starting from a population with 90 °/₀ of recessives, and assuming that different generations do not interbreed.

From Table II, when $y = \cdot 9$, $kn = -3\cdot 863$, ∴ $n = -94\cdot 2$. So 94·2 generations of selection will bring the recessives down to 25 °/₀. The remaining 5·8 generations give $kn = \cdot 238$, and from Table I by interpolation we find $y = \cdot 224$, i.e. only 22·4 °/₀ of recessives remain.

2. In the same case how many generations are needed to reduce the number of recessives to 1 °/₀? $y_n = \cdot 01$, hence, from Table II, $kn = 10\cdot 197$, ∴ $n = 248\cdot 7$. So 248·7 generations after 25 °/₀ is reached, or 343 in all, will be required.

3. The dominant melanic form *doubledayaria* of the peppered moth *Amphidasys betularia* first appeared at Manchester in 1848. Some time before 1901 when Barrett (13) described the case, it had completely ousted the recessive variety in Manchester. It is required to find the least intensity of natural selection which will account for this fact.

TABLE II.

Effect of slow selection on an autosomal Mendelian character.

°/₀ of favoured type	·0001	·001	·01	·05	·1	·2	·5
kn when dominants are favoured	−15·51	− 13·21	− 10·90	− 9·294	− 8·600	− 7·905	− 6·996
kn „ recessives „ „	− 1005	−320·0	−102·60	−45·50	−33·04	−23·42	−14·72

°/₀ of favoured type	1	2	3	5	10	15	20	25	30	35	40
kn (dominants)	− 6·286	−5·580	−5·161	−4·624	−3·863	−3·290	−2·979	−2·712	−2·439	−2·180	−1·964
kn (recessives)	−10·197	−6·875	−4·976	−3·717	−1·933	−1·041	− ·448	0	+ ·366	+ ·681	+ ·962

°/₀ of favoured type	45	50	55	60	65	70	75	80	85	90	95
kn (dominants)	−1·708	−1·467	−1·220	− ·962	− ·681	− ·366	0	+ ·448	+1·041	+1·933	+3·717
kn (recessives)	+1·220	+1·467	+1·708	+1·964	+2·180	+2·439	+2·712	+2·979	+3·290	+3·863	+4·620

°/₀ of favoured type	97	98	99	99·5	99·8	99·9	99·95	99·99	99·999	99·9999
kn (dominants)	+4·976	+6·875	+10·197	+14·72	+23·42	+33·04	+45·50	+102·60	+320·0	+1005
kn (recessives)	+5·161	+5·580	+ 6·286	+ 6·996	+ 7·905	+ 8·600	+ 9·294	+ 10·90	+ 13·21	+15·51

Assuming that there were not more than 1 °/₀ of dominants in Manchester in 1848, nor less than 99 °/₀ in 1898, we have, from Table II, $kn = 16\cdot 58$ as a minimum. But $n = 50$, since this moth usually has one brood per year. ∴ $k = \cdot 332$ at least, i.e. at least 3 dominants must survive for every 2 recessives, and probably more; or the fertility of the dominants must be 50 °/₀ greater than that of the recessives. Direct calculation step by step from equation (2·1) shows that 48 generations are needed for the change if $k = \cdot\dot{3}$. Hence the table is sufficiently accurate. After only 13 generations the dominants would be in a majority. It is perhaps instructive, in view of the fact that attempts have been made to explain such cases by epidemics of mutation due either to the environment or to unknown causes, to note that in such a case one recessive in every five would have to mutate to a dominant. Hence it would be impossible to obtain true breeding recessives as was done by Bate (14). Another possible explanation would be a large excess of dominants begotten in mixed families, as occurs in human night-blindness according to Bateson (15). But this again does not agree with the facts, and the only probable explanation is the not very intense degree of natural selection postulated above.

FAMILIAL SELECTION OF A SIMPLE MENDELIAN CHARACTER.

Consider the case of a factor A whose presence gives any embryo possessing it an advantage measured by k over those members of the same family which do not possess it. In this case the Pearson-Hardy law does not hold in the population. Each family may have both parents in common, as in mammals, or only the mother, as in cross-pollinated seed-plants. In the first case let the population consist of

$$p_n AA : 2q_n Aa : r_n aa, \text{ where } p_n + 2q_n + r_n = 1.$$

Then in a mixed family where equality was to be expected the ratio of dominants to recessives will be $1 : 1 - k$. But since the total is unaltered, the actual number of dominants will be to the expected as $2 : 2 - k$, of recessives as $2 - 2k : 2 - k$, and similarly for a family where a $3 : 1$ ratio was to be expected. The nth generation mating at random will therefore produce surviving offspring in the following proportions:

	AA	Aa	aa
From mating $AA \times AA$ ...	p_n^2	0	0
" " $AA \times Aa$ ...	$2p_n q_n$	$2p_n q_n$	0
" " $AA \times aa$ ...	0	$2p_n r_n$	0
" " $Aa \times Aa$ ...	$\frac{4q_n^2}{4-k}$	$\frac{8q_n^2}{4-k}$	$\frac{(4-4k)\,q_n^2}{4-k}$
" " $Aa \times aa$ ...	0	$\frac{2q_n r_n}{2-k}$	$\frac{(2-2k)\,q_n r_n}{2-k}$
" " $aa \times aa$ ...	0	0	r_n^2

$$\left.\begin{aligned} \therefore\ [AA] &= p_{n+1} = (p_n + q_n)^2 + \frac{kq_n^2}{4-k} \\ \tfrac{1}{2}[Aa] &= q_{n+1} = (p_n + q_n)(q_n + r_n) + kq_n\left(\frac{q_n}{4-k} + \frac{r_n}{2-k}\right) \\ [aa] &= r_{n+1} = (q_n + r_n)^2 - kq_n\left(\frac{3q_n}{4-k} + \frac{2r_n}{2-k}\right) \end{aligned}\right\} \quad \text{............(3·0)}$$

With complete selection, when $k = 1$, we have $r_{n+1} = r_n^2$, so the proportion of recessives, starting from $\frac{1}{4}$, will be $\frac{1}{16}$, $\frac{1}{64}$, etc., in successive generations, provided of course that all-recessive families survive, as in *Oenothera*. So recessives are eliminated far more quickly than in the ordinary type of selection. Clearly however dominants are not eliminated at once when $k = -\infty$ (provided that they survive in all-dominant families), for

$$p_{n+1} = p_n(1 - r_n) = p_n p_{n-1}(2 - p_{n-1}).$$

Starting from the standard population, successive proportions of recessives are 25 %, 56·25 %, 66·02 %, 84·25 %, etc.

In the more interesting case when k is small we can solve approximately, as follows. From equation (3·0) we see that $q^2_{n+1} - p_{n+1}r_{n+1}$ and hence $q_n^2 - p_n r_n$ is a small quantity of the order kq_n^2, i.e. is less than k. Hence if we write $u_n = \frac{p_n + q_n}{q_n + r_n}$, then q_n only differs from $\frac{u_n}{(1 + u_n)^2}$ by a small quantity of the order of k.

Now
$$u_{n+1} = \frac{p_{n+1} + q_{n+1}}{q_{n+1} + r_{n+1}}$$
$$= \frac{p_n + q_n + kq_n\left(\frac{2q_n}{4-k} + \frac{r_n}{2-k}\right)}{q_n + r_n - kq_n\left(\frac{2q_n}{4-k} + \frac{r_n}{2-k}\right)}$$
$$= \frac{p_n + q_n + \frac{1}{2}kq_n(q_n + r_n)}{q_n + r_n - \frac{1}{2}kq_n(q_n + r_n)} \text{ approximately}$$
$$= \frac{u_n + \frac{1}{2}kq_n}{1 - \frac{1}{2}kq_n}$$
$$= u_n + \frac{1}{2}kq_n(1 + u_n) \text{ approximately}$$
$$= u_n + \frac{ku_n}{2(1 + u_n)}. \quad \text{.....(3·1)}$$

Solving as for equation (2·1) we find
$$\tfrac{1}{2}kn = u_n + \log_e u_n - 1. \quad \text{.....(3·2)}$$

And since as above r_n (the proportion of recessives) $= (1 + u_n)^{-2}$, it follows that the species changes its composition at half the rate at which it would change if selection worked on the species as a whole, and not within families only.

If each family has its mother only in common, but the fathers are a random sample of the population, we assume the nth generation to consist of
$$p_n AA : 2q_n Aa : r_n aa, \text{ where } p_n + 2q_n + r_n = 1.$$

Let $u_n = \dfrac{p_n + q_n}{q_n + r_n}$, hence $p_n + q_n = \dfrac{u_n}{1 + u_n}$, $q_n + r_n = \dfrac{1}{1 + u_n}$.

Then families will be begotten as follows:

	AA	Aa	aa
From AA females ...	$\frac{p_n u_n}{1+u_n}$	$\frac{p_n}{1+u_n}$	0
„ Aa „ ...	$\frac{q_n u_n}{1+u_n}$	q_n	$\frac{q_n}{1+u_n}$
„ aa „ ...	0	$\frac{r_n u_n}{1+u_n}$	$\frac{r_n}{1+u_n}$

After selection and replacement the proportions will be:

	AA	Aa	aa
From AA females ...	$\frac{p_n u_n}{1+u_n}$	$\frac{p_n}{1+u_n}$	0
„ Aa „ ...	$\frac{q_n u_n}{1+u_n-\frac{1}{2}k}$	$\frac{q_n(1+u_n)}{1+u_n-\frac{1}{2}k}$	$\frac{q_n(1-k)}{1+u_n-\frac{1}{2}k}$
„ aa „ ...	0	$\frac{r_n u_n}{1+u_n-k}$	$\frac{r_n(1-k)}{1+u_n-k}$

With complete selection, where $k = 1$, recessives are eliminated at once, provided families are large enough. Where $k = -\infty$, dominants are not eliminated at once if pure dominant families survive, since $p_{n+1} = \frac{p_n(1-p_n)}{1+p_n}$. Starting from the standard population, successive values of r_n are 25 °/₀, 75 °/₀, 87·5 °/₀, 99·7 °/₀, etc. Where k is small we obtain approximate equations analogous to (3·0) whose solution is

$$\tfrac{3}{4}kn = u_n + \log_e u_n - 1. \quad \text{.......(3·3)}$$

Thus selection proceeds at $\frac{3}{4}$ of the rate given by equation (2·4).

Sex-limited characters and unisexual selection.

We have next to deal with characters which only appear in one sex, for example milk yield or other secondary sexual characters; or on which selection at least is unisexual, as for example in Darwinian sexual selection. Let the $(n-1)$th generation form spermatozoa in the ratio $u_n A : 1a$, eggs in the ratio $v_n A : 1a$. Then the nth generation consists of zygotes in the ratios

$$u_n v_n AA : (u_n + v_n) Aa : 1aa,$$

$$\therefore \; y_n = (1 + u_n)^{-1}(1 + v_n)^{-1}. \quad \text{.......(4·0)}$$

If only $1-k$ recessive ♀ survives for every dominant ♀, whilst ♂'s are not affected by selection, we have

$$\left.\begin{aligned} u_{n+1} &= \frac{2u_n v_n + u_n + v_n}{u_n + v_n + 2} \\ v_{n+1} &= \frac{2u_n v_n + u_n + v_n}{u_n + v_n + 2 - 2k} \end{aligned}\right\}. \quad \text{.......(4·1)}$$

With complete selection, when $k = -\infty$, and all dominants of one sex are weeded out, we have $v_n = 0$, and $u_{n+1} = \frac{u_n}{2 + u_n}$.

$$\therefore \; u_n = \left[2^{n-1}\left(1 + \frac{1}{u_0}\right) - 1\right]^{-1},$$

and

$$y_n = 1 + 2^{1-n}(y_0^{\frac{1}{2}} - 1). \quad \text{.......(4·2)}$$

Hence the proportion of dominants is halved in every successive generation. When $k = 1$, and all the recessives of one sex die childless, the proportions of recessives in successive generations, starting from the standard population, are 25 °/₀, 16·7 °/₀, 12·5 °/₀, 9·56 °/₀, 7·94 °/₀, and so on.

When k is small, since

$$v_{n+1} - u_{n+1} = \frac{2k(2u_n v_n + u_n + v_n)}{(u_n + v_n + 2)(u_n + v_n + 2 - 2k)}$$

and

$$\Delta u_n = \frac{(1 + u_n)(v_n - u_n)}{u_n + v_n + 2},$$

and hence the differences between u_n, u_{n+1}, v_n, v_{n+1} may be neglected in comparison with themselves;

$$\therefore \; v_n - u_n = \frac{ku_n}{1 + u_n} \text{ approximately,}$$

and

$$\Delta u_n = \frac{ku_n}{2(1 + u_n)} \text{ approximately.}$$

$$\therefore \; \tfrac{1}{2}kn = u_n + \log_e u_n - 1, \quad \text{.......(4·3)}$$

and selection proceeds at half the rate given by equation (2·4), a result stated by Punnett (3).

4—2

SELECTION OF AN ALTERNATIVELY DOMINANT CHARACTER.

A few factors, such as that determining the presence or absence of horns in Dorset and Suffolk sheep, according to Wood (16) are dominant in one sex, recessive in the other. Consider a factor dominant in the male sex, recessive in the female. Let the nth generation be produced by

$$\text{spermatozoa } u_n A : 1a, \quad \text{eggs } v_n A : 1a.$$

It consists of

$$\text{zygotes} : u_n v_n AA : (u_n + v_n) Aa : 1aa,$$

and the survivors after selection are in the ratios

$$♂ \quad u_n v_n AA : (u_n + v_n) Aa : (1-k) aa,$$

$$♀ \quad \frac{u_n v_n}{1-k} AA : (u_n + v_n) Aa : 1aa.$$

$$\left.\begin{aligned} \therefore u_{n+1} &= \frac{2u_n v_n + u_n + v_n}{u_n + v_n + 2 - 2k} \\ v_{n+1} &= \frac{\frac{2}{1-k} u_n v_n + u_n + v_n}{u_n + v_n + 2} \end{aligned}\right\} \quad \text{.............................(5·0)}$$

Whilst $$y_n \text{ (for males)} = (1+u_n)^{-1}(1+v_n)^{-1}. \quad \text{..........................(5·1)}$$

With complete selection, when all members of the type dominant in the female sex are weeded out, $k = 1$.

$$\therefore v_{n+1} = \infty, \text{ and } u_{n+1} = 1 + 2u_n, \text{ after the first generation.}$$

$$\therefore 1 + u_n = 2^{n-1}(1 + u_1),$$

and if z_n be the proportion of the weeded out type occurring in the female sex,

$$\left.\begin{aligned} \therefore y_n &= 0 \\ z_n &= 2^{1-n} z_1 \end{aligned}\right\}. \quad \text{.......................................(5·2)}$$

So this type disappears in the male sex, and is halved in successive female generations. If $k = \infty$ the type recessive in the female sex disappears in that sex and is halved in successive male generations.

When k is small,

$$\therefore v_{n+1} - u_{n+1} = \frac{k u_n (u_n - 1)}{1 + u_n} \text{ approximately,}$$

and $\Delta u_n = \frac{1}{2} k u_n$.

$$\therefore kn = 2 \log_e u_n \quad \text{.......................................(5·3)}$$

if $u_0 = 1$, so selection occurs on the whole more rapidly than by equation (2·4). (See Table V.) y_n is given by equation (2·0).

SEX-LINKED CHARACTERS UNDER NO SELECTION.

The events in an unselected population whose members differ with regard to a sex-linked factor have been considered by Jennings (17) but can be treated more simply. We suppose the male to be heterozygous for sex, but the argument is the same where the female is heterozygous. Consider a fully dominant factor A, such that the female may be AA, Aa, or aa, the male Aa or aa (or in Morgan's notation, which will be adopted, A or a). As Jennings showed, a population with

$$♀\text{'s } u^2 AA : 2u Aa : 1aa; \quad ♂\text{'s } uA : 1a$$

is stable during random mating, and other populations approach it asymptotically. In any population let the eggs of the $(n-1)$th generation be $u_n A : 1a$, the ♀-producing spermatozoa $v_n A : 1a$. Then the nth generation will be: ♀'s $u_n v_n AA : (u_n + v_n) Aa : 1aa$; ♂'s $u_n A : 1a$.

$$\left.\begin{aligned} \therefore\ u_{n+1} &= \frac{2u_n v_n + u_n + v_n}{2 + u_n + v_n} \\ v_{n+1} &= u_n \end{aligned}\right\}, \quad \text{.......(6·0)}$$

and if y_n be the proportion of recessive ♀'s, z_n of recessive ♂'s,

$$\left.\begin{aligned} y_n &= (1 + u_n)^{-1}(1 + v_n)^{-1} \\ z_n &= (1 + u_n)^{-1} \end{aligned}\right\}, \quad \text{.......(6·1)}$$

$$\therefore\ z^{-1}{}_{n+1} = 1 + u_{n+1} = \frac{2(1 + u_n)(1 + u_{n-1})}{2 + u_n + u_{n-1}} = \frac{2}{z_n + z_{n-1}}.$$

$$\therefore\ 2z_n = z_{n-1} + z_{n-2}.$$

Solving as usual for recurring series, we have

$$\left.\begin{aligned} 3z_n &= z_0 + 2z_1 + (-\tfrac{1}{2})^n (2z_0 - 2z_1) \\ y_n &= z_n z_{n-1} \end{aligned}\right\}, \quad \text{.......(6·2)}$$

$$\left.\begin{aligned} \therefore\ z_\infty &= \tfrac{1}{3}(z_0 + 2z_1) = \tfrac{1}{3}(z_n + 2z_{n+1}) \\ y_\infty &= z_\infty^2 \end{aligned}\right\}. \quad \text{.......(6·3)}$$

Hence from the proportion of males in two successive generations, or both sexes in one, we can calculate the final values. Successive values of y_n and z_n oscillate alternately above and below their final values, but converge rapidly towards them.

Bisexual selection of a sex-linked character.

If the conditions are as above, except that in each generation one dominant survives for every $(1 - k)$ recessives in each sex, then

$$\left.\begin{aligned} u_{n+1} &= \frac{2u_n v_n + u_n + v_n}{u_n + v_n + 2 - 2k} \\ v_{n+1} &= \frac{u_n}{1 - k} \end{aligned}\right\}, \quad \text{.......(7·0)}$$

and

$$\left.\begin{aligned} y_n &= (1 + u_n)^{-1}(1 + v_n)^{-1} \\ z_n &= (1 + u_n)^{-1} \end{aligned}\right\}. \quad \text{.......(6·1)}$$

With complete selection if $k = -\infty$, and no dominants survive to breed, selection is complete in one generation. If $k = 1$, and no recessives survive to breed,

$$u_{n+1} = 1 + 2u_n, \text{ and } v_n = \infty.$$

$$\therefore\ 1 + u_n = 2^n(1 + u_0),$$

and

$$\left.\begin{aligned} z_n &= 2^{-n} z_0 \\ y_n &= 0 \end{aligned}\right\}. \quad \text{.......(7·1)}$$

So no recessive females are produced and the number of recessive males is halved in each generation. Selection is therefore vastly more effective than on an autosomal character. If colour-blind or haemophilic persons were prevented from breeding, these conditions could be almost abolished in a few generations, which is not the case with feeble-mindedness. If selection is slow we solve as for equations (4·1), and find approximately

$$v_n - u_n = \frac{2ku_n^2}{3 + 3u_n},$$

$$\Delta u_n = \frac{ku_n(3 + u_n)}{3 + 3u_n}.$$

So, reckoning generations to or from a standard population where $u_0 = 1$, and 50 °/₀ of the males and 25 °/₀ of the females are recessives,

$$kn = \log_e u_n + 2 \log \left(\frac{3 + u_n}{4} \right), \quad \text{.............................(7·2)}$$

$$\left. \begin{array}{l} y_n = (1 + u_n)^{-2} \\ z_n = (1 + u_n)^{-1} \end{array} \right\}. \quad \text{...(7·3)}$$

Table III and Fig. 3 are calculated from these equations. Within the limits covered by the figure selection acts more rapidly on a sex-linked character in the homozygous sex than on an

TABLE III.

Effect of slow selection in both sexes on a sex-linked character, dominants being favoured.

°/₀ of recessives of homozygous sex	99·998	99·98	99·80	99·60	99·00	98·01
" " " heterozygous sex	99·999	99·99	99·9	99·8	99·5	99
kn (number of generations × k) ...	– 12·09	– 9·786	– 7·481	– 6·787	– 5·866	– 5·164

°/₀ of recessives of homozygous sex	96·04	90·25	81	64	49	36	25	16	10	6·25	4
" " " heterozygous sex	9·8	95	90	80	70	60	50	40	31·62	25	20
kn (number of generations × k) ...	– 4·454	– 3·485	– 2·700	– 1·802	– 1·156	– ·580	0	619	1·282	1·910	2·506

°/₀ of recessives of homozygous sex	2	1	·5	·25	·1	·01	·001	·0001	$\cdot 0_5 1$	$\cdot 0_7 1$
" " " heterozygous sex	14·14	10	7·071	5	3·162	1	·3162	·1	·01	·001
kn (number of generations × k) ...	3·441	4·394	5·366	6·353	7·679	11·07	14·50	17·95	24·86	31·76

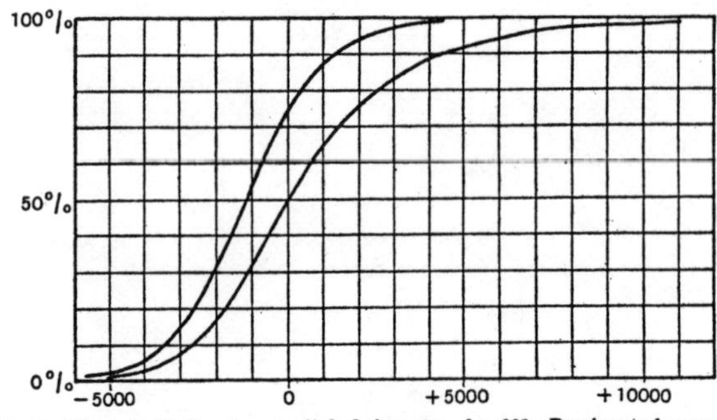

Fig. 3. Effect of selection on a sex-linked character. $k = \cdot 001$. Dominants favoured. Upper curve, homozygous sex; lower curve, heterozygous sex. Abscissa = generations. Ordinate = percentage of sex with the favoured character.

autosomal character. In the heterozygous sex selection occurs at about the same rate in the two cases. However, as appears from Table V, sex-linked recessive characters increase far more rapidly in the early stages, and sex-linked dominants in the late stages of selection, the autosomal characters.

Table III is not quite accurate unless selection is very slow, the error being of the order of k. Thus when $k = 0{\cdot}2$ the error in n is nearly 10 °/₀. Still even for these large values it furnishes a useful first approximation.

BISEXUAL FAMILIAL SELECTION OF A SEX-LINKED CHARACTER.

Here we need only consider the case where the family within which selection occurs has both parents in common. Sex-linkage of the animal type is rare in plants, and families with many fathers per mother are rare in animals. Let the nth generation be

$$♀\ p_n AA : 2q_n Aa : r_n aa; \quad ♂\ s_n A : t_n a,$$

where $p_n + 2q_n + r_n = s_n + t_n = 1$. Let the dominants have an advantage of $1 : 1 - k$ over the recessives in the mixed families. Then the $(n+1)$th generation occurs in the following proportions, after selection has operated:

From mating	AA ♀	Aa ♀	aa ♀	A ♂	a ♂
$AA \times A$ ...	$p_n s_n$	0	0	$p_n s_n$	0
$Aa \times A$ ...	$q_n s_n$	$q_n s_n$	0	$\frac{2q_n s_n}{2-k}$	$\frac{2(1-k) q_n s_n}{2-k}$
$aa \times A$ ...	0	$r_n s_n$	0	0	$r_n s_n$
$AA \times a$ ...	0	$p_n t_n$	0	$p_n t_n$	0
$Aa \times a$ ...	0	$\frac{2q_n t_n}{2-k}$	$\frac{2(1-k) q_n t_n}{2-k}$	$\frac{2q_n t_n}{2-k}$	$\frac{2(1-k) q_n t_n}{2-k}$
$aa \times a$ ...	0	0	$r_n t_n$	0	$r_n t_n$

Hence, writing $k' = \frac{k}{2-k}$,

$$\left.\begin{aligned} p_{n+1} &= (p_n + q_n) s_n \\ 2q_{n+1} &= (p_n + q_n) t_n + (q_n + r_n) s_n + k' q_n t_n \\ r_{n+1} &= (q_n + r_n) t_n - k' q_n t_n \\ s_{n+1} &= p_n + q_n + k' q_n \\ t_{n+1} &= q_n + r_n - k' q_n \end{aligned}\right\} \quad \text{.........................(8·0)}$$

With complete selection, when the recessives are eliminated, $k' = k = 1$, and

$$r_{n+1} = r_n t_n,$$
$$t_{n+1} = r_n,$$
$$\therefore\ r_n = r_0^{\phi(n+1)} t_0^{\phi(n)},$$

where

$$\phi(n) = \frac{2^{-n}}{\sqrt{5}} [(1+\sqrt{5})^n - (1-\sqrt{5})^n], \quad \text{..............................(8·1)}$$

i.e. $\phi(n)$ is the nth term of Fibonnacci's series 1, 2, 3, 5, 8, 13, 21, etc. So the recessives disappear very fast. When dominants are eliminated $k' = -1$, $k = -\infty$, and the equations are less tractable. The percentages of recessives in succeeding generations, starting from a standard population, are:

♀	25	37·5	56·25	66·80	82·97	etc.
♂	50	75	75	89·06	93·16	etc.

When k is small we solve as in equations (3·0), and find

$$\Delta u_n = \frac{k' u (2+u)}{3+3u} \text{ approximately.}$$

$$\left.\begin{aligned} \therefore\ kn &= 3\log_e\left[\frac{u_n(2+u_n)}{3}\right] \\ r_n &= (1+u_n)^{-2} \\ t_n &= (1+u_n)^{-1} \end{aligned}\right\}, \qquad \text{(8·2)}$$

starting from the standard population, and p_n, q_n have very nearly the values for a population in equilibrium. Selection therefore proceeds much as in racial selection but at from a half to a third of the rate. Some figures are given in Table V.

SELECTION OF A SEX-LINKED CHARACTER IN THE HOMOZYGOUS SEX ONLY.

Several sex-linked factors are known which have a much more marked effect on the homozygous than the heterozygous sex. Thus in *Drosophila melanogaster* "fused" females are sterile, males fertile; whilst the character "dot" occurs in 8 °/₀ of the genetically recessive females, but only 0·8 °/₀ of the males. [Morgan and Bridges (18).] But the chief evolutionary importance of this type of selection must have been in the Hymenoptera and other groups where the males are haploid and all amphimictic inheritance sex-linked. The characters of the diploid females and neuters are generally more important (especially in the social species) than those of the males. On the other hand it must be remembered that for a few drone characters selection must be very intense, and largely familial. Using the usual notation

$$\left.\begin{aligned} u_{n+1} &= \frac{2u_n v_n + u_n + v_n}{u_n + v_n + 2 - 2k} \\ v_{n+1} &= u_n \end{aligned}\right\}. \qquad \text{(9·0)}$$

With complete selection, if all dominants are eliminated and $k = -\infty$, all dominants disappear in two generations. If all recessives are eliminated $k = 1$, and starting with a standard population the percentages of recessives in successive generations are:

♂ (heterozygous sex)	50	33·3	30	23·2	21·4	18·6
♀ (homozygous sex)	25	16·7	10	6·96	5·14	3·95

So elimination is vastly slower than when selection occurs in both sexes (equation (7·1)). When k is small we solve as in (7·0), and find

$$3\Delta u_n = \frac{2ku_n}{1+u_n} \text{ approximately.}$$

$$\therefore\ \tfrac{2}{3}kn = u_n + \log_e u_n - 1, \qquad \text{(9·1)}$$

$$\left.\begin{aligned} y_n &= (1+u_n)^{-2} \\ z_n &= (1+u_n)^{-1} \end{aligned}\right\}. \qquad \text{(7·3)}$$

So selection of the homozygotes proceeds as in Fig. 2 and Tables I and II, but 1·5 times as many generations are needed for a given change. The heterozygous sex changes rather more slowly.

SELECTION OF A SEX-LINKED CHARACTER IN THE HETEROZYGOUS SEX ONLY.

In certain cases sex-linked factors appear only or mainly in the heterozygous sex. Thus in *Drosophila melanogaster* "eosin" eye-colour is far more marked in the male than the female, and the sex-linked fertility factor L_2 postulated by Pearl (19) in poultry can only show in the female sex. If selection is limited to the heterozygous sex,

$$\left.\begin{aligned} u_{n+1} &= \frac{2u_n v_n + u_n + v_n}{u_n + v_n + 2} \\ v_{n+1} &= \frac{u_n}{1-k} \end{aligned}\right\}. \qquad \text{(10·0)}$$

With complete selection, if all dominants are eliminated, $k = -\infty$, and

$$u_{n+1} = \frac{u_n}{2 + u_n} \text{ (after the second generation),}$$

$$v_n = 0.$$

$$\therefore\ u_n = \left[2^{n-1}\left(1 + \frac{1}{u_1}\right) - 1\right]^{-1}, \text{ but } u_1 = u_0;$$

$$\therefore\ y_n = z_n = 1 - 2^{1-n} z_0. \qquad (10\cdot1)$$

So the number of dominants is halved in each generation after the second. If recessives are eliminated, $k = 1$, and

$$u_{n+1} = 1 + 2u_n \text{ (after the second generation),}$$

$$v_{n+1} = \infty.$$

$$\therefore\ u_n = 2^{n-1}(1 + u_1) - 1;$$

$$\therefore\ \left.\begin{aligned} y_n &= 0 \\ z_n &= 2^{1-n} z_0 \end{aligned}\right\}, \qquad (10\cdot2)$$

the proportion of recessives being halved in each generation. If selection is slow

$$\Delta u_n = \tfrac{1}{3} k u_n \text{ approximately;}$$

$$\therefore\ kn = 3 \log u_n \qquad (10\cdot3)$$

if $u_0 = 1$; and y_n, z_n are given by equations (7·3). Hence selection in the heterozygous sex proceeds as in Fig. 1, but at one-third of the pace, whilst selection in the homozygous sex is slightly faster.

Certation, or gametic selection of an autosomal character.

The work of Renner (6) and Heribert-Nilsson (20) shows that gametes or gametophytes may be selected according to what factors they carry. The field of such selection may be wide, as in wind pollination, but is more often restricted, and mainly familial, i.e. among the gametes of the same individual. Except in homosporous plants the intensity must be different in gametes of different genders, and we shall here only consider the case where one is affected. Let the nth generation be formed from gametes carrying $u_n A : 1a$, this proportion being reduced by selection in one gender to $u_n A : (1-k)\, a$, the selection being general and not familial. Then the nth generation will be $u_n^2 AA : (2-k)\, u_n Aa : (1-k)\, aa$.

$$\therefore\ u_{n+1} = \frac{u_n(2u_n + 2 - k)}{(2-k)\, u_n + 2 - 2k}. \qquad (11\cdot0)$$

With complete selection, if all dominant-carrying gametes are eliminated, $k = -\infty$, and

$$u_{n+1} = \frac{u_n}{2 + u_n};$$

$$\therefore\ y_n = \frac{1}{1 + u_n} = 1 - 2^{1-n}(1 - y_1). \qquad (11\cdot1)$$

So the proportion of dominants is halved in each generation. If recessive-carriers are eliminated, no recessive zygotes appear, and the proportion of heterozygotes is halved in each generation. If selection is slow,

$$\Delta u_n = \tfrac{1}{2} k u_n \text{ approximately;}$$

$$\therefore\ kn = 2 \log_e u_n, \qquad (11\cdot2)$$

$$y_n = (1 + u_n)^{-2}. \qquad (2\cdot0)$$

If the gametes are replaced in heterozygous organisms, as must happen in a large batch of pollen-grains or spermatozoa from the same source, let the nth zygotic generation be formed from unselected gametes (say megagametes) $u_n A : 1a$, and selected gametes (say microgametes) $v_n A : 1a$, so its proportions are $u_n v_n AA : (u_n + v_n) Aa : 1aa$.

$$\left.\begin{aligned} \therefore\ u_{n+1} &= \frac{2u_n v_n + u_n + v_n}{u_n + v_n + 2} \\ v_{n+1} &= \frac{2u_n v_n + u_n + v_n + k'(u_n + v_n)}{u_n + v_n + 2 - k'(u_n + v_n)} \end{aligned}\right\}, \qquad \text{(11·3)}$$

where $k' = \frac{k}{2-k}$, as in equation (8·0). With complete selection (when for example there is a very great disparity between growth-rates of pollen-tubes, though both types are viable), if dominant gametes are eliminated, $k' = -1$, and the percentages of recessive zygotes in successive generations, starting from a standard population, are:

25, 37·5, 54·69, 71·48, 84·16, 91·83, etc.

If recessive gametes are eliminated, $k' = 1$, and the percentages of recessive zygotes in successive generations are:

25, 12·5, 4·56, 1·14, 0·17, ·014, etc.

When selection is slow, $\Delta = \frac{1}{4} k u_n$ approximately.

$$\therefore\ kn = 4 \log_e u_n \qquad \text{(11·4)}$$

if $u_0 = 1$, so selection proceeds at half the rate given by equation (11·2), y_n being given by (2·0).

Gametic selection of a sex-linked character.

This is not known to occur, and at all complete gametic selection is very unlikely in animals, so we need only consider slow selection. Let selection occur among the gametes of the homozygous sex, with no replacement within heterozygous organisms. Let the nth generation be formed from eggs in the ratio $u_n A : 1a$ before, or $u_n A : (1-k)\, a$ after selection, and ♀-producing spermatozoa in the ratio $v_n A : 1a$. Then the nth generation is

$$♀\ u_n v_n AA : (u_n + v_n - kv_n)\, Aa : (1-k)\, aa;\quad ♂\ u_n A : (1-k)\, a.$$

$$\left.\begin{aligned} \therefore\ u_{n+1} &= \frac{2u_n v_n + u_n + v_n - kv_n}{u_n + v_n + 2 - kv_n - 2k} \\ v_{n+1} &= \frac{u_n}{1-k} \end{aligned}\right\}; \qquad \text{(12·0)}$$

$$\therefore\ \Delta u_n = \tfrac{2}{3} k u_n \text{ approximately.}$$

$$\therefore\ \tfrac{2}{3} kn = \log_e u_n, \qquad \text{(12·1)}$$

whilst y_n and z_n are given by equation (7·3), so selection proceeds twice as fast as in equation (10·3). In the more important case of familial selection (replacement in heterozygous individuals), if $k' = \frac{k}{2-k}$, then

$$\left.\begin{aligned} u_{n+1} &= \frac{2u_n v_n + u_n + v_n + k'(u_n + v_n)}{u_n + v_n + 2 - k'(u_n + v_n)} \\ v_{n+1} &= u_n \end{aligned}\right\}, \qquad \text{(12·2)}$$

u_n being here the gametic ratio after selection.

$$\therefore\ \Delta u_n = \tfrac{2}{3} k' u_n = \tfrac{1}{3} k' v_n;$$

$$\therefore\ \tfrac{1}{3} kn = \log_e u_n, \qquad \text{(12·3)}$$

so selection proceeds as in equation (10·3).

If selection occurs among the gametes of the heterozygous sex there is clearly no effect if they are replaced, whilst otherwise the effects are the same as those of zygotic selection, and are given by equation (10·3).

COMPARATIVE RESULTS OF COMPLETE (ARTIFICIAL) SELECTION.

The results of complete selection in the more important cases are summarized in Table IV. In every case the field of selection considered is the whole population. Complete familial selection occasionally occurs through natural causes, but never through human agency. Column 3 gives the sex to which the numbers in columns 4 and 5 refer. Selection is supposed to begin on

TABLE IV.

Effects of complete selection.

Character eliminated	Type of selection	Sex	°/₀ after 5 generations from 50 °/₀	°/₀ after 10 generations from 50 °/₀	Equation
Non-amphimictic ...	Any	Both	0	0	—
Autosomal dominant	Bisexual	„	0	0	—
„ recessive	„	„	2·44	0·768	2·2
„ dominant	Unisexual	„	1·83	0·0572	4·2
„ recessive	„	„	8·88	3·27	—
Sex-linked dominant	Bisexual	Homozygous	0	0	—
		Heterozygous	0	0	—
„ recessive	„	Homozygous	0	0	7·1
		Heterozygous	1·56	0·0484	7·1
„ dominant	In homozygous sex	Homozygous	0	0	—
		Heterozygous	0	0	—
„ recessive	„ „ „	Homozygous	5·34	1·74	—
		Heterozygous	18·5	13·28	—
„ dominant	In heterozygous sex	Homozygous	1·83	0·0572	10·1
		Heterozygous	3·125	0·0977	10·1
„ recessive	„ „ „	Homozygous	0	0	—
		Heterozygous	3·125	0·0977	10·2
Autosomal dominant	Gametic unisexual	Both	1·83	0·0572	11·1
„ recessive	„ „ „	„	0	0	—

a population in equilibrium containing equal numbers of dominants and recessives of the sex considered. It is worth noting that in the case of sex-linked characters, and autosomal recessives when selection is gametic, individuals of types which have wholly disappeared reappear if selection ceases. With many types of heredity dominants are eliminated in one or two generations, and where this is not the case they generally decrease more rapidly than recessives.

APPLICATIONS TO SLOW SELECTION.

With the exception of (1·1) the equations found for the rate of slow selection are not rigorously accurate. n is in general a higher transcendental function of u, but of what nature is not clear. It will be shown later that the finite difference equations found in this paper are special cases of integral equations which may possibly prove more tractable. The values for kn found in terms of u all have inexactitudes of the order k^2n. Thus if one type has an advantage of 1 °/₀, the number of generations required for a given change can also be found within about 1 °/₀.

Table V shows the effect of slow selection in the various cases considered. The third column gives the sex to which the subsequent figures apply. Selection is throughout supposed to give the favoured type an advantage of $\frac{1}{1000}$, i.e. 1000 of this type survive for 999 of the other. If the advantage is $\frac{1}{100}$, one-tenth of the number of generations is required for a given change, and so on, but when selection is very rapid the numbers are somewhat inaccurate.

It will at once be seen that selection is most rapid when amphimixis is avoided by any of the means cited above. Moreover selection is ineffective on recessive characters when these are rare, except in the case of sex-linked factors, when selection is effective in the heterozygous sex, and in gametic selection. It seems therefore very doubtful whether natural selection in random-mating

TABLE V.

Generations required for a given change with various types of slow selection. $k = \cdot 001$.

Dominant factor favoured	Type of selection	Sex	·001—1 °/₀	1—50 °/₀	50—99 °/₀	99—99·999 °/₀	Equations
Non-amphimictic	Bisexual racial	Both ...	6,921	4,592	4,592	6,921	1·1
Autosomal ...	" "	" ...	6,920	4,819	11,664	309,780	2·0, 2·4
" ...	Unisexual racial / Bisexual familial	" ...	13,841	9,638	23,328	619,560	2·0, 3·2, 4·3
" ...	" familial*	" ...	9,227	6,425	15,522	413,040	2·0, 3·3
" † ...	" racial	♂	13,831	8,819	6,157	7,112	2·0, 5·3
Sex-linked ...	" "	Homozygous	6,916	4,668	5,593	10,106	7·2, 7·3
		Heterozygous	6,928	5,164	11,070	20,693	
" ...	" familial	Homozygous	20,753	13,785	13,785	20,753	8·2
		Heterozygous	20,768	14,987	24,332	41,450	
" ...	Racial of homozygous sex	Homozygous	10,380	7,228	17,496	464,670	7·3, 9·1
		Heterozygous	10,392	8,378	153,893	149,860,377	
" ...	Racial of heterozygous sex	Homozygous	20,746	13,228	9,236	10,668	7·3, 10·3
		Heterozygous	20,753	13,785	13,785	20,753	
Autosomal ...	Unisexual gametic	Both ...	13,831	8,819	6,157	7,112	2·0, 11·2
" ...	" " ‡	" ...	27,661	17,638	12,314	14,224	2·0, 11·4
Sex-linked ...	Gametic of homozygous sex	Homozygous	10,373	6,619	4,618	5,334	7·3, 12·1
		Heterozygous	10,377	6,892	6,892	10,377	
" ...	Gametic of homozygous sex‡	Homozygous	20,746	13,228	9,236	10,668	7·3, 12·3
		Heterozygous	20,753	13,785	13,785	20,753	

* The families have only one parent in common.

† Dominant in ♂, recessive in ♀.

‡ In heterozygous individuals gametes are replaced (as zygotes in familial selection).

The effect of selection on recessive characters may be found by inverting the order of the four numerical columns. Thus 309,780 generations are needed for an autosomal recessive to increase from ·001 °/₀ to 1 °/₀, 11,664 generations to increase from 1 °/₀ to 50 °/₀, and so on.

organisms can cause the spread of autosomal recessive characters unless they are extraordinarily valuable to their possessors. Such characters appear far more frequently than dominant mutations, but in their early stages are selected infinitely more slowly. It is thus intelligible that none of the melanic varieties of Lepidoptera which are known to have spread should be recessive.

There are at least four ways out of this impasse:

A. In a species which adopts self-fertilization or very close inbreeding advantageous autosomal recessive characters can spread rapidly. Thus supposing that in each of two otherwise similar species, one of which is mainly self-fertilizing, an advantageous recessive mutation occurs, it will spread far more quickly in the self-fertilizing species, and this species will tend to replace the other. This fact may well account for the widespread presence of self-fertilization and close inbreeding, in spite of the fact that they seem often to be physiologically harmful, and must certainly check the combination of useful variations which have arisen independently.

B. Recessives may be helped to spread by assortative fertilization. This may take place in the following ways:

1. Psychological isolation. Thus Pearson and Lee (21) have shown that a tall man is more likely to marry a tall woman than a short woman if presented with equal numbers of each. Of course the recessives must not be so repulsive to the dominants as to escape mating altogether at first. In plants psychological isolation may be due to the psychology of the insect or other pollinating organism. Thus a mutant plant with a new colour, scent, or shape may be isolated because it attracts a different insect from the type plant.

2. Anatomical isolation. Pearl (22) and Crozier and Snyder (23) have shown that in *Paramoecium* and *Gammarus* there is a strong tendency for organisms of like size to mate. This will be effective provided mutations are not so great as to leave the first mutants unmated.

3. Temporal isolation. If the recessive factor causes (or is very closely coupled with a factor causing) a change in the breeding or flowering time, this will serve as an effective barrier against crossing.

4. Spatial isolation. If the recessive has a different habitat, e.g. a different range of soil or temperature conditions to which it is adapted, some of its individuals will be spatially isolated from the dominants.

5. Selective fertilization. If the results of Jones (24) are due to this cause, as seems almost certain, we have here a *vera causa*, though it must be remembered that he did not work with single factor-differences. He found that when either of two races of maize is fertilized with a mixture of pollen the proportion of hybrids was less than was to be expected from random fertilization. This does not seem to have been due to inviability of the hybrids, which were more vigorous and fertile than the parent races. Clearly if the hybrid zygotes are inviable or sterile the rarer form of the species will be weeded out whether it is dominant or recessive, weak or vigorous. But if there is selective fertilization due for example to increased activity of pollen-tubes in tissue of the type which produced them, the increase of the rare form, especially if it is recessive, will be facilitated.

All these types of isolation, then, will favour the replacement of a type species by a recessive mutant. May it not be that in many cases mutual infertility is the cause and not the effect of specific differences? A new mutant form arises within a species. If it crosses freely with the type we call it a variety, and a moderately advantageous recessive variety will only spread very slowly indeed. But if it does not cross freely we call it a new species, and it is much more likely to establish itself. Possibly then interspecific sterility is partly to be explained by its having a selective value.

C. The increase of recessives is greatly facilitated, as will be shown later, by incomplete dominance. Thus if there is only one recessive in a million, and the recessives have an advantage of ·001, their rate of increase will be speeded up elevenfold if the heterozygotes have an advantage of ·00001 over the pure dominants.

D. If heterozygotes have any advantage *as such* this will tend to favour any new factors so long as they are rare. But no "stimulus of heterozygosis" has yet been demonstrated in cases of single factor-differences.

Whether the isolation of small communities, or what comes to much the same thing, great immobility of individuals at all stages of their lives, will help or hinder the spread of a new recessive type in the species as a whole is a nice question. It will certainly slow the spread of a dominant.

At first sight the selection of dominant factors would not seem to be a probable cause of the origin of species rather than new varieties. But it must be remembered that dominant mutations are very often lethal in the homozygous condition. Under certain circumstances, to be discussed later, their selection may lead to the establishment of a system of balanced lethals, and a probable change in the chronosome number.

The theory so far developed gives a quantitative account of the spread of a new advantageous type within a population under certain simple conditions, and demonstrates that inbreeding, homogamy, and inter-varietal sterility may sometimes be of selective value, and therefore preserved by natural selection. It is proposed in later papers to discuss the selection of semi-dominant, multiple, linked, and lethal factors, partial inbreeding and homogamy, overlapping generations, and other complications.

SUMMARY.

Mathematical expressions are found for the effect of selection on simple Mendelian populations mating at random. Selection of a given intensity is most effective when amphimixis does not affect the character selected, e.g. in complete inbreeding or homogamy. Selection is very ineffective on autosomal recessive characters so long as they are rare.

REFERENCES.

1. Pearson. *Proc. Roy. Soc.* 54—72.
2. Warren. *Genetics*, 2, p. 305, 1917.
3. Punnett. *Mimicry in Butterflies*, p. 154.
4. Ibsen and Steigleder. *Am. Nat.* 51, p. 740, 1917.
5. Durham. *Journ. Genetics*, 1, p. 107, 1911.
6. Renner. *Zeit. Ind. Abst. u. Ver.* 18, p. 121, 1917.
7. Schmidt. *Comptes rendus trav. Lab. Carlsberg*, 14, 1920.
8. Goldschmidt. *Zeit. Ind. Abst. u. Ver.* 23, p. 1, 1920.
9. Pearson. *Phil. Trans. Roy. Soc.* A, 203, p. 53, 1904.
10. Hardy. *Science*, 28, p. 49, 1908.
11. Punnett. *Journ. Hered.* 8, p. 464, 1917.
12. Detlefsen. *Genetics*, 3, p. 573, 1918.
13. Barrett. *Lepidoptera of the British Islands*, 7, p. 130.
14. Bate. *Ent. Rec.* 1895, p. 27.
15. Bateson. *Mendel's Principles of Heredity*, p. 221.
16. Wood. *Journ. Agric. Science*, 1, p. 364.
17. Jennings. *Genetics*, 1, p. 53, 1915.
18. Morgan and Bridges. *Carn. Inst. Wash. Pub.* 237, 1916.
19. Pearl. *Am. Nat.* 46, p. 130.
20. Heribert-Nilsson. *Hereditas*, 1, p. 41, 1920.
21. Pearson and Lee. *Biometrika*, 2, p. 371, 1903.
22. Pearl. *Biometrika*, 5, p. 213, 1907.
23. Crozier and Snyder. *Proc. Soc. Exp. Biol. & Med.* 19, p. 327, 1922.
24. Jones. *Biol. Bull.* 38, p. 251, 1920.

16

Reprinted from *Cambridge Phil. Soc. Proc. Biology* 1:158–163 (1924)

A MATHEMATICAL THEORY OF NATURAL AND ARTIFICIAL SELECTION. PART II

THE INFLUENCE OF PARTIAL SELF-FERTILISATION, INBREEDING, ASSORTATIVE MATING, AND SELECTIVE FERTILISATION ON THE COMPOSITION OF MENDELIAN POPULATIONS, AND ON NATURAL SELECTION.

BY J. B. S. HALDANE, M.A.

IN the first paper (1) of this series expressions were found for the effect of natural selection of small and constant intensity on Mendelian populations whose generations do not overlap; either during random mating, or when all zygotes are self-fertilised. An intermediate condition as regards mating may arise when there is a tendency to self-fertilisation, to mating between relatives, or to unions between similar but not necessarily related zygotes or gametes. We consider a population whose mth generation consists of $p_m AA : 2q_m Aa : r_m aa$, where A is a completely dominant Mendelian factor, and $p_m + 2q_m + r_m = 1$. When such a population is subjected to any system of mating it falls rapidly or instantly into a new equilibrium. During this process it will be shown that the gametic ratio $u_m = \frac{p_m + q_m}{q_m + r_m}$ is unaltered. When equilibrium is reached under the given mating system we find p, q, r in terms of u.

We now suppose selection to take place at such a rate that $(1 - k)$ recessives survive for every dominant, and so slowly that the population is always very nearly in equilibrium under the mating system. If this condition were not fulfilled we should have to investigate the problem by the method of Lotka (2), which in this case presents considerable difficulties. During selection we have

$$u_{n+1} = \frac{p_n + q_n}{q_n + r_n - kr_n},$$

$$\therefore \Delta u_n = \frac{kr_n u_n (1 + u_n)}{1 - kr_n (1 + u_n)}.$$

Since k is small, and $r_n (1 + u_n) = \frac{r_n}{q_n + r_n}$ and is therefore less than unity,

$$\therefore \frac{du}{dn} = \Delta u_n = kr_n u_n (1 + u_n) \text{ approximately.}$$

$$\therefore kn = \int_1^{u_n} \frac{du}{ru(1+u)} \qquad \ldots(1),$$

putting $u_0 = 1$, as in Part I. This can be evaluated as r is a known function of u. Under random mating when recessives are few $r_n = \frac{1}{(1 + kn)^2}$ approximately, so selection is very slow. It will be shown that with some systems of mating successive small values of r_n approximate to a geometrical series, so that selection is vastly more rapid.

PARTIAL SELF-FERTILISATION.

Let a proportion l of the population be self-fertilised, $(1 - l)$ mated at random, where l may have any value from 0 to 1 inclusive.

$$\therefore\ p_{m+1} = l\,(p_m + \tfrac{1}{2}q_m) + (1 - l)\,(p_m + q_m)^2,$$
$$q_{m+1} = \tfrac{1}{2}lq_m + (1 - l)\,(p_m + q_m)\,(q_m + r_m),$$
$$r_{m+1} = l\,(\tfrac{1}{2}q_m + r_m) + (1 - l)\,(q_m + r_m)^2.$$

Clearly $u_{m+1} = u_m$, and

$$q_m = \frac{2\,(1 - l)\,u}{(2 - l)\,(1 + u)^2} + \left(\frac{l}{2}\right)^m \left[q_0 - \frac{2\,(1 - l)\,u}{(2 - l)\,(1 + u)^2}\right].$$

So there is a rapid approach to equilibrium, when

$$\left.\begin{aligned} p &= \frac{u\,(l + 2u - lu)}{(2 - l)\,(1 + u)^2} \\ q &= \frac{2\,(1 - l)\,u}{(2 - l)\,(1 + u)^2} \\ r &= \frac{2 - l + lu}{(2 - l)\,(1 + u)^2} \end{aligned}\right\} \qquad \ldots(2{\cdot}1).$$

During selection

$$\left.\begin{aligned} kn &= \int_1^{u_n} \frac{(2 - l)\,(1 + u)\,du}{u\,(2 - l + lu)} \\ &= \log_e u_n + \frac{2}{l}\log_e\left(\frac{2 - l + lu_n}{2}\right) \\ r_n &= \frac{2 - l + lu_n}{(2 - l)\,(1 + u_n)^2} \end{aligned}\right\} \qquad \ldots(2{\cdot}2),$$

unless $l = 0$, when $kn = u_n + \log_e u_n - 1$.
When recessives are sufficiently few

$$(2 - l)\,r_n = \frac{l}{u_n} = le^{-\frac{lkn}{2+l}} \text{ approximately,}$$

so $\dfrac{r_n}{r_{n+1}} = 1 + \dfrac{lk}{2 + l}$ approximately, and selection is rapid.

PARTIAL INBREEDING.

Let a proportion l of the population be mated to whole brothers or sisters, $(1 - l)$ mated at random. Let matings occur in the following proportions:

Mating	Proportion	Producing offspring	Matings of inbred offspring
$AA \times AA$	α_m	$\alpha_m\,AA$	$l\alpha_m\,(AA \times AA)$
$AA \times Aa$	$4\beta_m$	$2\beta_m\,(AA + Aa)$	$l\beta_m\,(AA \times AA + 2AA \times Aa + Aa \times Aa)$
$AA \times aa$	$2\gamma_m$	$2\gamma_m\,Aa$	$2l\gamma_m\,(Aa \times Aa)$
$Aa \times Aa$	$16\delta_m$	$4\delta_m\,(AA + 2Aa + aa)$	$l\delta_m\,(AA \times AA + 4AA \times Aa + 2AA \times aa + 4Aa \times Aa + 4Aa \times aa + aa \times aa)$
$Aa \times aa$	$4\epsilon_m$	$2\epsilon_m\,(Aa + aa)$	$l\epsilon_m\,(Aa \times Aa + 2Aa \times aa + aa \times aa)$
$aa \times aa$	ζ_m	$\zeta_m\,aa$	$l\zeta_m\,(aa \times aa)$

Where $\alpha_m + 4\beta_m + 2\gamma_m + 16\delta_m + 4\epsilon_m + \zeta_m = 1$,

$$\therefore\ p_m = \alpha_m + 2\beta_m + \gamma_m,\ p_{m+1} = \alpha_m + 2\beta_m + 4\delta_m,$$
$$q_m = \beta_m + 8\delta_m + \epsilon_m,\ q_{m+1} = \beta_m + \gamma_m + 4\delta_m + \epsilon_m,$$
$$r_m = \gamma_m + 2\epsilon_m + \zeta_m,\ r_{m+1} = 4\delta_m + 2\epsilon_m + \zeta_m,$$
$$\alpha_{m+1} = (1-l)\,p_{m+1}^2 + l\,(\alpha_m + \beta_m + \delta_m),$$
$$4\beta_{m+1} = 4\,(1-l)\,p_{m+1}q_{m+1} + 2l\,(\beta_m + 2\delta_m),$$
$$2\gamma_{m+1} = 2\,(1-l)\,p_{m+1}r_{m+1} + 2l\delta_m,$$
$$16\delta_{m+1} = 4\,(1-l)\,q_{m+1}^2 + l\,(\beta_m + 2\gamma_m + 4\delta_m + \epsilon_m),$$
$$4\epsilon_{m+1} = 4\,(1-l)\,q_{m+1}r_{m+1} + 2l\,(2\delta_m + \epsilon_m),$$
$$\zeta_{m+1} = (1-l)\,r_{m+1}^2 + l\,(\delta_m + \epsilon_m + \zeta_m),$$
$$\therefore\ u_{m+1} = u_m.$$

When equilibrium is reached we can suppress suffixes in the above, and find $\gamma = 4\delta$.

$$\therefore\ (1-l)\,pr = (4-l)\,\delta = (1-l)\,q^2 + \frac{lq}{4}.$$

But
$$pr = \left(\frac{u}{1+u} - q\right)\left(\frac{1}{1+u} - q\right).$$

$$\therefore\ lq = 4\,(1-l)\left(\frac{u}{(1+u)^2} - q\right).$$

$$\left.\begin{aligned} \therefore\ p &= \frac{l + (4-3l)\,u}{(4-3l)\,(1+u)^2} \\ q &= \frac{4\,(1-l)\,u}{(4-3l)\,(1+u)^2} \\ r &= \frac{4-3l+lu}{(4-3l)\,(1+u)^2} \end{aligned}\right\} \quad \text{...(3·1)},$$

$$\left.\begin{aligned} \therefore\ kn &= \int_1^{u_n} \frac{(4-3l)(1+u)\,du}{u\,(4-3l+lu)} \\ &= \log_e u_n + \frac{4\,(1-l)}{l}\log_e\left(\frac{4-3l+lu_n}{4-2l}\right) \\ r_n &= \frac{4-3l+lu_n}{(4-3l)\,(1+u_n)^2} \end{aligned}\right\} \quad \text{...(3·2)},$$

unless $l = 0$, when $kn = u_n + \log_e u_n - 1$.

When recessives are very few,

$$(4-3l)\,r_n = \frac{l}{u_n} = le^{-\frac{lkn}{3l-4}} \text{ approximately,}$$

so $\dfrac{r_n}{r_{n+1}} = 1 + \dfrac{lk}{4-3l}$ approximately, and selection is rapid.

PARTIAL ASSORTATIVE MATING.

We consider a population containing a proportion r of recessives, the sexes being in equal numbers and mating so conducted that while each zygote is mated once and only once in a given period, the probability of a recessive mating with a given recessive is greater than that of its mating with a given dominant, and similarly for dominants. Let θ be the proportion of dominant $\times$ recessive and recessive $\times$ dominant matings, then that of matings between two dominants is $1 - r - \theta$, between two recessives $r - \theta$,

$$\therefore (r - \theta)(1 - r - \theta) = (1 + \lambda)\theta^2,$$

where λ is positive. In general λ is a function of r, but since $\frac{\lambda}{1 + 2\lambda}$ is the coefficent of association as defined by Yule (3), between the phenotypic characters of spouses, and such coefficients are found to be valuable even when the proportions of the different classes vary greatly, it is probable that λ varies rather little with changes in the population. In a case of human assortative mating given by Yule $\lambda = 0{\cdot}18$.

$$\theta = \frac{\sqrt{1 + 4\lambda r(1 - r)} - 1}{2\lambda},$$

$$\therefore p_{m+1} = \frac{(p_m + q_m)^2}{1 - r_m} - \theta_m \left(\frac{p_m + q_m}{1 - r_m}\right)^2,$$

$$q_{m+1} = \frac{q_m(p_m + q_m)}{1 - r_m} + \theta_m \left(\frac{p_m + q_m}{1 - r_m}\right)^2,$$

$$r_{m+1} = r_m + \frac{q_m^2}{(1 - r_m)} - \theta_m \left(\frac{p_m + q_m}{1 - r_m}\right)^2.$$

$\therefore u_{m+1} = u_m$; and, at equilibrium,

$$\left.\begin{aligned} p &= \frac{u}{1 + u} - q \\ \lambda(1 + u)^4 q^4 &+ u^2(1 + u)^2 q - u^3 = 0 \\ r &= \frac{1}{1 + u} - q \end{aligned}\right\} \quad \ldots(4{\cdot}1).$$

During selection

$$\frac{du_n}{dn} = kr_n u_n(1 + u_n),$$

$$\therefore \lambda\left(1 - \frac{1}{ku_n}\frac{du_n}{dn}\right)^4 - \frac{u_n(1 + u_n)}{k}\frac{du_n}{dx} + u_n^2 = 0.$$

$$\left.\begin{aligned} \therefore kn &= \int_1^{u_n} \frac{du}{u - uf(\lambda, u)} \\ r_n &= \frac{1 - f(\lambda, u_n)}{1 + u_n} \end{aligned}\right\} \quad \ldots(4{\cdot}2),$$

where $f(\lambda, u)$ is the real positive root of

$$\lambda x^4 + u^2(1 + u)x - u^3 = 0.$$

Clearly

$$\frac{u}{1 + u} > f(\lambda, u) > 0,$$

$$\therefore |u_n + \log_e u_n - 1| > |kn| > |\log_e u_n|$$

and

$$(1 + u_n)^{-1} > r_n > (1 + u_n)^{-2}.$$

Hence selection proceeds at a rate intermediate between those of equations (1·2) and (2·3) of Part I. When recessives are few, so that u_n^3 is large compared with λ, $f(\lambda, u) = \frac{u}{1+u}$, approximately, and selection proceeds according to equation (2·3) of Part I. Hence the effect of partial assortative mating in speeding up selection is unimportant.

SELECTIVE FERTILISATION.

If λ has the same meaning as above, except that it applies to unions between gametes and not zygotes, as in Jones' (4) case, where λ was generally less that 100, though in one experiment it exceeded 10,000, equilibrium is reached in one generation, and $pr = (1 + \lambda) q^2$, .

$$\left.\begin{aligned} \therefore\ p &= \frac{2\lambda + (1 + 2\lambda) u - \sqrt{(1 + u)^2 + 4\lambda u}}{2\lambda (1 + u)} \\ q &= \frac{\sqrt{(1 + u)^2 + 4\lambda u} - 1 - u}{2\lambda (1 + u)} \\ r &= \frac{1 + 2\lambda + u - \sqrt{(1 + u)^2 + 4\lambda u}}{2\lambda (1 + u)} \end{aligned}\right\} \quad \ldots(5\cdot1).$$

During selection,

$$\begin{aligned} kn &= \int_1^{u_n} \frac{2\lambda du}{u [1 + 2\lambda + u - \sqrt{(1 + u)^2 + 4\lambda u}]} \\ &= \int_1^{u_n} \frac{1 + 2\lambda + u + \sqrt{(1 + u)^2 + 4\lambda u}}{2 (1 + \lambda) u} du, \end{aligned}$$

and, if λ be constant,

$$\left.\begin{aligned} kn = \log_e u_n + \frac{1}{2 + 2\lambda} [u_n &- 1 + \sqrt{(1 + u_n)^2 + 4\lambda u_n} - 2\sqrt{1 + \lambda} \\ &+ (1 + 2\lambda) \log_e (1 + 2\lambda + u_n \\ &+ \sqrt{(1 + u_n)^2 + 4\lambda u_n}) - \log_e (1 + u_n \\ &+ 2\lambda u_n + \sqrt{(1 + u_n)^2 + 4\lambda u_n}) \\ &- 2\lambda \log_e 2 (1 + \lambda + \sqrt{1 + \lambda})] \end{aligned}\right\} \quad \ldots(5\cdot2),$$

$$r_n = \frac{1 + 2\lambda + u_n - \sqrt{(1 + u_n)^2 + 4\lambda u_n}}{2\lambda (1 + u_n)}.$$

Here again selection occurs at a rate intermediate between that of equations (1·2) and (2·3) of Part I, and when recessives are few $r_n = \frac{1 + \lambda}{u_n^2} = \frac{1}{(1 + \lambda) k^2 n^2}$ approximately, so selection is only very slightly more rapid than during random mating.

DISCUSSION.

Effects similar to those produced by partial brother-sister mating may be expected from less drastic types of inbreeding, *e.g.* mating of cousins. Such moderate degrees of inbreeding must occur in any population where neither zygotes nor gametes of both genders are very mobile. When recessives are sufficiently rare any cause which promotes inbreeding, even of distant relatives, will enormously increase their number for a given gametic ratio, and will make u_n tend to vary as r_n^{-1} rather than $r_n^{-\frac{1}{2}}$, making u an exponential function of kn instead of being proportional to it. Assortative mating will have little effect. Thus, if recessives number one in a million, and if only one mating in a hundred is between whole brothers and sisters, more than one recessive in 400 will mate with another recessive. To attain a like result by assortative mating a recessive must be more than 2500 times as likely to mate with a recessive as a dominant. This would imply such obstacles to mating with a dominant that the first recessive to appear could never mate at all. Probabilities of this order may, however, occur in selective fertilisation. Hence inbreeding or self-fertilisation appears to be necessary in the early stages of selection of a recessive character if this process is to be fast enough to be an effective cause of evolution. They cannot be replaced by moderate degrees of selective mating or fertilisation.

SUMMARY.

Expressions (2·1), (3·1), (4·1), (5·1) are found for the composition of Mendelian populations subjected to partial self-fertilisation, inbreeding, assortative mating, or selective fertilisation, and equations (2·2), (3·2), (4·2), (5·2) derived for the effect of selection on such populations. The effect of selection is greatly increased by inbreeding and self-fertilisation.

REFERENCES.

(1) Haldane (1924). *Trans. Camb. Phil. Soc.* **23**, 19.
(2) Lotka (1921). *Proc. Nat. Acad. Sci.* **7**, 168.
(3) Yule (1919). *An Introduction to the Theory of Statistics*, p. 38.
(4) Jones (1920). *Biol. Bull.* **38**, 251.

17

Reprinted from *Cambridge Phil. Soc. Proc.* **23**:363–372 (1926)

A MATHEMATICAL THEORY OF NATURAL AND ARTIFICIAL SELECTION. PART III

J. B. S. Haldane

In this part the cases of a single but incompletely dominant factor, and of several interacting factors are considered. Mating is supposed to be at random, populations to be very large, and generations not to overlap. The notation is, so far as possible, that of Part I (1).

Selection of an incompletely dominant autosomal character.

Let the nth generation be formed from female gametes in the ratio $u_n A : 1a$, male gametes in the ratio $v_n A : 1a$. The nth generation is therefore in the proportions $u_n v_n AA : (u_n + v_n) Aa : 1aa$. Let the ratios after selection has occurred be:

$$♂\ u_n v_n AA : (1 - K_m)(u_n + v_n) Aa : (1 - k_m) aa,$$
$$♀\ u_n v_n AA : (1 - K_f)(u_n + v_n) Aa : (1 - k_f) aa,$$

where K_m, K_f, k_m, k_f are small.

$$\therefore\quad u_{n+1} = \frac{2u_n v_n + (1 - K_f)(u_n + v_n)}{(1 - K_f)(u_n + v_n) + 2 - 2k_f}.$$

Hence, since $\dfrac{u_n - v_n}{u_n}$ is clearly small,

$$\Delta u_n = \frac{v_n - u_n}{2} + \frac{u_n (K_f u_n - K_f + k_f)}{1 + u_n}, \text{ approximately,}$$

and $$\Delta v_n = \frac{u_n - v_n}{2} + \frac{u_n (K_m u_n - K_m + k_m)}{1 + u_n}, \text{ approximately.}$$

Δu_n and Δv_n can be shewn to differ by a small quantity of the second order.

$$\therefore\quad \Delta u_n = \frac{u_n (K u_n - K + k)}{1 + u_n} \quad\ldots\ldots\ldots\ldots\ldots(1\cdot0)$$

where $$K = \tfrac{1}{2}(K_f + K_m);\ k = \tfrac{1}{2}(k_f + k_m).$$

Equilibrium can only occur when $\Delta u_n = 0$, i.e. u_n tends either to zero, infinity, or to $1 - \dfrac{k}{K}$. Hence for equilibrium to be possible $\dfrac{k}{K} < 1$. If K be positive, i.e. heterozygotes are at a disadvantage compared with pure dominants, then $\Delta u_n \gtrless 0$ according as $u_n \gtrless 1 - \dfrac{k}{K}$. Hence the equilibrium is unstable if it exists. If K

be negative the equilibrium is stable if it exists. We have thus three cases to consider. In each

$$\frac{du_n}{dn} = \frac{u_n(Ku_n - K + k)}{1 + u_n} \text{ approximately,}$$

and the proportion of recessives $y_n = (1 + u_n)^{-2}$.

(*a*) No equilibrium, $\frac{k}{K} > 1$.

$$\therefore \quad (k - K)\,n = \log_e u_n + \frac{k - 2K}{K} \log_e \left(\frac{Ku_n - K + k}{k}\right) \quad \ldots(1\cdot1)$$

making the usual convention that $u_0 = 1$.

Hence the values of u_n lie between two geometrical series, and selection is therefore vastly more efficacious on recessives than when dominance is complete, as in equations 2·4 and 4·3 of Part I.

(*b*) Stable equilibrium, $k > K$, $0 > K$.

$$\therefore \quad (K - k)\,n = \log_e \left(\frac{u_n}{u_0}\right) + \frac{2K - k}{K} \log_e \left(\frac{Ku_n - K + k}{Ku_0 - K + k}\right) \quad \ldots(1\cdot2).$$

We must take $u_0 \gtrless 1 - \frac{k}{K}$ according as $u_n \gtrless 1 - \frac{k}{K}$.

Here again successive values of u_n lie between two geometrical series, so that the population proceeds fairly rapidly towards equilibrium. As Fisher (2) has pointed out, such cases probably occur in nature in connexion with factors governing size, where the heterozygote is at an advantage as compared with either type of homozygote.

(*c*) Unstable equilibrium, $K > 0$, $K > k$.

The population proceeds towards homozygosis in one direction or the other. This case can hardly occur in nature, as any mutants, either in an AA or an aa population, would be weeded out while still few in number.

Selection of an incompletely dominant sex-linked character.

The female sex is throughout supposed to be homogametic; if the male is homogametic the argument is the same *mutatis mutandis*. Let the nth generation be formed from ova in the ratio $u_nA : 1a$, female-producing spermatozoa in the ratio $v_nA : 1a$. Let the ratios of the nth generation after selection be:

$$♀ \quad u_nv_nAA : (1 - K)(u_n + v_n)\,Aa : (1 - k)\,aa,$$
$$♂ \quad u_nA : (1 - k')\,a,$$

where K, k and k' are small.

$$\therefore \quad u_{n+1} = \frac{2u_n v_n + (1-K)(u_n + v_n)}{(1-K)(u_n+v_n) + 2 - 2k},$$

$$v_{n+1} = \frac{u_n}{1-k'}.$$

$$\therefore \quad \Delta u_n = \Delta v_n = \frac{u_n}{3+3u_n}[(2K+k')u_n - 2K + 2k + k'],$$

approximately, ...(2·0)

and $u_n = v_n$, approximately.

Hence u_n tends to zero, infinity, or $\frac{2K-2k-k'}{2K+k'}$. Equilibrium is possible if $\frac{k+k'}{2K+k'} < \frac{1}{2}$. It is stable if $2K+k'$ be negative, unstable if this quantity be positive. In each case

$$\frac{du_n}{dn} = \frac{u_n(2Ku_n + k'u_n - 2K + 2k + k')}{3(1+u_n)}, \text{ approximately,}$$

and the proportion of recessive females is $(1+u_n)^{-2}$, of recessive males $(1+u_n)^{-1}$. Three cases occur.

(*a*) No equilibrium, $\frac{k+k'}{2K+k'} > \frac{1}{2}$.

$$\therefore \quad \frac{2k+k'-2K}{3} n = \log_e u_n + \frac{2k-4K}{2K+k'} \log_e \left(\frac{2Ku_n + k'u_n - 2K + 2k + k'}{2k+2k'} \right)$$

......(2·1),

putting $u_0 = 1$. Selection therefore proceeds much as according to equation 7·2 of Part I.

(*b*) Stable equilibrium, $0 > 2K + k'$, $2k + k' > 2K$.

$$\therefore \quad \frac{2K-2k-k'}{3} n = \log_e \left(\frac{u_n}{u_0} \right) + \frac{2k-4K}{2K+k'} \log_e \left(\frac{2Ku_n + k'u_n - 2K + 2k + k'}{2Ku_0 + k'u_0 - 2K + 2k + k'} \right)$$

......(2·2)

where $u_0 \gtrless u_\infty$ according as $u_n \gtrless u_\infty$.

The results of Robertson (3) suggest that milk-yield in cattle depends on one or more sex-linked factors which act most effectively when heterozygous, besides autosomal factors. If so human effort in this case has given K a negative value, while k and k' are nearly zero. Hence an equilibrium should be reached.

(*c*) Unstable equilibrium, $2K + k' > 0$, $2K > 2k + k'$.

The population proceeds in one direction or the other to homozygosis. This case can hardly occur in nature.

Multiple factors.

Many cases exist in nature where several factors are needed to ensure the appearance of a character. Thus in wheat Nilsson-Ehle (4) found that any one of three dominant factors will produce redness, that is to say a white plant must be a triple recessive. On the other hand Saunders (5) found that in Matthiola three dominant factors are needed for slight hoariness, four for complete hoariness, so that a hoary plant is a multiple dominant. In other cases the effects of factors are merely additive, and selection will act on each independently of the others. It will be shown later that linkage between factors, unless very strong, is unimportant. We shall therefore at first consider unlinked factors, and shall confine ourselves to the case of complete dominance.

Selection of a multiple autosomal recessive character.

Let $A_1, A_2, \ldots A_r, \ldots A_m$ be m autosomal dominant factors, each of which produces the same effect, so that the multiple recessive alone competes with the other genotypes. Let $1-k$ of this type survive for every one of the others. Let y_n be the proportion of multiple recessives in the nth generation, formed from gametes in the proportion ${}_ru_n A_r : 1 a_r$, and similarly for the other factor pairs. Then the nth generation consists of zygotes in the ratios:

$${}_ru_n{}^2 A_r A_r : 2{}_ru_n A_r a_r : 1 a_r a_r, \text{ etc.}$$

$$\therefore \quad y_n = \prod_{r=1}^{m} (1 + {}_ru_n)^{-2}.$$

Of the $a_r a_r$ zygotes only $y_n (1 + {}_ru_n)^2$ are multiple recessives. Hence

$${}_ru_{n+1} = \frac{{}_ru_n (1 + {}_ru_n)}{1 + {}_ru_n - k y_n (1 + {}_ru_n)^2}.$$

$\therefore \quad \Delta {}_ru_n = k y_n \, {}_ru_n (1 + {}_ru_n)$, approximately, if k be small.

Putting $x = kn$ we have approximately:

$$\left.\begin{array}{l} \dfrac{d\,{}_1u_n}{dx} = y_n \, {}_1u_n (1 + {}_1u_n) \\ \dots\dots\dots\dots\dots\dots\dots\dots \\ \dots\dots\dots\dots\dots\dots\dots\dots \\ \dfrac{d\,{}_ru_n}{dx} = y_n \, {}_ru_n (1 + {}_ru_n) \\ \dots\dots\dots\dots\dots\dots\dots\dots \\ \dots\dots\dots\dots\dots\dots\dots\dots \end{array}\right\}$$

To eliminate the u's from these m equations put $s_r = \dfrac{{}_ru_n}{1 + {}_ru_n}$.

$$\therefore \quad \frac{ds_r}{dx} = y_n s_r. \qquad \therefore \quad \log s_r = \int y_n dx + \log a_r.$$

$\therefore\ s_r = a_r s$, where a_r is an integration constant independent of n and given by the initial state of the population.

$$\therefore\ \left.\begin{aligned} y_n &= \prod_{r=1}^{m} (1 - a_r s)^2 \\ kn = x &= \int \frac{ds}{s y_n} \end{aligned}\right\} \qquad \text{...............(3·0).}$$

The latter equation is integrable, and the elimination of s gives the required relation between y_n and kn.

$$\frac{dy_n}{dx} = -\,2y_n{}^2 \sum_{r=0}^{m} {}_r u_n,$$

whereas if only one factor is concerned,

$$\frac{dy_n}{dx} = -\,2y_n{}^2 u_n.$$

Now comparing these rates for equal values of y_n in the two cases, we note that since $y_n{}^{-\frac{1}{2}} = 1 + u_n = \prod_{r=1}^{m} (1 + {}_r u_n)\ \therefore\ u_n > \sum_{r=1}^{m} {}_r u_n$. Hence selection is slower than in the case of a character determined by one factor only. When however dominants are very rare, or when one a_r greatly exceeds the rest, i.e. one recessive factor is far commoner than the others, selection proceeds at about the same rate in the two cases. It is slowest when all the a_r's are equal.

Selection of a multiple sex-linked recessive character.

If $A_1, A_2, \ldots A_r, \ldots A_m$ are sex-linked (the female being homogametic) the nth generation formed from eggs in the ratios ${}_r u_n A : 1a$, etc., and female-producing spermatozoa in the ratios ${}_r v_n A : 1a$, etc., while z_n is the proportion of multiple recessive males, y_n of such females, and k is the coefficient of selection.

$$\therefore\ {}_r u_{n+1} = \frac{2\,{}_r u_n\,{}_r v_n + {}_r u_n + {}_r v_n}{{}_r u_n + {}_r v_n + 2 - 2k y_n (1 + {}_r u_n)}.$$

$$.\,{}_r v_{n+1} = \frac{{}_r u_n}{1 - k z_n (1 + {}_r u_n)}.$$

$$y_n = \prod_{r=1}^{m} (1 + {}_r u_n)^{-1} (1 + {}_r v_n)^{-1}.$$

$$z_n = \prod_{r=1}^{m} (1 + {}_r u_n)^{-1}.$$

$\therefore$ Approximately ${}_r u_n = {}_r v_n,\ y_n = z_n{}^2$.

$$\therefore\ 3\Delta\,{}_r u_n = k\,{}_r u_n (1 + {}_r u_n)(2z_n{}^2 + z_n).$$

As above, putting $kn = x$, $\frac{{}_r u_n}{1 + {}_r u_n} = a_r s$, we have a_r constant, and

$$\left. \begin{aligned} z_n &= \prod_{r=1}^{m} (1 - a_r s) \\ kn &= x = 3 \int \frac{ds}{s z_n (2 z_n + 1)} \end{aligned} \right\} \quad \ldots\ldots\ldots\ldots\ldots(4{\cdot}0).$$

This again is soluble in finite terms by the elimination of s.

$$\frac{dz_n}{dx} = -\tfrac{1}{3} z_n^2 (2z_n + 1) \sum_{r=1}^{m} {}_r u_n,$$

while in the single factor case

$$\frac{dz_n}{dx} = -\tfrac{1}{3} z_n^2 (2z_n + 1) u_n, \text{ where } u_n = z_n^{-1} - 1.$$

Comparing these rates for equal values of z_n, we find as above $u_n > \Sigma\, {}_r u_n$. Hence selection proceeds more slowly with many factors than with one. When, however, dominants are very rare or one a_r much larger than the rest, selection proceeds as with one factor.

Selection of a multiple autosomal dominant character.

When each of m autosomal dominant factors is needed to produce a character, we find, using the same notation as above except that y_n is the proportion of dominants,

$$y_n = \prod_{r=1}^{m} [1 - (1 + {}_r u_n)^{-2}],$$

$$\frac{d\, {}_r u_n}{dx} = \frac{y_n (1 + {}_r u_n)}{2 + {}_r u_n}, \text{ with } m - 1 \text{ similar equations.}$$

$$\therefore \quad (1 + {}_r u_n)\, e^{1 + {}_r u_n} = e^{\int y_n dx} = a_r s.$$

Hence the problem can be reduced to the elimination of s between:

$$\left. \begin{aligned} y_n &= \prod_{r=1}^{m} [1 - \{\phi(a_r s)\}^{-2}] \\ kn &= x = \int \frac{ds}{sy} \end{aligned} \right\} \quad \ldots\ldots\ldots\ldots\ldots(5{\cdot}0)$$

where ϕ is defined by the equation $t = \phi(t)\, e^{\phi(t)}$.

Numerical integration would be possible for known values of a_r,

$$\frac{dy_n}{dx} = 2y_n^2 \sum_{r=1}^{m} \frac{1}{{}_r u_n (2 + {}_r u_n)^2},$$

while in the single factor case

$$\frac{dy_n}{dx} = \frac{2y_n^2}{u_n (2 + u_n)^2}.$$

Now when one a_r is very much smaller than the rest these two rates are nearly equal for equal values of y_n. When all the a_r's are equal,

$$\frac{dy_n}{dx} = my_n^2\left[\left(1-y_n^{\frac{1}{m}}\right)^{-\frac{1}{2}}+1\right]^{-2}\left[\left(1-y_n^{\frac{1}{m}}\right)^{-2}-1\right]^{-1}.$$

The ratio of this rate to the rate with a single factor (putting $t^m = y_n$) is

$$\frac{mt^{m-1}(1+\sqrt{1-t^m})(1-t)^{\frac{3}{2}}}{(1+\sqrt{1-t})(1-t^m)^{\frac{3}{2}}}.$$

When t is small this tends to the small mt^{m-1}; when t is nearly unity, to $m^{-\frac{1}{2}}$ which is also small. The ratio when all the a_r's are equal is, by Purkiss' theorem, the minimum value. Hence it would seem that in general natural selection acts more slowly on a multiple dominant than a single dominant. The case of a multiple sex-linked dominant and various more complicated cases present still greater difficulties to analysis, though of course individual cases could always be solved numerically.

Linkage.

Consider two autosomal factors A, B, linked with such intensity that the cross-over value is $100l$ in the female, $100l'$ in the male sex. Let the nth generation be formed from:—

$$\text{eggs} \quad p_nAB : q_nAb : r_naB : s_nab,$$
$$\text{spermatozoa} \quad p_n'AB : q_n'Ab : r_n'aB : s_n'ab,$$

where $p_n + q_n + r_n + s_n = p_n' + q_n' + r_n' + s_n' = 1$.

The nth generation therefore consists of:—

$$p_np_n'ABAB : (p_nq_n' + p_n'q_n)ABAb : (p_nr_n' + p_n'r_n)ABaB$$
$$: q_nq_n'AbAb : (p_ns_n' + p_n's_n)AB.ab : (q_nr_n' + q_n'r_n)Ab.aB$$
$$: r_nr_n'aB.aB : (q_ns_n' + q_n's_n)Abab : (r_ns_n' + r_n's_n)aBab$$
$$: s_ns_n'abab.$$

If no selection occurs they produce gametes in the proportions:

$$2p_{n+1} = p_n + p_n' + l(q_nr_n' + q_n'r_n - p_ns_n' - p_n's_n)$$
$$2q_{n+1} = q_n + q_n' - l(q_nr_n' + q_n'r_n - p_ns_n' - p_n's_n)$$
$$2r_{n+1} = r_n + r_n' - l(q_nr_n' + q_n'r_n - p_ns_n' - p_n's_n)$$
$$2s_{n+1} = s_n + s_n' + l(q_nr_n' + q_n'r_n - p_ns_n' - p_n's_n),$$

whilst the values of p_{n+1}', etc., are given by the same expressions with l' substituted for l. Hence after one generation

$$\frac{p_n + q_n}{r_n + s_n} \text{ and } \frac{p_n' + q_n'}{r_n' + s_n'}$$

have the same constant value u, while

$$\frac{p_n + r_n}{q_n + s_n} = \frac{p_n' + r_n'}{q_n' + s_n'} = v.$$

We may therefore write:

$$p_n = \frac{uv}{(1+u)(1+v)} + x_n; \quad q_n = \frac{u}{(1+u)(1+v)} - x_n;$$

$$r_n = \frac{v}{(1+u)(1+v)} - x_n; \quad s_n = \frac{1}{(1+u)(1+v)} + x_n.$$

$$p_n' = \frac{uv}{(1+u)(1+v)} + x_n'; \quad q_n' = \frac{u}{(1+u)(1+v)} - x_n';$$

$$r_n' = \frac{v}{(1+u)(1+v)} - x_n'; \quad s_n' = \frac{1}{(1+u)(1+v)} + x_n'.$$

$$\therefore \quad q_n r_n' + q_n' r_n - p_n s_n' - p_n' s_n = -2(x_n + x_n').$$

$$\therefore \quad 2x_{n+1} = (1-2l)(x_n + x_n'); \quad 2x_{n+1}' = (1-2l')(x_n + x_n').$$

Hence if $x_0 + x_0' = c$,

$$\therefore \quad \left. \begin{aligned} x_n &= (\tfrac{1}{2} - l)(1-l-l')^{n-1}c \\ x_n' &= (\tfrac{1}{2} - l')(1-l-l')^{n-1}c \end{aligned} \right\} \quad \text{............(6·0).}$$

Hence the proportions of the various types of gamete approach asymptotically those which would be reached in one generation without linkage, the ratio of successive differences from the final values being $1-l-l'$. Hence if either l or l' is larger than k the effects of linkage are unimportant. A similar proof holds for a pair of sex-linked factors.

Selection in a tetraploid organism.

In a tetraploid race which is stable, i.e. yields only diploid gametes, five types of zygote and three of gamete exist. Gregory (6) and Blakeslee, Belling and Farnham (7) have shown that zygotes produce gametes as follows:

Zygotes	Gametes
$AAAA$	AA
$AAAa$	$1AA : 1Aa$
$AAaa$	$1AA : 4Aa : 1aa$
$Aaaa$	$1Aa : 1aa$
$aaaa$	aa

Gregory thought that $AAaa$ gave $1AA : 2Aa : 1aa$, but his results, as well as theory, suggest the above ratio. As in Part II we first consider tetraploidy without selection, and then the process of selection in a population which would be in equilibrium but for that selection. Let the mth generation be formed from gametes in the ratios $p_m AA : 2q_m Aa : r_m aa$, where $p_m + 2q_m + r_m = 1$, and $u_m = \frac{p_m + q_m}{q_m + r_m}$. They form zygotes in the ratios:

$$p_m^2 AAAA : 4p_m q_m AAAa : (4q_m^2 + 2p_m r_m) AAaa$$
$$: 4q_m r_m Aaaa : r_m^2 aaaa.$$

$$\therefore \quad p_{m+1} = p_m + \tfrac{2}{3}(q_m^2 - p_m r_m)$$
$$q_{m+1} = q_m - \tfrac{2}{3}(q_m^2 - p_m r_m)$$
$$r_{m+1} = r_m + \tfrac{2}{3}(q_m^2 - p_m r_m).$$

Hence $u_{m+1} = u_m = u$, and when equilibrium is reached $q_\infty^2 = p_\infty r_\infty$.

Hence $p_\infty = \dfrac{u^2}{(1+u)^2}$, $q_\infty = \dfrac{u}{(1+u)^2}$, $r_\infty = \dfrac{1}{(1+u)^2}$,

and the population in equilibrium is in the ratios:

$$\frac{u^4}{(1+u)^4} AAAA : \frac{4u^3}{(1+u)^4} AAAa : \frac{6u^2}{(1+u)^4} AAaa$$
$$: \frac{4u}{(1+u)^4} Aaaa : \frac{1}{(1+u)^4} aaaa.$$

Putting $\theta_m = q_m^2 - p_m r_m$, we find $\theta_{m+1} = \frac{1}{3}\theta_m$, $\therefore\ \theta_m = 3^{-m}\theta_0$.

$$\therefore \quad p_m = p_{m-1} + \theta_{m-1}$$
$$= p_0 + \sum_{r=0}^{m-1} \theta_r$$
$$= p_0 + \tfrac{3}{2}(1 - 3^{-m})\,\theta_0$$
$$= p_\infty - \frac{3^{1-m}}{2}\theta_0 \quad \text{......................(7·0).}$$

Hence the ratios of the different classes converge very rapidly to their final values. Under selection of a population which has reached such an equilibrium, if A is completely dominant,

$$u_{n+1} = \frac{p_n + q_n}{q_n + r_n - kr_n^2}$$
$$= \frac{u_n}{1 - k(1+u_n)^{-3}},$$

$$\therefore \quad \frac{du_n}{dn} = \Delta u_n = \frac{ku_n}{(1+u_n)^3}, \text{ approximately, if } k \text{ be small.}$$

$$\therefore \text{ if } u_0 = 1,\ kn = \log_e u_n + 3u_n + \tfrac{3}{2}u_n^2 + \tfrac{1}{3}u_n^3 - 4\tfrac{5}{6} \text{...(7·1),}$$

Hence when dominants are few u_n changes at the same rate as in a diploid organism; when they are many, much more slowly. To compare the change in the number y_n of recessives we find

$$kn = \log_e(1 - y_n^{\frac{1}{4}}) - \tfrac{1}{4}\log_e y_n + y_n^{-\frac{1}{4}} + \tfrac{1}{2}y_n^{-\frac{1}{2}} + \tfrac{1}{3}y_n^{-\frac{3}{4}} - 6\tfrac{2}{3} \text{...(7·2),}$$

$$\therefore \quad \frac{dy_n}{dn} = -4ky_n^2(y_n^{-\frac{1}{4}} - 1),$$

while in a diploid population

$$\frac{dy_n}{dn} = -2ky_n^2(y_n^{-\frac{1}{2}} - 1).$$

Hence here too the rate is always slower in the tetraploids, though not much so when recessives are few.

If dominance is incomplete, as is usual in tetraploid organisms, and after selection the zygotes are in the ratios:

$$u_n^4 AAAA : 4(1-k_1)\,u_n^3 AAAa : 6(1-k_2)\,u_n^2 AAaa$$
$$: 4(1-k_3)\,u_n Aaaa : (1-k_4)\,aaaa,$$

$$\therefore \quad \Delta u_n = \frac{k_1 u_n^4 + 3(k_2-k_1)\,u_n^3 + 3(k_3-k_2)\,u_n^2 + (k_4-k_3)\,u_n}{(1+u_n)^3},$$

approximately, if the coefficients are small. The possible equilibria, if any, are given by the roots of

$$k_1 u_\infty^3 + 3(k_2-k_1)\,u_\infty^2 + 3(k_3-k_2)\,u_\infty + k_4 - k_3 = 0.$$

The various possible cases, and their stability, could easily be investigated. If the advantage of the various genotypes increases or decreases with the number of dominant factors they contain, so that $k_4 > k_3 > k_2 > k_1 > 0$, or $0 > k_1 > k_2 > k_3 > k_4$, no equilibrium is possible,

$$\therefore \quad n = \int \frac{(1+u_n)^3\,du_n}{k_1 u_n^4 + 3(k_2-k_1)\,u_n^3 + 3(k_3-k_2)\,u_n^2 + (k_4-k_3)\,u_n} \quad \ldots(7{\cdot}3).$$

If $k_1 = 0$ this contains a term proportional to u_n or u_n^2. If $k_1 \neq 0$ all the terms are logarithmic and selection is always rapid. But $AAAa$ is more likely to resemble $AAAA$ than Aa to resemble AA. Hence polyploidy diminishes the probability of a rapid selection in populations where recessives are few. Since stable polyploidy is only known in hermaphrodite plants there is no need to discuss cases of sex-linkage or different intensities of selection in the two sexes. The theory can readily be extended to the higher forms of polyploidy.

Summary.

Expressions are found for the changes caused by slow selection in populations whose characters are determined by incompletely dominant, multiple, or polyploid factors, and for the equilibria attained in certain of these cases.

REFERENCES.

1. J. B. S. Haldane. *Trans. Camb. Phil. Soc.*, vol. 22, p. 19, 1924.
2. R. A. Fisher. *Proc. Roy. Soc. Edin.*, vol. 42, p. 321, 1922.
3. E. Robertson. *Journ. Genetics*, vol. 11, p. 79, 1921.
4. H. Nilsson-Ehle. *Lund's Univ. Arsskrift.*, p. 69, 1909.
5. E. R. Saunders. *Journ. Genetics*, vol. 10, p. 149, 1920.
6. Gregory. *Proc. Roy. Soc.*, B, vol. 87, p. 484, 1914.
7. Blakeslee, Belling and Farnham. *Science*, vol. 52, p. 388, 1920.

18

Reprinted from *Amer. Phil. Soc. Proc.* **105**(2):167–195 (1961)

ON CERTAIN ASPECTS OF THE EVOLUTIONARY PROCESS FROM THE STANDPOINT OF MODERN GENETICS *

S. S. CHETVERIKOV

Institute of Experimental Biology, Moscow, U.S.S.R.

Translated by

MALINA BARKER

Edited by

I. MICHAEL LERNER

University of California, Berkeley

INTRODUCTORY NOTE

A TRANSLATION into English of the classical paper by S. S. Chetverikov is long overdue. This work has been much cited as one of the cornerstones of experimental population genetics, but, even in the original Russian version, it is a bibliographic rarity difficult of access. Recently Th. Dobzhansky published translated excerpts as an appendix to his 1959 address at the Cold Spring Harbor Symposium on Quantitative Biology. Now, Malina Barker's translation of the full text of the article makes it possible for biologists unfamiliar with Russian to obtain first-hand acquaintance with the work which had exercised such a profound influence on, among others, Th. Dobzhansky, N. V. Timofeeff-Ressovsky, and N. P. Dubinin.

For a man who has been instrumental in initiating a whole trend of thought and experiment, which has now in the main withstood a test of a third of a century, Chetverikov remained until his death last year an unknown personality outside of the immediate circle of friends and students. For most of the following material I am indebted to Dr. Raisa L. Berg of the Department of Darwinism, Leningrad State University. She, in turn, drew for her information on Chetverikov's autobiographical note prepared about a year before he died.†

Sergei Sergeevich Chetverikov was born on May 6, 1880. He graduated from the University of Moscow in 1906. From 1909 to 1919 he lectured on general entomology at the Higher School for Women in Moscow. When, after the Soviet revolution, this institution was absorbed by the University, Chetverikov continued as a University lecturer in entomology and in systematics. Principles of both genetics and biometry were covered by him in the latter course. In 1924 the two subjects acquired independent status as courses in General Heredity and Introductory Biometry, for which he remained responsible until 1929. It was in this period that on the invitation of N. K. Koltzov, Chetverikov organized and headed the Department of Genetics in the former's Institute of Experimental Biology. After having spent three years in the city of Vladimir, Chetverikov in 1935 accepted the call to the chair of genetics at the University in Gorkiy. Shortly thereafter he became Dean of the Faculty of Biology and Director of the Saturnid Silkworm Experimental Station.

He virtually retired from active work in 1948, though as late as 1956 he described a new butterfly species from the southern Ural foothills. He spent the remaining years of his life as an invalid, having suffered a series of heart attacks and nearly complete loss of sight. The only surviving relative, his brother Nicholas (who himself belonged to the Moscow Society of Naturalists, of which Chetverikov was a prominent member), cared for him. S. S. Chetverikov died on July 2, 1959.

His published work is not extensive. Many of his early writings were notes on various species of Lepidoptera. He was an avid and expert collector, and in the course of a lifetime acquired a tremendous collection of butterflies, which even before his death was transferred to the Zoological Museum of the U.S.S.R. Academy of Sciences in Leningrad. A number of Chetverikov's publications, including his very first one in 1902, were

* Originally published in Russian in *Zhurnal Eksperimental'noi Biologii* **A2**: 3–54, 1926.

† Since this was written the German text of this note together with a photograph appeared in the *Nova Acta Leopoldina*, n.s., no. 143: 308–310, 1960.

guides for collectors and museum preparators. Others (of a total of twenty-six titles) included articles on butterflies and on biometry in the Great Soviet Encyclopedia, an anatomical study of the crustacean, *Asellus aquaticus,* and an account of selection in the silkworm for monovoltinism. One of his papers, dealing with the significance of the exoskeleton in insect evolution, was singled out for translation into English and published in the 1918 *Annual Report* of the Smithsonian Institution, where it rubs shoulders with contributions by E. W. Berry, Prince Kropotkin and Sir Ray Lankester. A somewhat earlier one (1905), entitled "The Waves of Life," first drew attention to the evolutionary importance of fluctuations in population size (population waves).

But the two publications for which Chetverikov will be longest remembered are an abstract presented to the Fifth International Congress of Genetics in Berlin in 1927, and the essay to which this note is intended as an introduction. The former dealt with an attempt at experimental verification of the deductive conclusions reached in the latter, regarding the extent of cryptic genetic variability in natural populations.

His experimental technique is, of course, today a commonplace one, but in 1926 it was a trail-blazing undertaking. Having concluded that the apparent phenotypic uniformity of many populations must conceal behind it a great amount of genetic variability, Chetverikov trapped 239 wild *Drosophila melanogaster* females and proceeded to examine the offspring of their immediate progeny, mated brother × sister. As many as 32 different segregating loci were found in this small population, giving rise to future work of this kind on many species by a great number of investigators. One need only to refer to Dobzhansky's *Genetics and the Origin of Species,* or to the Symposia of the Cold Spring Harbor Biological Laboratory dedicated to evolutionary topics (1955, 1959) to appreciate the far-reaching growth of experimental population genetics since the modest beginning of Chetverikov's experiment.

Perhaps Chetverikov's most wide-reaching influence has been as a teacher. He conducted a seminar on evolutionary problems which met under his guidance hundreds of times. He initiated into research such luminaries as N. V. Timofeeff-Ressovsky, B. L. Astaurov, N. P. Dubinin and D. D. Romashov. P. V. Terentjev, now Professor of Vertebrate Zoology at Leningrad, and H. A. Timofeeff-Ressovsky were also his students. It must have been a source of great satisfaction to Chetverikov to have both Timofeeff-Ressovskys visit him shortly before his death. A number of long-overdue signs of recognition reached him also at that time. They included honorary membership in the Russian Entomological Society, a plaque and diploma of the German Academy of Naturalists, and, perhaps, most meaningful to himself, a dedication to "My dear teacher and friend, Sergei Sergeevich Chetverikov" of an extensive review of the current status of microevolution published by N. V. Timofeeff-Ressovsky in the 1957 *Botanichesky Zhurnal.*

His best monument, of course, will be the vast field of population genetics developed from the point of departure of his 1926 essay. Many of the views expressed in it obviously are no longer tenable. For example, Chetverikov's arguments regarding the failure of artificial production of mutations were being rendered obsolete by Muller's work probably at the very time that they were made. Similarly, he was unaware of Castle's change of opinion on the effects of selection expressed in the 1919 paper in the *American Naturalist* (vol. 53). But these and other lapses were, needless to say, trivial, compared with the general clarity of his insight into the evolutionary process.

Stylistically, the essay is not well written. It is repetitive in many places, abounds in somewhat inexact metaphors, and many of the tortuous paragraph-long periods would be the despair of the most competent translator. But through the work, there shines a comprehension and vision of the synthetic theory of evolution, which most certainly antedated other ventures into the new systematics, and which make it of first-rate historical importance.

In editing the attached translation, a big share of the work consisted of trying to bring the bibliography and text citations into conformity with current standards of stylistic uniformity and accuracy. A number of uncorrected errors, especially in citations of nineteenth-century literature, may have escaped detection. Russian surnames and journal titles have been rendered in an English transliteration.

Some inconsistencies may be found by purists in the present English version. Most of them represent calculated risks in order to improve readability, or to up-date discarded terms. Thus, following Dobzhansky's lead, *mutation* is employed throughout instead of *genovariation.* In current usage mutation in its broad sense covers all of the types of genetic changes that Chetverikov's genovariation subsumed (see his first footnote). Confusion with Waagen's definition of mutation, as a phylogenetic displacement of one form by another, is avoided by using *Waagen's mutation* to describe this process. Similarly, I have accepted Dobzhansky's suggestion that *free crossing* represents the phenomenon discussed better than the translator's *random mating*, though this term conveys Chetverikov's meaning in a majority of cases. Finally, I have added some editorial footnotes giving modern names of several varieties and species discussed in the text.

I. Michael Lerner
University of California,
Berkeley

Hardly any other field of biological knowledge can look back on the course traversed in the last twenty-five year period with such satisfaction as one of its youngest branches—genetics. One feels in the significant words concerning genetics by the greatest American paleontologist, Professor Henry Osborn (1912), written already at the beginning of 1912: "Genetics is the most positive, permanent and triumphant branch of modern biology. Its contributions to heredity are epoch-making," not only acknowledgment of those enormous attainments of genetics in the beginning of the present century, but also the prescience of that colossal development which it achieved in the decade that followed, owing particularly to the work of the American geneticists, at the head of which, without doubt, we must place the name of Professor T. H. Morgan.

And if but ten or fifteen years ago it seemed to the great majority of biologists that the Mendelian laws of heredity, which lie at the basis of modern genetics, were only particular, special, cases of the transmission of heritable properties, then nowadays one can hardly find even a few individuals who do not recognize the validity of the words of East and Jones, who stated in 1919: "Mendelian heredity has proved to be the heredity of sexual reproduction: the heredity of sexual reproduction is Mendelian" (East and Jones, 1919: 50).

But then, sexual reproduction, in one or another form, is a basic type of multiplication in the plant as well as in the animal kingdom, and, therefore, it becomes at once evident what enormous significance Mendelism acquires in a whole series of general biological phenomena, and how deeply the basic aspects of genetics, of this "most positive, triumphant branch of biology," must penetrate the depths of the various departments of our science. And genetics, which is constantly developing theoretically and perfecting its methodology, is encompassing more and more new fields, of organisms as well as of the phenomena of heredity. It is not necessary to be a great fanatic about it to foresee the day when all aspects of heredity for the whole organic world will converge into one channel under the direction of a single and universal law of genetics. Nevertheless, one is often obliged to encounter points of view and opinions, which, if not directly hostile to genetics, then, at any rate, are characteristic of an extremely reserved and distrustful attitude toward it on the part of scientists expressing such an opinion. What is the cause of this distrust?

It seems to me that one should look for the cause in the fact that genetics in its conclusions touches too sharply and definitely upon certain long-established general theoretical views; that it demolishes too harshly conventional, deeply-rooted notions, while our theoretical outlook changes unwillingly from the well-marked ruts of conventional, logical generalizations to the uneven road of new constructs, though they might correspond more closely to our modern understanding.

Genetics is in a similar contradiction with conventional views of general evolutionary concepts, and in this, undoubtedly, lies the reason that Mendelism was greeted with such hostility on the part of many outstanding evolutionists, both here and abroad. The present article sets itself the goal of clarifying certain questions on evolution in connection with our current genetic concepts.

How can one link evolution with genetics, and bring our current genetic notions and concepts within the range of those ideas which encompass this basic biological problem? Would it be possible to approach the questions of variability, the struggle for existence, selection, in other words, Darwinism, starting not from these completely amorphous, indistinct, indefinite opinions on heredity, which existed at the time of Darwin and his immediate followers, but from the firm laws of genetics?

As yet we do not have an orderly elaborated system of evolutionary knowledge based on a contemporary genetic foundation, while the still-remembered, recent, extremely unsuccessful attempt of Lotsy (1916) to provide a new scheme of evolution, supposedly based on purely genetic notions, rather promoted the reinforcement of the distrust of skeptics than served to bring actually sound ideas into the general consciousness.

Naturally, in my undertaking, it is not possible to inspect the question at hand from all possible angles. I propose to dwell only upon certain aspects, which, in my opinion, are especially important for a correct appraisal of the role of our genetic ideas in the general structure of the theory of evolution. I will enumerate three such aspects: (1) the origin of mutations (or, as I shall call them, henceforth, "genovariations" in nature, (2) the role of free crossing in Mendelian heredity, (3) the significance of selection under these conditions.

I. THE ORIGIN OF MUTATIONS IN NATURE[1]

The opinion that the numerous mutations studied in chickens and mice, in maize and peas, and, finally, in various species of fruit flies (*Drosophila*), and in a whole series of other laboratory and domestic animals and plants are a result of the influence of man, is often encountered. It seems to many, consciously or sometimes subconsciously, that the origin of all these numerous hereditary changes is an effect of the artificial environment of domestication or of laboratory conditions. And it is, therefore, precisely as a result of such influences, that the majority of newly arisen variations are more or less definitely expressed monstrosities, not encountered in nature and not capable of having any significance in the process of evolution.

Such a view is absolutely false, although to refute it by experimental means is rather difficult because of the vicious circle into which the question falls. For to prove that a given variation encountered in nature is inheritable, that is, is genotypic, it must be submitted to genetic analysis, and for this it is necessary to keep it at least two generations under laboratory environment, that is, to expose it again to the action of artificial conditions.

But even granting that it is possible to prove convincingly that there is a certain number of mutations in a wild population in its natural surroundings, still it is never possible to determine whether they are actually new structures, i.e., mutations, or whether they appeared only as a result of recombination of structures already existing (from the beginning, according to Lotsy) in the population of genes; that is, it is not possible to demonstrate the very fact of *origin* of new genes in nature.

There is, nevertheless, a whole series of indirect considerations which logically leads us to acknowledge the existence of the process of origin of mutations in nature of precisely the same order that is found under our artificial conditions.

First of all, the generally sad fact must be recognized, *that up to the present we are not only completely unable to produce desirable mutations artificially, but are even unable to influence the frequency of their occurrence.* Not only ordinary conditions of the laboratory environment, but even the application of factors, so strongly acting on the organism, as unusual temperatures, radium, X-rays, alcohol, ether, abnormal barometric pressure, and, finally, hybridization, and a whole series of still other more or less strong influences, have not so far led to the positive result desired. If the origin of new mutations depended on the conditions of a laboratory experiment, then, by changing them in one or another way, we could

[1] I prefer to use the term "genovariation" in place of the customary word "mutation," since this latter term began to be used in paleontology (Waagen, 1869) much earlier than in genetics (de Vries, 1901, 1903) for the designation of a concept also pertaining to the phenomenon of variability, but essentially different. To wit, "mutation," in paleontology, refers to the alteration of the organism *in time,* in passing from one stratum to another. Moreover, the term "mutation" has several different meanings even in genetics, the "mutation" of de Vries by no means corresponding to the "mutation" of Morgan.

Therefore, I prefer to use the term "genovariation," meaning under it every newly developed inheritable change, each new form of "genotypic variant"; that is, in the sense similar to the "mutation" of Morgan.

The concept of genovariation includes both the hereditary changes of genotype dependent on changes in the chromosome complex itself (tetraploidy, trisomics, nondisjunction, etc.) or changes of whole sections of the chromosome (deficiency, translocation), and the very much more frequent changes within individual chromosomes, existence of which is manifested only by the changes of external characters conditioned by the corresponding genes. These are the changes which have lately been designated "gene mutation" by American authors (Sturtevant, 1925). But from the point of view of their general biological significance, in particular, in the question of the interrelationships between genetics and evolution, in so far as we may judge at the present time, there is no basic difference between these kinds of genotypic changes, so that it is possible to unite them all under one general term, *"genovariation."*

But from these genovariations all changes of genotype that are the result of the recombination of genes in crosses are naturally excluded, since no new structures arise in this way.

Editor's note: The term *mutation* has now acquired the broad meaning given by Chetverikov to *genovariation.* It is, therefore, used throughout the text. The footnote is retained in the translation merely for the record.

have an opportunity to control their origin. Since this is not so, then evidently no influence of laboratory environment exists.

The old experiments of Standfuss (1906), Fischer (1901, 1902), Tower (1906), and Kammerer (1907, 1908, 1913), which the defenders of the influence of external conditions on genotypic variability love to cite, lose more and more of their charm daily. Some of these have already received a more appropriate genetic interpretation, and in particular, the work of Kammerer with spotted salamanders, which resounded so much, has received a genetically very clear, though unexpected, explanation in the studies of Herbst (1919, 1924). Another part of the literature, in particular, the data of Tower, has now lost general confidence in his native country, America, and insistently calls for review and verification by means of more perfected genetic methods.

The situation is not better with the more recent studies in that direction, of which the work of Guyer and Smith (1920, 1924) on heritable changes in the eye, produced by injection of certain antibodies into the blood, and the article of Academician I. P. Pavlov concerning the inheritance of acquired conditioned reflexes, have attracted most attention. They not only did not strengthen the point of view defended, but, on the contrary, called for extremely thorough and harsh criticism (Koltzov, 1924; Morgan, 1924) showing the complete lack of genetic foundation for such kind of conclusions.

I repeat, that by means of laboratory treatments we cannot produce mutations, and, consequently, their appearance is not dependent on artificial conditions of investigation.

I consciously omitted for the present the cases of artificial changes in the number of chromosomes (resulting in tetraploidy and other forms of polyploidy) indicated in the works of Winkler (1916), the Marchal brothers (1906), von Wettstein (1924), and others. Here we are dealing with an exceptional type of genotypic variability, not with changes in single genes, but merely with the increase in the number of their groups. In several specific instances of such kind, we are in a position to produce mutations artificially but the very exceptional nature of this form of genotypic variability places the cited instances apart from the general and principal mass of mutations, conditioned by changes in single genes. From the point of view of the origin of mutations in nature, all these cases are interesting because of the simplicity of the stimuli producing these genotypic changes—injury to the vegetative parts of plants —which gives us a full right to admit similar changes under natural conditions. The existence of polyploid forms in the wild state, further strengthens our conviction that the origin of this type of mutation is not solely dependent on exclusively laboratory conditions.[2]

But still the question remains, why, in such a case, in the presence of thousands of mutations among laboratory and domestic animals and among cultivated plants, we know so little about their existence in nature; why, in the fruit fly (*Drosophila*) where we now number more than four hundred mutations in our culture jars, we know almost nothing about the same process under natural conditions of existence.

Undoubtedly, their often observed lowered viability is one of the important reasons for the rare occurrence of mutations in nature. In this connection, the abundant material collected on *Drosophila* yields such a quantity of exactly verified and analyzed facts that no other material is comparable to it.

We have here all the most imperceptible transitions from such mutations which have completely normal viability (for example, in *D. ampelophila*,[3] *black*, *vermilion*, a series of changes in wing venation, etc.; here also belongs *radius incompletus* of *D. funebris* Meig, described by Timofeeff-Ressovsky, 1925), through those having decreased viability to a greater or lesser degree (for example, *yellow*, *rudimentary*), to mutations which have exceedingly low viability, such as branched legs (*reduplicated*), wingless (*apterous*), and such like, which are directly contiguous with the so-called lethal mutations.

It is obvious that in the severe struggle for survival, which reigns in nature, the majority of these less viable mutations, originating among normal individuals, must perish very quickly, usually not leaving any descendants.

It stands to reason, that it is extremely difficult to find by chance such a mutation in nature, and, in general, it may be said that the number of mutations with decreased viability is significantly greater than that of those the viability of which does not suffer.

Here we have arrived directly at the other side

[2] Perhaps the observations made by Kosminsky (1924) on polyploidy in the spermatogenesis of the silkworm under the influence of raised temperature belong to the same category.

[3] *D. melanogaster* in current usage. Ed.

of the question of the origin of mutations in nature and their role in the process of evolution. Often the opinion is encountered that the whole process of the origin of mutations is, essentially, a teratological one, that all the changes occurring in genes are really only more or less clearly apparent "freaks," because of which they cannot have any significance in the process of directional evolution.

Such an opinion could be merely the result of an insufficiently profound study of the phenomenon of mutation. It is true that the number of "freakish" mutations harmful to the organism is incomparably greater than that of the harmless ones (not to mention beneficial ones), but such a ratio is entirely to be expected because of the nature of the currently perfect adaptation of the organisms to their environment.

The living organism in its normal habitat represents an extremely fine, complex and perfect mechanism, adjusted to all the varied requirements which are demanded from it by this habitat. To "injure" such a mechanism is very much easier than to "improve" it.

If someone should set himself the goal of inventing chance variations of the organism, and would thereupon go on to classify these variations under the categories of "harmful," "neutral," and "beneficial," then it certainly can be said that the first group would be many times larger than both the others taken together, and the last group of "beneficial" changes would seem completely negligible. These three groups occur with respect to mutational variability in approximately the same ratio, and this fact merely serves as an extra confirmation of the randomness of the changes that arise.

But, along with "harmful" freakish mutations, there is also a series of "neutral" ones, not having any biological significances and therefore not subject to selection (see below), that can be ascertained by anyone who had worked in genetics. Examples of such mutations in *Drosophila* were mentioned above.

It is also particularly interesting and important to note the fact that some of these "biologically neutral" mutations occurring randomly in the normal populations of some species or other sometimes correspond to the "normal" features of neighboring species *or even genera and families.* Thus the darkening of the coloration of the body characteristic for the *"black"* mutation in *D. ampelophila,* is a normal feature of the species *D. funebris;* the lack of the crossvein in the wing (*"crossveinless"*) serves as a characteristic distinguishing feature of the family closely related to *Drosophila, Asteidae;* the curving downwards of the wings, encountered in a series of mutations (*"depressed," "curved"*), appears at the same time as a diagnostic trait of the genus *Stegana* (fam. *Drosophilidae*), etc. I myself had the opportunity to see H. G. Shaposhnikov's photographs of a whole brood of *Saturnia pyri* Schiff., which had on the outer margin of the front wings a halfmoon-shaped scallop, a trait which in this form is characteristic of a series of genera of Lepidoptera (for instance, *Drepana, Macaria,* and others).

It will hardly occur to anyone to see deformities in these characteristically "normal" features of the genera and species noted, and consequently there is no basis for considering them as freaks when they appear unexpectedly as mutations in species not normally having them. On the contrary, facts of this kind can only strengthen the conviction that features, such as those specified above, could have originated in the corresponding families, genera and species in exactly the way in which they occur at the present time before our very eyes in the species under investigation—that is, mutationally.

Nevertheless, it would be completely erroneous to think that the viability of mutations depends on the intensity of the morphological change of a character. We see very often that genes, at first glance affecting completely "neutral" characters, in fact exert a marked influence on viability. Thus, while the darkening of the body under the action of the gene *"black"* does not affect the viability of the fly *D. ampelophila,* its lightening under the action of the gene *"yellow"* proves to be coupled with some weakening of the organism.

Moreover, it should not be forgotten that we now know a whole series of physiological genes, not expressing themselves in any morphological way, but undoubtedly playing a tremendous role in the life processes of the organisms. As examples of the action of genes of such a kind one may point to the early-ripening varieties of cultivated plants, to frost-resistant or frost-susceptible varieties of cereal grasses, to the natural immunity of various plant as well as animal organisms, to the univoltine and bivoltine varieties of the silkworm, to the unequal egg production of various breeds of chickens, and to many others. It is evident that, among these physiological genes, not a few will be found which would either affect the organism very harmfully, or would, on the contrary, under specific environmental conditions, prove to be very beneficial (for instance, frost-resistant varieties in the North), although, externally, that is, morphologically, these mutations would not be distinguishable. Thus, no direct corrélation between the intensity of morphological change produced by a mutation and its viability exists.

We have already seen in several examples mentioned above that mutational variability touches upon various characters greatly differing in significance. Alongside of the least salient traits, such as the color of the body, such important characters of *Drosophila* are changed as venation, wing structure, etc., which are fundamental in the modern systematics of insects for distinguishing the higher systematic categories. Consequently, it is necessary to acknowledge as completely erroneous the idea which some express, that mutations deal in only a superficial way with species traits, being characteristic of differences between varieties, while, apart from these small aberrations, there is a "basic substance" of organisms which is not subject to mutational changes. Because of this, it is argued, the process of evolution, the process of the transformation of whole organisms into others, could not be achieved by means of mutations. Speaking figuratively, it is claimed that with all mutations, a fly always remains a fly, and a rat a rat, and never does the latter produce deviations in the direction of a rabbit or a dog.

But here two concepts are confused: the *diversity* of characters subject to genotypic variability, and the scope of variability, i.e., the *amplitude of deviation*. Actually, we see that *absolutely all* parts of the organism are subject to genotypic variation. But, although genotypic variability is discontinuous, its leaps, naturally, cannot be infinitely large, so that the amplitude of aberration is limited, and the limit is determined by the structure of the genes themselves. Abrupt and profound changes of the organism are possible only by means of a *prolonged accumulation* of mutational changes, of long-termed stratification of one deviation upon another.

But in mutational variability we did not recognize anything new that was not there before; it is not a *new means of producing variation,* but rather the most fundamental, timeless means of evolution by which the organic world has gone on and developed from its very appearance on earth to our own days. And if we only see *sequential* changes on this way, only see the slow transformation of some forms into others, this comes about exactly because mutational variability cannot transform one organism into another suddenly in one leap, but only gradually by means of accumulation of individual mutations.

On the contrary, it would be fatal to our conception of the role of mutations in the course of the evolutionary process if we actually had cases of such sudden transformation of some organisms into others. For we know that the course of evolution was slow and gradual, and that such sudden transformations would only attest that mutations are apart from the evolutionary path, that between them and evolution there is an impassable gulf. Fortunately we know that such a gulf does not exist in reality.

In the discussion of the question of the origin of mutation in nature, it is necessary to have in view at all times the tremendous role of free crossing, the engulfing influence of which must strongly affect the frequency of detection of mutations under natural conditions. The next section will be dedicated to an analysis of the role of free crossing, while here it is relevant to mention only the fact that owing to the effect of such crossing, each newly originating recessive mutation, through crossing with the normal forms, would be dissolved, so to say, in the latter, and disappear (practically, within the limits of our investigation, forever) from morphological detection.

Thus, it would be possible to observe the origin of a new mutation in the majority of cases only at the very moment of its origin, when it either is not yet destroyed by natural selection, in the case of decreased viability, or while its morphological expression has still not been dissolved in the surrounding normal forms, or swallowed up by the effects of free crossing.

In the laboratory environment, where the conditions of existence and direction of crossing are controlled and regulated by the conscious will of man, both of the preceding powerful factors, effecting the decrease in the number of observed mutations in nature, are removed, and therefore frequency of detection of new mutations is immeasurably increased.

Finally, in the discussion of the question of the origin of mutations in nature it is necessary to take into consideration also the following thought: as the data of Morgan and his school on *D. ampelophila* (*melanogaster*) have shown, one mutation occurs in approximately 10,000 flies observed. If, lacking data for other animals or plants, we accept this ratio as the usual one between changed and original organisms, then granting the same intensity of the process of appearance of mutations in nature, one can expect to come across one case of the origin of a mutation in 10,000 normal specimens. But it is doubtful that there exist many collections where one may carry out a survey of 10,000 specimens of one species in order to detect one mutation.

However, as we will immediately see, this discussion has only conditional importance, and, as I shall try to show in the following sections, there are in nature at all times two opposing processes: one, the accumulation of mutations; the other, their elimination. The presence of a greater or smaller number of mutations in nature depends on the balance between them.

Thus, we ought to expect *a priori* that mutations in nature will come to our attention immeasurably less frequently than in the laboratory environment or in cultivated plants or domestic animals. The correspondence of the actual state of affairs with theoretical expectations is the best proof of the validity of the premises stated.

And thus, we do not as yet have any basis for seeing in the process of origin of mutations the result of the artificial influence of man. On the contrary, the whole presently available stock of facts indicates that this process goes on just as regularly under natural conditions as in the laboratory conditions of an experiment, but that there is a whole series of weighty reasons removing a vast number of cases of such origin of mutations in nature from our observation.

II. MUTATIONS UNDER CONDITIONS OF FREE CROSSING

I have already briefly pointed out in the preceding section the enormous significance which free crossing has on the fate of newly arising mutations in nature. In the present section I shall attempt to analyze this significance in greater detail.

The mechanisms operating in the inheritance of various mutations in different forms of mating, the most simple of which were already established by Mendel in 1865, served many times as the theme for mathematical studies by a series of investigators. Beginning with the remarkable work of Pearson (1904) dating as early as 1904, a whole series of individuals—Hardy (1908), Jennings (1912, 1914, 1916, 1917), Pearl (1913, 1914*a*, 1914*b*), Fish (1914), Robbins (1917-1918), Wentworth and Remick (1916), Wright (1921*a* and *b*), Philipchenko (1919, 1924), Tietze (1923), and Romashov (1926)—subjected to mathematical analysis the results of the inheritance of Mendelizing factors under all imaginable combinations of matings. But from this long list of works only those which investigated the results of mating and the fate of individual features under conditions of free crossing have significance for an accurate understanding of the role of genotypic variation in the process of evolution, since the natural state of a species pre-supposes precisely a state of a "freely crossing community"—"*Paarungsgemeinschaft*" of the German authors.

The definition of species, as an aggregate of individuals constituting a single freely intercrossing complex, corresponds most closely to our genetic and systematic ideas. Naturally, the freedom of crossing in a whole series of cases depends on many internal as well as external causes, sometimes reinforcing, sometimes diminishing its significance. But *potentially,* all the individuals of a *single species* are able to cross *freely,* without encountering any hindrance either in the process of fertilization itself or in the viability or fertility of their offspring.

Naturally, in so far as there are not and cannot be boundaries between the definitively formed "good" species and sharply expressed, strongly differentiated varieties, the above criterion for a species cannot always be unconditionally applicable. But in any case, if two generally acknowledged species are known to be freely crossing with one another, both as pure "species" themselves and as hybrids, and crossing is, indeed, free, without any direct or indirect influence of man, then from the genetical point of view such "species" actually belong to *one* species. And, on the contrary, if two varieties of one species begin to show some reproductive isolation, expressed either in the area of instincts, or in the physiology of fertilization, or, finally, in the viability or fertility of their offspring, then even if we cannot in this case speak of two distinct "good" species, all the same, we undoubtedly are dealing with the beginning of an as yet incomplete process of speciation, with a new species *in statu nascendi.*

Naturally, such a definition of *species* should not be understood in the sense that one or another form of reproductive isolation is the initial step of the process of speciation. Undoubtedly, an accumulation of a greater or lesser number of morphological or physiological differences always precedes the appearance of reproductive isolation, but the very fact of existence of these differences, in whatever form they should appear, is not a sufficient basis for the formation of a new species. And from our genetic experiments on the most diverse animal and plant organisms, we now know that it is possible to create two groups of such organisms, which will differ from each other by a perfectly concrete complex (theoretically speaking, as large as one wishes) of morphological characteristics, which are not connected by intermediate forms; that is, having a so-called morphological *hiatus*, but at the same time belonging genetically to the same species.

Thus, free crossing is a characteristic condition of the enormous majority of natural species of animals as well as plants, and the significance of this factor in the process of evolution must not be underestimated.

Actually, the extent of significance which the greatest theoreticians of evolution attributed to this factor is seen, if in nothing else than in the fact that, until the very end of his life, Darwin considered the most substantial objection to his theory not the critical remarks of even the greatest biologists, but that which was given in 1867 by a certain engineer, Professor Jenkin, who demonstrated by a simple arithmetical calculation that as a result of free crossing every accidentally originating deviation, be it even beneficial, must be very quickly dissolved among the normal individuals. Thus, crossing exerts a dissolving ("swamping"—Darwin) influence on every newly arising single deviation, even if it is a very marked one.

These considerations pressed Darwin in subsequent revisions of his theory further and further away from current genetic views of the role and significance of single deviations, mutations (in Darwin's terminology, "sports"), in the process of evolutionary development of the organic world, and caused him to attribute a greater and greater role in this process to the small but mass individual deviations, now included under the term "fluctuations."

As we shall see further, the considerations which so strongly disturbed Darwin are not in conformity with our current concepts of the role of crossing, but still it is important to note the fact that this nonconformity deflected the theoretical thought of Darwin from his *earlier and more accurate* concepts in the direction of acceptance of views closer to current Neo-Lamarckism.

And the other founder of current evolutionary theory, that is, A. R. Wallace, occupies exactly the same position on this question, emphasizing in several places in his book *Darwinism* the idea that individual, accidental changes of organisms (in our sense, mutations) are in no case a source of the evolutionary process, which rather lies in insignificant but *mass* variability, which is characteristic of the whole organic world (see particularly the section: "The swamping affects of intercrossing," p. 210, 1898 edition).

In precisely this manner all the following evolutionary theoreticians, among whom Wagner (1868, 1889), Weismann (1872, 1904), Romanes (1926), Petersen (1903), Pearson (1900), and Platé (1913), must be mentioned, attributed an extremely important, sometimes directly decisive, role to free crossing in the process of evolution.

But, at the present time, now that genetics imperatively dominates the whole field of sexual reproduction, and with it, naturally, the process of free crossing, a pressing need is felt for a review of the conclusions at which the previous investigators arrived, to bring them into conformity with our current genetic concepts.

Of all the numerous works cited at the beginning of the present section, the investigations of K. Pearson and G. H. Hardy have the greatest interest and significance for us. The latter, in a short work of only two pages, established the extremely important law, characterizing the condition of equilibrium existing under Mendelian laws of heredity and free crossing. One may term this the *law of equilibrium under free crossing* or *Hardy's law*. This law may be briefly formulated as follows: the relative frequencies of homozygous (dominant as well as recessive) and heterozygous individuals, under conditions of free crossing and in the absence of any kind of selection, remain constant, provided that the product of the frequencies of homozygous individuals (dominant and recessive) is equal to the square of half of the frequency of heterozygous forms.

Expressing this law as a genetic formula and representing the composition of the population analyzed by the expression $pAA + 2qaA + raa$, where p, $2q$ and r denote the frequencies of the respective groups of homozygous and heterozygous individuals, we may define the condition of equilibrium of such a freely crossing population by the condition:

$$pr = q^2.$$

From this law the extremely important conclusion follows: since for any value of p and r a value of $2q$ may be found to satisfy the equation $pr = q^2$, a freely crossing population may be in a state of equilibrium with any proportions of homozygous dominant and recessive forms.

Thus, in a freely crossing population, there can be preserved from generation to generation not only the classical Mendelian ratio 1:2:1 (phenotypically, 3:1) but the numbers of one of the homozygous forms (regardless of whether it is dominant or recessive) may exceed the numbers of the other by 1, 2, 3 . . . times, and the population will still remain in a state of equilibrium, so long as the fundamental condition, $pr = q^2$, is satisfied.

Directly related to the law just established, is another very important law, which is also concerned with the state of equilibrium within a freely crossing population, and may be termed *the law of stabilizing crossing.* First established by K. Pearson, as early as 1904, it remained completely unnoticed for a long time, owing to the extremely abstract formulation, inaccessible to the great majority of biologists, which was initially given it.

This law was proven anew by Hardy (1908) in the above-mentioned article, and was subsequently established again several times in various formulations and on the basis of various mathematical and biological considerations by a series of investigators (Jennings, 1916; Wentworth and Remick, 1916; Tietze, 1923).

Briefly, this second law of free crossing, or as we shall call it, the *law of stabilizing crossing* (*Pearson's law*) may be formulated thus: under conditions of free crossing with any initial ratio of frequencies of homozygous and heterozygous parental forms, a state of equilibrium will be established in the population as a consequence of the very first generation of free crossing.

Thus, should a state of equilibrium in a freely crossing population be disturbed from without, as a result of the very first subsequent crossing, which we will call *stabilizing crossing,* a new state of equilibrium is established within the population, at which the given population will remain until some new external force removes it from this state.

Proceeding again to a genetic notation, we may formulate the specified law thus: if we have a certain freely crossing population,

$$xAA + 2yAa + zaa, \qquad (P)$$

displaced from the state of equilibrium, that is, where

$$xz \neq y^2,$$

then as a result of the first *stabilizing crossing,* the respective frequencies of homozygous and heterozygous forms are expressed by the formula

$$x_1{}^2AA + 2x_1z_1Aa + z_1{}^2aa, \qquad (F_1)$$

so that

$$x_1{}^2z_1{}^2 = (x_1z_1)^2$$

and our population will be in a state of stable equilibrium (according to Hardy's law).

Thus, the apparatus which stabilizes the frequencies of the components of a given population lies in the very mechanism of free crossing. Any change of the ratio of these frequencies is possible only by action from without and is possible only so long as the external force which disturbs the equilibrium is acting.

In the present study we shall concern ourselves with only two of these external forces: selection, in the broadest sense of the term, and the origin of new genotypic changes—mutations. We shall now turn to the investigation of the role of these under conditions of free crossing.

In the preceding section I tried to show that we have no basis whatsoever for denying the existence under natural conditions of a continuous process of origin of new mutations. As all the data on *Drosophila,* which was most thoroughly investigated, show, the number of newly originating mutations appears to be increasing without limit. At the same time some of the mutations in certain cases tend to arise repeatedly and more or less frequently (for example, "*white*" and "*Notch*"). In other cases, the same gene changes in a different fashion, giving a series of numerous alleles such as are characteristic of the same "*white*" or, for instance, of "*Truncate*" and "*dumpy,*" etc. The greatest majority of mutations, however, arose but once, and, for the present, no limits to the torrent of these singly occurring changes and to the variety of their expression can be seen.

What is the fate of these individual mutations, these "sports," in Darwin's terminology? Are they actually condemned to disappearance without a trace, to be dissolved in the sea of normal individuals, in no way influencing the subsequent fate of the species, the process of its evolution?

Let us begin the analysis with a case of the not infrequent appearance in nature of a recessive homozygous mutation (*aa*). What shall be the fate of such a new gene? The appearance of such mutation disrupts the state of equilibrium of the freely crossing species at that moment. If this mutation is not immediately destroyed by natural selection because of its lesser viability or lack of adaptability, then it will cross with a normal form (*AA*). Here we know, on the basis of the law of stabilizing crossing, that in the immediately following generation the equilibrium will be restored, while our recessive gene will become the heterozygous (*Aa*).

Assuming that the pair of parent individuals, $aa \times AA$, with a constant population size of a species, must leave one pair of offspring, we must conclude that two individuals which are phenotypically normal, but genotypically heterozygous (*Aa*), will enter into the composition of the population. After this, because of the state of equilibrium, this composition of the population (species) will continue from generation to generation.

A short and simple calculation shows that the probability of two such heterozygous individuals meeting, as a result of which a homozygous recessive form (*aa*) could appear anew (that is, that the initial homozygous form of the mutation will be again manifested) is equal to one over the number of individuals in the whole population (species) minus one.

Translating this into symbolic language, and taking the number of individuals in a population as $N + 1$, we can express the probability (p) of two such heterozygous individuals meeting by the equation

$$p = 1/N.$$

This equation means that in N consecutive generations one such meeting of heterozygous individuals can occur.

Let us imagine some concrete example as an illustration. Assume that there are now in the whole of Northern Eurasia 1,000,000 plus 1 gray crows. Suppose that among these a recessive mutation, an albino, appears suddenly. If it is not destroyed and crosses with a normal individual, then in the generation following, the state of equilibrium of the species will be restored, so that in the general mass of gray crows there will be preserved from generation to generation a pair of individuals which are externally normal, but are heterozygous for color.

The probability that these two heterozygous individuals will meet and produce offspring (in the ideal conditions of free crossing) is equal to $1/N = 1/1{,}000{,}000$, that is, the renewed appearance of a white (homozygous) individual owing to the recombination of genes may be expected once in 1,000,000 consecutive matings. Practically speaking, this probability is entirely negligible, and, in fact, our albino mutation will be absorbed, "swallowed up" by free crossing.

But its fate will be completely different from what former evolutionists imagined. The mutation will not be destroyed, will not be dissolved in the mass of normal individuals. It will exist in the heterozygous state remaining hidden from the eye, but in the form of a specific inheritable genotype generation after generation.

These considerations make it possible for us to examine more deeply and more clearly the genetic structure of a freely crossing aggregate, a species.

To what extent is a species-population homogeneous in its hereditary properties? And if we still admit a share of it to be heterogeneous, then how can we explain in that case that great uniformity, the "monotypism" of natural wild species, which so characteristically distinguishes them from domestic breeds?

We have just seen that each newly arising natural mutation is swallowed up by the basic mass of the species, though it is not destroyed but preserved in the heterozygous condition in the "depths" of the species. In the first section I tried to show that the process of origin of new mutations must be viewed not as an accidental occurrence, but as entirely normal, systematic one. Thus, in instances of repeated origin of natural mutations, they will again and again be swallowed up by the original species. A new phenomenon now arises, providing us once more with an opportunity to examine several interesting questions.

Assume that a certain freely crossing population (an aggregate) consists of $N + 1$ individuals. The probability of reappearance among them of the earlier appearing mutation a (as a result of the meeting of heterozygous individuals) will be, as we have seen, $1/N$, which for a large value of N is a negligibly small quantity.

But let us imagine that in that very same aggregate there should arise another independent individual variation (*bb*), which would also pass into the heterozygous state. The probability of its second appearance as a result of recombination will also be $1/N$; for a third mutation (*cc*) the probability will be the same, as it will be for the fourth (*dd*), for the fifth (*ee*), etc., etc.

All these mutations, originating within a "normal" species, pass, as a result of crossing, into the heterozygous state, and are thus swallowed up, absorbed by the species, remaining in it in the form of isolated individuals. As a result, we arrive at the conclusion that a *species, like a sponge, soaks up heterozygous mutations, while remaining from first to last externally (phenotypically) homogeneous.* That such a conclusion about the genotypic structure of a species corresponds to reality is confirmed by the results of an as yet unfinished analysis of the constitution of the wild species of the genus *Drosophila,* undertaken by the genetical laboratory of the Institute of Experimental Biology last summer (1925).[4]

The probability ($1/N$) of a meeting of two like heterozygotes, in a least bit numerous species (N), is so small that it can practically be disregarded. But under consecutive appearances of new mutations (independent of each other), the probability of appearance of some one of them will obviously become greater and greater, being determined by the law of the *summation* of independent, equal and compatible probabilities.

Thus, with two concealed mutations the probability of a second appearance of either one will already be almost twice as great:

$$p = 2/N - 1/N^2.$$

For three mutations, it will be equal to $3/N - (3/N^2 - 1/N^3)$, and, in general, for m mutations absorbed by the population, the probability (p) of appearance of some one of them as a result of the recombination under free crossing is expressed by the formula:

$$p = 1 - (N\text{-}1)^m/N$$

in which the exponent m, equal to the number of independent probabilities, that is, to the number of mutations that have arisen and been swallowed up by the species, can obviously increase indefinitely.

It is evident that with the increase in this number, that is, with the increase of the exponent m, this probability can prove to be very high, and a given species will then disclose now one, and now another, of the mutations contained within it.

Here we approach an important and interesting question. We have just seen that a species, like a sponge, soaks up more and more mutations, remaining all the time externally monotypic. But as the accumulation within the species of a greater and greater number of such concealed mutations proceeds, one or another of them will appear with increasing frequency in the homozygous state, leading to the external manifestation by the species of a greater and greater genotypic variability.[5]

Thus, we form a concept that is exactly the contrary of the accepted opinion about the correlation between the age of a species and its variability. Ordinarily, it is claimed that a young species has not yet been stabilized, that its characteristics are just formed, and consequently are strongly fluctuating, and the whole species possesses a large degree of variability of instability. It is only gradually, as the species "ages," that its characteristics are fixed more and more, "are fixed hereditarily," and the species becomes stable, monomorphic.

In reality, as we have seen, the situation is exactly opposite. The older the species, the more mutations are accumulated within it, the more frequently is one or another of them disclosed in the homozygous state, and the more the species becomes externally genetically variable. Generally speaking, all other conditions being equal, *genotypic variability of a species increases proportionally to its age.*

In the foregoing analysis of the genotypic structure of a freely crossing population (species) we touched upon another very important question, namely, upon the importance of population size in the manifestation of its genotypic variability. Here we are dealing with two opposing tendencies: on the one hand, the more *numerous* the population, the greater are the chances for *origin* of new mutations in it. Thus, the frequency of origin of variability is directly proportional to the size of the freely crossing population.

On the other hand, the *less numerous* the population (that is, the smaller the value of N), the greater is the probability of *manifestation* in it in homozygous form of mutations absorbed by it earlier. In other words, the frequency of the manifestation of mutations is inversely proportional to population size.

These conditions ordinarily counterbalance each other, and what is lost by the less numerous species through the infrequency of the appearance of mutations, is gained by them through the frequency of manifestation of the mutations absorbed, and *vice versa.* But in some cases this equilib-

[4] An abstract dealing with these results appears in the Proceedings of the 5th International Congress of Genetics, 2: 1499–1500, Berlin, 1927. Ed.

[5] It is necessary to point out here that this statement refers only to genotypic, inheritable variation. Fluctuating phenotypic variation, as a reaction of the organism to the influence of various external conditions, complies with its own laws. This variability is only a manifestation of some "reactive irritability" of the organism, and is not related to inheritable variability or to the problem of evolution and speciation in general.

rium is disturbed, that is, when, because of some cause or other, the freedom of crossing within the limits of the species is disturbed. If we imagine that the total number of individuals of a given species, *N*, is subdivided into a series of *isolated* colonies, then the frequency of origin of new mutations within the limits of the entire species will not suffer, but the probability of reappearance of each such mutation will be once more considerably increased, depending on the reduced size (*n*) of the colony, within which it originally arose.

Thus, we approach a more profound understanding of the enormous role which the factor of *isolation* plays in the origin of species variability. At first glance, it may appear that the very fact of isolation, taken by itself cannot play any role in the process of evolution (Jordan, 1905). As much as identical groups are isolated from each other, they remain identical to each other. But this is exactly the point: a species, as we tried to show above, represents limitless diversity of genotypic combinations, and each isolation creates in it at once the exceptionally favorable conditions for the manifestation of heritable variations, either already existing within the species prior to isolation (given an initially unequal distribution of them), or originating after the separation of the isolated colonies which do not cross with each other.

Thus, isolation entirely automatically leads to a *differentiation* within a species, to the fact the colonies of one species, isolated from each other, begin, with time, to manifest differences in individual characters, which may be detected either by direct morphological study, or by biometric evaluation of their means and variabilities. *And so, isolation, under the conditions of a process of continuous accumulation of mutations becomes, by itself, a cause of intraspecific (and consequently, eventually, also of interspecific) differentiation.*

Of all the factors contributing to the break-up of a species into separate non-interbreeding colonies, it is necessary, naturally, to put spatial, *geographical isolation* in the first place as the most powerful and common factor of intraspecific differentiation. The colossal number of geographical races already reported and the many times greater number still to be reported (sub-species and "nations" in the sense of A. P. Semenov-Tyanshansky, 1910) is the best illustration and proof of the power of this isolating and differentiating factor.

But in no case is it to be imagined that geographical isolation is the *exclusive* factor in intraspecific differentiation. Although they occur incomparably rarer, other forms of isolation leading to the same results, that is, to the formation of separate non-interbreeding colonies, are also encountered. Thus, undoubtedly, there exists *isolation in time*, that is, the subdivision of a species into a series of colonies living sympatrically but isolated from each other owing to the non-concurrence of reproductive periods within each. In this connection, the best studied example is our common herring (*Clupea harengus* L.) which is subdivided into several colonies living in one place but separated from each other by difference in time of their egg-laying (fall and spring-spawning herrings). As the classical investigations of Heincke (1898) have shown, these separate colonies, isolated in time, vary among themselves in mean values of a whole series of characters, and making use of the method of "combined deviations," developed by the same author, it is possible to assign each separately caught specimen, entirely by its morphological characters, to one or another of these "seasonal" races with a high degree of probability.

There is a perfectly analogous example in the vegetable world, namely, the Linnaean species *Euphrasia officianalis* L., which at the present time is split by botanists into a whole series of "elementary species," differing from one another not only in their time of flowering, but also in several small and not entirely stable morphological characters.

Finally, the possibility is not excluded of having *ecological isolation*, when a species is subdivided into separate colonies, existing in one and the same geographical area, reproducing at one and the same time, but isolated from one another by the conditions of their habitat. Theoretically, it is very easy to imagine such a possibility, but the actual existence of such ecological colonies, differing from one another in hereditary, genotypic properties, requires still further investigation and verification. But ecological types of certain ants, ecological types (for instance, the marsh forms) of certain insects, hint at the existence of such kind of isolation. Yet, until now the nature of such ecological types remains entirely unexplained, and, in particular, it is not at all known what proportion of these "ecological traits" is actually inheritable (if, in general, some is), and what is simply a phenotypic reaction of the organism to changed external conditions in which they develop and exist.

Instances of different biotypes constituting the

population of the species reacting in varying degrees to the unfavorable conditions of existence encountered by the species on the margins of its range, must be considered also under the subject of isolation. In this marginal zone (providing it is not a simple mechanical barrier impeding the further spread of the species), a more intense struggle for existence must obviously take place, a struggle both against environmental inorganic nature, and against individual organisms hindering the further spread of the species (biological barriers). There is no doubt whatsoever that individual biotypes of the species may resist the unfavorable conditions encountered in different ways, and, because of this, a part of them will spread further, whereas the remaining bulk of biotypes will stop earlier without passing across the barrier encountered.

Because of this, the fraction of the biotypes which proved to be more resistant and has penetrated further beyond the barrier encountered will be *isolated* in this marginal area of the range of the species from the remaining mass and will give rise to a separate colony, where, by virtue of the causes indicated above, the remaining biotypes will not penetrate.

The more frost-resistant biotypes of various animals and plants in the northern margin of their ranges, or, for instance, the drought-resistant biotypes on the borders of steppes and deserts, and, finally, cases of natural immunity to specific diseases and parasites, etc., may serve as examples of such ecological isolation.

Be this as it may, all the factors indicated seem to be *external* in relation to the organism and are important only in so far as they do not meet any resistance from the structure of the organism itself. The influence of spatial isolation is paralyzed by all possible means of active or passive movement; isolation in time is prevented by lengthening of periods of reproduction, and finally ecological isolation conflicts with possession of wide adaptability by organisms.

As a result of the interaction of all these external and internal factors, a definite state of equilibrium is created, which determines the degree of freedom of crossing, characteristic of each species. The stronger act the isolating, dissociative factors, and the stronger is intraspecific variability expressed, the more often must be manifested in separate colonies the genotypic differentiation concealed within the species. Thus a law may be established: ***all other conditions being equal, the degree of differentiation within a species is directly proportional to the degree of isolation of its separate parts.***

This law in its purest form should be expressed, for instance, in insular faunas, where often colonies isolated on neighboring islands display a varying degree of differentiation of heritable characters and where this differentiation could most probably be attributed precisely to isolation.

But exactly the same results are obtained when we proceed to the detailed study of variability in organisms which possess only a very low ability for self-dissemination, and the well-known work of Coutagne (1895) on the variability of different colonies of terrestrial mollusks of France provides a brilliant example of this.

But this regularity appears most clearly when the external factors of isolation are combined in the most complete manner with lack of mobility of the forms studied. The classical studies of Gulick (1872, 1888) on the variability of the family *Achatinellidae* on the Sandwich Islands and the analogous work of Garrett (1884), Mayer (1902), and Crampton (1916) on the spread and variability of the genus *Partula* on the Society Islands, yield the most brilliant illustrations of the above law, and at the same time indicate with extreme clarity the degree of differentiation that species may attain, if there is at the same time subdivision and low mobility, which determine the simultaneous existence of a large number of small non-interbreeding colonies.

The factual side of the above law is formulated in sharp relief by Crampton (1925) in the following words, referring to a species of the genus *Partula* on the islands of the Pacific Ocean:

> In general, each group of islands bears species which occur in no other group; there is one known exception which serves to emphasize the observed rule Within the confines of a given group the several islands bear their own distinctive forms, again with rare exceptions. Finally, the several valleys of one and the same island constitute the homes of colonies which, in correlation with their nearer situations, are more alike than are any two associations living in separated islands (p. 6).

Examples, in which the noted regularity of differentiation in connection with isolation would be exhibited more distinctly, could hardly be found.

All of the above analysis of the significance of free crossing was constructed on the hypothesis of the origin of homozygous and recessive mutations in nature. But dominant mutations can arise in exactly the same way, and furthermore, it is highly probable that mutations originate as heterozygotes.

The analysis of all these situations remains essentially the same. The only difference is in the fact that the probability of repeated manifestation of a mutation subsequent to the meeting of two heterozygous individuals in the third of these cases is greatly lowered, and in the last analysis, depends on the frequency of repeated occurrences of identical mutations and on some variability of our statistical means.

In the case of a dominant mutation, it will, naturally, not disappear as a result of the first stabilizing crossing, but will exist in the population in its expressed form, representing in certain instances a more or less noticeable morphological change or physiological deviation from type, scarcely perceptible, because of its rarity, among the millions of normal individuals. The main difference here will be in that such a manifested, dominant form in a stabilized population will be subjected continuously generation by generation to the action of natural selection, which in many cases will lead either to its final extinction or, on the contrary, to its greater and greater diffusion. In this lies the essential difference between dominant and recessive mutations, since the latter, as a result of the very first stabilizing crossing, passes into a cryptic, unexpressed condition, and thereby at once escapes the powers of selection, regardless of whether its qualities might be favorable or harmful to the organism.

In this manner, we now, naturally, approach the question of the role of selection under conditions of free crossing with genotypical variability.

III. NATURAL SELECTION

In the foregoing analysis of free crossing, we tried to establish its role as a factor stabilizing a given population. In its very essence it is a conservative factor, preserving the genotypic composition of the species in the condition in which it is found at a given moment.

Natural selection (and, in general, selection in any form) is, in this connection, its direct antagonist. If free crossing stabilizes the population, then selection, on the contrary, all the time displaces the equilibrium state, and, if in this sense we may call free crossing a conservative principle, then selection, undoubtedly, is the dynamic principle, leading ceaselessly to modification of the species.

It is perfectly obvious that if selection will favor one of the alleles (it is completely irrelevant whether it is dominant or recessive, or whether it is basic "wild-type" or the newly arisen mutation), its action, directed toward the preservation of one form to the detriment of the other, will, without fail, disrupt the basic regularity underlying free crossing and thereby will all the time keep moving the population out of the equilibrium state. In this way, two opposing forces will constantly struggle against each other with respect to the frequencies of the different pairs of alleles: one, the force of selection, which disrupts the existing state of equilibrium to the advantage of the selected gene; the other, the stabilizing action of free crossing, striving in the very first generation to re-establish order and equilibrium, which will again and again be disturbed by the action of selection. As a result, the population will continuously pass from one equilibrium state to another, and this process will last until the operation of selection ceases.

It is evident that the *rate* of this process is directly proportional to the *intensity*, or *force*, of selection, and here it is necessary to clarify a concept which plays a very great role in the action and significance of selection, and to which, nonetheless, too little attention had been paid in questions on the course of evolution.

The basic, fundamental differences between artificial selection and all kinds of natural selection is that the first is personal and purposeful, whereas the second is automatic. This difference is apparent first of all in those factors which determine degree of intensity of selection itself. If in artificial selection the *person* controlling selection is the foremost of these factors, and the intensity of selection is determined by the *goal* which this person sets for his activity, then in natural selection its force or intensity is determined, in the final analysis, by the importance of the selected traits in the struggle for existence and survival of the most adapted types. While in the first case, the intensity of selection may easily be, and, actually often is, brought to 100 per cent, that is, only the individuals having the trait selected are permitted to reproduce, under natural conditions one may say with full confidence that the force of selection does not even remotely approach this limit. Even when the selected trait has very great importance in the struggle for existence, under natural selection in the course of a series of generations there will exist and propagate side by side individuals both having and lacking this trait. In the final analysis, survival of an individual in the struggle for existence depends on such a complex interrela-

TABLE 1

NORTON'S TABLE

Percentage of total population formed by old variety	Percentage of total population formed by the hybrids	Percentage of total population formed by the new variety	Number of generations taken to pass from one position to another as indicated in the percentages of different individuals in left-hand column							
			A. Where the new variety is dominant				B. Where the new variety is recessive			
			100/50	100/75	100/90	100/99	100/50	100/75	100/90	100/99
99.9	.09	.000								
98.0	1.96	.008	4	10	28	300	1920	5740	17,200	189,092
90.7	9.0	.03	2	5	15	165	85	250	744	8,160
69.0	27.7	2.8	2	4	14	153	18	51	149	1,615
44.4	44.4	11.1	2	4	12	121	5	13	36	389
25	50	25	2	4	12	119	2	6	16	169
11.1	44.4	44.4	4	8	18	171	2	4	11	118
2.8	27.7	69.0	10	17	40	393	2	4	11	120
.03	9.0	90.7	36	68	166	1,632	2	6	14	152
.008	1.96	98.0	170	333	827	8,243	2	6	16	165
.000	.09	99.9	3840	7653	19,111	191,002	4	10	28	299

tion of causes and effects, that the importance of one or another single beneficial trait is, generally speaking, a matter of chance.

We shall call the intensity of selection in a given trait as being 10 per cent, 20 per cent, etc., when the probability of survival in the struggle for existence of individuals not having a favorable trait is 10 per cent, 20 per cent, etc., lower than individuals having it. Thus, with an intensity of selection of 10 per cent, of 100 surviving individuals having the selected trait, 10 individuals survive precisely because they have it, that is, they had the opportunity to utilize the advantage conferred by it in the struggle for life.

Naturally, the actual computation of selection intensity is at present an unapproachable problem. But in this case it is the principle that is more important; it is important to establish the concept of intensity of selection itself and its quantitative measurement.

Basing himself on the above concept of selection intensity, the mathematician, H. T. J. Norton, calculated the table (table 1) which was published by Punnett (1915) in his book, *Mimicry in Butterflies.*

It shows a tabulation of the number of generations [6] in the course of which a population existing under conditions of free crossing passes from one equilibrium state to another under various intensities of *positive* selection, both with dominance as well as recessivity of the selected trait.

In the three columns on the left are, in accordance with Hardy's formula, the percentages of frequencies in the population of individuals in various equilibrium states satisfying the condition that the product of the frequency of homozygotes be equal to the square of half of the frequency of heterozygotes, $pr = q^2$.

In the right part of the table, in the vertical columns, the number of generations, which must elapse for a given population under a particular intensity of selection to pass from one equilibrium state to another, is shown. Under A the figures are given for the case in which the trait positively selected is dominant, while under B the figures stand for that case when it is recessive.

This table merits more careful inspection. First of all, attention is attracted to the fact that in both cases—when the selected trait is dominant as well as when it is recessive—the process of the *transformation of the species,* that is, of the *complete replacement* of a former unadapted form by the more adapted one, proceeds, practically speaking, to an *end.* This process of complete replacement of one form by another goes on even under the very weakest intensity of selection of 1 per cent, the only contrast being in *different rates of the process.*

Considering the transformation process as a whole, its duration under a low intensity of selection (as probably happens most often in nature) proceeds in both cases at approximately

[6] For the sake of simplicity it is assumed here that we are dealing with organisms which reproduce themselves once in their lifetime (insects, annual plants, etc.).

the same rate. Only with the increase of the intensity of selection (to 25 per cent) does the difference in the rate of the transformation of the species between the cases of dominance and recessivity of the selected trait become noticeable. And with an intensity of selection of 50 per cent, the transformation of the species proceeds twice as fast in the case of the positive selection of a recessive mutation than of a dominant one.

It is also characteristic that the process of transformation proceeds at the *most rapid rate* under an intermediate equilibrium, that is to say, when the numbers of dominant and recessive homozygotes are nearly equal to each other (the classical Mendelian ratio 1:2:1). Here we see that for an intensity of selection of 10 per cent, $16 + 11 = 27$ generations in all are sufficient (should the selected mutation be recessive) in order to pass from a state of equilibrium in which the number of recessive homozygotes was four times less than the number of dominant homozygotes (44.4 *AA* and 11.1 *aa*) to a condition with the reverse ratio of frequencies (11.1 *AA* and 44.4 *aa*). This transformation proceeds almost as rapidly when the selected trait is dominant ($12 + 18 = 30$ generations).

The differences in the course of the transformation of the species between the cases of selection of recessive or dominant mutations manifests itself only when we come to the examination of the extreme phases (beginning and end) of the process. Thus, should the favorably selected trait be dominant, the process of replacement of the less adapted form by the more adapted one proceeds very rapidly from the very beginning. For instance, with an intensity of selection of 10 per cent, 305 generations are sufficient for a freely crossing population to pass from a condition in which 99.9 per cent lack the selected character to a condition in which, on the contrary, 90.7 per cent will possess this character; that is, in practical terms, the whole population is completely altered in the direction of better adaptation in the struggle for life. Even with a negligible intensity of selection of 1 per cent, this process for an almost complete transformation is realized in the relatively brief time of approximately 3,000 generations.

On the contrary, should the newly appearing favorably selected form be recessive, the beginning of the process of transformation of the population proceeds extremely slowly. At this same intensity of selection of 10 per cent, after almost 18,000 generations, more than 90.7 per cent of the population will still consist of forms not displaying the favorable variation; that is, once again, in practical terms, the population will not manifest noticeable evidences of transformation. Naturally, with even weaker degrees of selection, this process must be protracted to hundreds of thousands of generations. And it is so even in the case where the initial point of our calculation is taken to be that 0.09 per cent of the population already represents the heterozygous form; that is, the probability of the meeting of two such heterozygous forms is almost 1/1000. And we have seen above that with single isolated mutations, the probability of a meeting of two heterozygous individuals depends on the size of the total freely crossing population, and must ordinarily be less than the one just indicated.

On the other hand, the picture changes abruptly if we turn to the end of the process, which concludes the complete transformation of the species. Should the selected character be recessive, this process of the complete and final extermination of the less favorable character proceeds and is completed extremely rapidly, in a few dozen generations, whereas, in the case of dominance of the selected character, the process of the final purification of the population from the recessive gene is protracted almost to infinity. Thus, with an intensity of selection of even 10 per cent almost 20,000 generations are required, just to increase the number of homozygous dominant forms from 90.7 to 99.9 per cent. The final eradication of all heterozygotes, that is, the complete disappearance of the recessive gene, is possible only after hundreds of thousands of generations.

To what conclusion does our examination of Norton's table lead? Many of these conclusions have a very great importance for a correct understanding of the evolutionary process.

First of all, we see that because of the effects of free crossing and selection, under the conditions of Mendelian heredity, every, even the slightest, improvement of the organism has a definite chance of spreading throughout the whole mass of individuals comprising the freely crossing population (species). Here Darwinism, in so far as natural selection and the struggle for existence are its characteristic features, received a completely unexpected and powerful ally in Mendelism.

One of the most substantial difficulties of Darwinism has always been the difficulty in imagining the process by which minute improvements of the organism, the importance of which for survival appeared, generally speaking, completely negligible. The living organism, in the course of its individual life, from the egg until its death, is exposed to such infinitely diverse influences of the surrounding environment, is faced with danger from the most varied causes so many times that it seems that the small advantage which a slight improvement of the organism can offer would be completely submerged in the thousands of deaths threatening it from all sides. Let this advance help it to escape a concrete danger *A;* it will all the same be destroyed by danger *B,* or *C,* or *D,* etc., and, thus, it would take an extremely advantageous concurrence of circumstances for the organism to survive and to pass on by inheritance its small advantage. And in later generations, the same struggle and accidental basis of survival will threaten its descendants.

Now, because of Mendelism, our understanding of this process has changed. By virtue of the properties of free crossing *nothing is lost of that which is acquired by the species.* No matter how small the newly arisen improvement be, hundreds of thousands of generations, perhaps, will pass, but finally, sooner or later, it will emerge and gradually *become a property of all the individuals of the species.*

Another important conclusion from the inspection of Norton's table is that transformational *change of a freely crossing population—species, the replacement of the less adapted form by the more adapted one, in a word, the process of adaptive evolution of the species, always proceeds to the end.* It is a matter of indifference whether the better adapted form is dominant or recessive, whether the intensity of selection is 50 per cent or one per cent. Once the transformation has started, once the species has moved from a dead stop, the process automatically proceeds farther, until *the whole species changes in toto,* or until selection ceases.

This conclusion is very important for an accurate understanding of the role of various features in the evolutionary process. *Under conditions of free crossing, that is, until there is isolation* (in one of the forms indicated above), the struggle for existence and natural selection can continuously alter the physiognomy of the species, can disseminate more and more new adaptive characters through the *whole mass* of individuals of the species, can perfect any features of its organization, *but never under these conditions does the species give rise to a new species, never will there be a subdivision of the species into two, never will speciation occur.*

The species as a whole, in its total mass, will change, will evolve, will become more and more perfect in its adaptation to the external environment. In the process of the historical development of a species we shall observe that a form, the less perfect one, will be *completely* replaced by another more perfect one; we shall observe the process of adaptive evolution, the process of the formation of true *Waagen mutations,* when one form is completely replaced by another in the phylogenetic evolutionary process.

Fortunately, this process of complete transformation is known to us not only in paleontological material, not only from the classic examples of the ammonites, which served Waagen (1869) to establish the concept of *mutation* itself, or from the paludine fresh-water deposits of the Danube Basin, which gave Neumayr and Paul (1875) the material for their famous study. This process occurs before our very eyes, in the modern epoch. From the data in the systematics of Lepidoptera, best known to me, I can cite two examples which indubitably certify to the continuation of the process of transmutational evolution at the present time, and systematists of various other groups of animals and plants could surely cite analogous instances in their fields.

The first of the examples concerns the peppered moth, *Amphidasis betularia* L., and its melanic variety, *doubledayaria* Mill.[7]

In 1866 Milliere, in his famous *Iconographie et Descriptions de Chenilles et Lépidoptères Inédits* described a remarkable form of *Amphidasis betularia* L. obtained by him from northern England as a single specimen and later designated by him *doubledayaria* Mill. While the typical form of the species is white with characteristic scattered black streaks (as in the markings of a birch), the new form in well-expressed specimens appears coal-black in color. As experiments in crossing these forms have shown, this variation is dependent on the presence of a single gene and is dominant to the normal white.

There are data indicating that the melanic form originated about 1850 in the Manchester area and at first was an extraordinary rarity. But gradually notices are increasing about its greater and greater spread towards the north and the south

[7] Chetverikov refers in this discussion to the peppered moth, now known as *Biston betularia,* and to its *carbonaria* variety. Ed.

of England, and everywhere the new variety crowds out and displaces the basic white form, so that at the present in the greater part of England, the predominant (and in some places the exclusive) form is the black *doubledayaria* Mill.

A completely analogous case was described recently in northwestern Germany, where in the vicinity of Hamburg, in 1904, a melanistic form of another moth *Cymatophora or* F.,[8] named *f. albigensis* Hsbr. also suddenly appeared. And here an extremely rapid replacement of the initial form by a new (melanistic and dominant) one is being observed, this process proceeding so energetically that after only eight years in some Hamburg suburbs up to 90 per cent of the population of this species belong to the variant form.

In both cases presented, a living page of the actual adaptive evolutionary process passes in front of us. Before our eyes, within a natural species population a dominant mutation suddenly appears. The conditions of free crossing do not stabilize its frequency: on the contrary, it quickly propagates itself, and becomes more and more numerous each year. Already within several decades a whole series of localities appears to be taken over by it, and in many of them it is becoming the predominant, prevalent form. In the case of *A. betularia doubledayaria* Mill., in many places in England the new form "completely" replaces the earlier basic form, which then becomes an uncommon, occasional phenomenon.

We see how an occasional, single deviation, an individual mutation, gradually takes on the character of the *proles,* in the sense of Korzhinsky (1892), that is, we have an example of a form with a certain area of habitation, which becomes an insular race, gradually supplanting the initial original form in the area of its diffusion. Before our very eyes the process of mutation in the sense of Waagen and the paleontologists unfolds.

Evidently, such replacement of one form by another could occur only by virtue of the dark form's being, in some unknown way, better adapted to the conditions of its existence than the original, basic form, and we can even attempt, using Norton's table, to determine the intensity of selection. Taking the case of *A. betularia* L., and assuming that in the neighborhood of Manchester, where this form was first observed seventy-five years ago, the whole aggregate of the species (99.9 per cent) took on the black exterior of *doubledayaria* Mill., then the intensity of selection must have been not less than 50 per cent. But later, the process of fixation of the English race will have to slow down very much, and for many centuries typical white *betularia* L. will still appear among the English *doubledayaria* Mill. from time to time through the meeting of two heterozygous forms (especially often in the presence of inbreeding).

The case of evolution described is, however, still not completed. In some accidental way, most probably by the agency of man (possibly through the efforts of amateur entomologists), a form of *doubledayaria* Mill. seems to have been brought into the European continent about 1888. Originally, it appeared in the corner of Germany nearest to England, and in Holland, but subsequently began, just as in England, to spread victoriously, farther and farther into the depths of the country. In Germany the very same picture was repeated as in England. Gradually, news of the appearance of the black *doubledayaria* Mill. arrives from more and more remote places, and the nearer to the Rhine, the more numerous the form becomes. In many places it has already become predominant, gradually crowding out and replacing the typical *A. betularia* L. At the present time it has been reported throughout almost the whole of Germany, up to its eastern borders.

We do not yet have information on its appearance in Poland, but there is no doubt that it will appear there too in the near future, if it has not already appeared there by now, and one may say with complete assurance that within several decades it will reach us.

We can even make an attempt to describe the further development of this process. It will proceed along the same way farther and farther through the whole enormous territory occupied by the species up to the maritime region in the Far East. Throughout this extensive area, the basic form of *A. betularia* L. will be replaced by its more viable black rival, thus visibly demonstrating the proposition that a species is actually a single freely-crossing population, *"Paarungsgemeinschaft."* And the time will come when the finding of the former basic white form will be considered a rare occurrence, just as until recently the black English form was considered to be a rarity. Only on the isolated islands of Japan, where *A. betularia* L. also exists, will the original white race survive (unless the newcomer is carried there accidentally), and then it will be considered in systematics an island subspecies, a

[8] The name currently used for this moth, the poplar lutestring, is *Tetlea or.* Schiff. Ed.

final relict of the once widely abundant basic form.

Analogous cases are known to us from systematics. Thus, for example, the ghost moth, *Hepialus humuli* L., in the Shetland Isles (northern Britain) has survived in the form having the name subsp. *thuleus* Crotch., which shows, in comparison with the continental form, undoubtedly, more primitive features (absence of sexual dimorphism).

The same Norton's table shows that not only centuries, but thousands of years, will elapse before the dominant *doubledayaria* Mill. will completely pass into the homozygous condition. A certain percentage of heterozygous individuals, similar in appearance to the homozygotes, will always be conserved, and crosses between them will give rise to a number of the basic white *A. betularia* L. For future systematists there will be instances of atavism or (if they do not know the history of this moth) cases of sudden appearance of a recessive mutation. If on the islands of Japan the basic white race escapes destruction, then such cases of the sudden appearance somewhere in Europe of a specimen identical with the Japanese race will present examples of individual recurrences of characteristics distinctive of a sometimes extremely remote geographical race. Modern systematics knows of hundreds of analogous examples. Thus, making use of the already cited example of the ghost moth, *H. humuli* L., and its island subspecies, *thuleus* Crotch., it may be noted that specimens, similar to the Scottish ones, are, as instances of individual variability, occasionally captured in Holland. Evidently heterozygous individuals are still to be found in Holland, and as a result of crosses between them, the more primitive, basic race reappears. It is likely that the diffusion of the contemporary strongly dimorphic form of *H. humuli* L. proceeded from the East, and therefore in the West the replacement of the original form is still less complete than in the East, so that a pair of heterozygous individuals is more likely to meet for crossing there.

The example cited reveals to us the whole picture of the evolutionary process as it proceeds under the influence of natural selection and the struggle for existence. Its most characteristic feature under ordinary conditions is its completeness, the total replacement of the former form by the new, more adapted one, i.e., the mutation process of Waagen.

But what would happen in a case when selection is interrupted before the completion of the process of transformation, if it is imagined that in the course of selection, for instance, the enemy, in the struggle with which the adaptation was specially selected, disappears? On the basis of the second (stabilizing) law of free crossing, we know that in the generation following, a condition of equilibrium is established and in all generations following thereafter, the relative frequencies of individuals carrying the trait and not carrying it will remain unchanged. The species will become subdivided, but not into two independent species or varieties, but into two forms: it will become *polymorphic*. As before, all the individuals of this species will compose one "Paarungsgemeinschaft," indiscriminately crossing among themselves, and whether the young are born monotypic, or whether individuals belonging to both forms will be encountered among them in various ratios, will depend on the genetic structure of the parents. It is entirely possible that in at least several of the cases of polymorphic species presently observed we are dealing with precisely this kind of cessation of the action of selection, although it is also possible that in other cases the observed polymorphism of the species is just a transient condition of transformation of the species under a relatively weak intensity of selection. As Norton's table demonstrates, under such conditions thousands of generations may be necessary in order to erradicate the recessive form completely, and the period of our observations is much too brief to detect a noticeable change in the relative frequencies of the two forms.

Finally, the third important conclusion which may be made on the basis of an analysis of Norton's table consists of the fact that natural selection, like free crossing, *promotes the accumulation of recessive genes in the population.*

In the analysis of the example of positive selection of the dominant trait of *A. betularia doubledayaria* Mill., we saw that an almost complete replacement of the recessive form by the dominant may occur relatively fast. But the examination of the appropriate part of Norton's table (A) shows that this process is completed very slowly, and that, even with relatively intensive selection, hundreds of thousands of generations are required in order to bring all the dominant forms into a homozygous condition. Until then, as a result of the action of selection on the population, a certain number of heterozygous forms will still remain, which under a favorable concatenation of circumstances may cross with each other and give origin to what may appear as newly arising atavistic recessive mutations.

If we measure the intensity of ***positive*** selection in per cent, then precisely in the same way can we measure the intensity of ***negative*** selection. All disadvantageous changes of the organism are by no means inevitably ruinous for it. Only in the case of extremely harmful deviations from the norm will the intensity of negative selection reach 100 per cent; in a very great majority of cases the intensity of negative selection will be less, and not uncommon will be the cases of 10 per cent or even one per cent intensity of negative selection. This means that the organisms possessing the disadvantageous trait will not be quickly removed from the arena of the battle of life, and tens and, sometimes, hundreds of generations will be necessary for selection to eliminate these forms.

And here again the difference between dominant and recessive characters becomes markedly telling, and once more in the direction of accumulation of the latter. As we saw above, in the case of recessivity of the selected characters, in other words, in the case of ***dominance of the trait subjected to negative selection,*** selection finally eliminates the less adapted form, and this proceeds comparatively rapidly. On the contrary, in the case of recessivity of the disadvantageous character in the first (stabilizing) cross after its origin, this trait will pass into the heterozygous, concealed state, will escape from the controlling action of natural selection, and, in this way, the whole population will arrive at an equilibrium, conserving in its depths the recessive gene for the less adapted trait in the heterozygous condition. As we see, also in this case, selection involves a relatively rapid removal of unfavorable dominant characters, and, on the contrary, an accumulation of recessive genes.

Finally, even in the case of appearance of advantageous recessive mutations, they pass into the concealed heterozygous condition immediately after the first cross with the normal form, and in this way also escape the action of selection; it is necessary to wait $1/N$ generations[9] (see above) for selection to be able again to act to bring about fixation of the favorable trait.

Thus, free crossing like natural selection, leads to the same final result, namely, to ***the accumulation in the population in the heterozygous state of recessive genes,*** of which a very significant proportion can be connected with characters disadvantageous to the organism.

This analysis leads us to an understanding of a fact which is extremely strange at first glance, namely, that the number of known recessive mutations is many times that of dominant ones. This is especially clearly evident in precisely those organisms, the material for a genetic analysis of which is obtained in the wild, that is, from a population which, in our concept, is saturated with recessive heterozygotes. Thus, in the latest summary of Morgan, Bridges, and Sturtevant (1925) devoted to the genetics of the genus *Drosophila,* the number of recessive genes exceeds the number of dominant genes by more than ***six*** times. There is no doubt that the enormous majority of these recessive genes was ***accumulated*** by the species in the course of its long specific life, and only became ***manifest*** subsequently in the laboratory with the aid of more or less prolonged inbreeding. At the same time, because of random assortment of pairs for mating the initial appearance of the cryptic characters could extend through a whole series of generations, and then to distinguish a case of a new mutation from a case of simple manifestation of one already existing in the heterozygous condition becomes actually impossible.

We noted above that the role of free crossing in the process of evolution is a conservative one, striving to maintain the ***status quo,*** whereas natural selection acts as an opposing, dynamic factor. But if we bring within the scope of our analysis the process of continuous origin of new mutations as well, then this concept needs to be both changed and supplemented. While free crossing, storing and preserving within the species all the newly arising mutations, gradually unfixes the characteristics of the species, makes it less stable, and produces intraspecific differentiation, natural selection, on the contrary, preserves the stability of the species, its monomorphism. Removing and gradually eliminating all mutations, which in the last analysis appear to be harmful, natural selection ***purifies*** the species of contamination by accumulated variations, and, in the case of favorable changes, spreads them to all the individuals of the species, thereby reimposing on it homogeneity.

In this way, within each species the struggle of two processes goes on: the process of accumulation of mutations and the process of their elimination, and, the genotypic composition of the

[9] Chetverikov's clear intent here is "*N generations.*" Ed.

species is in the long run determined by the interactions between them.

We have analyzed a case of adaptive evolution on the example of *A. betularia* L. But are, in general, non-adaptive evolutionary processes possible in nature? This is the question which up to the present has remained open and debatable.

Systematics knows thousands of examples where the species are distinguished not by adaptive but rather by neutral (in the biological sense) characters, and to try to ascribe adaptive significance to all of them is work which is as little productive as it is unrewarding, and in which one does not know at times whether to be more surprised by the boundless ingenuity of the authors themselves or by their faith in the limitless naïveté of their readers.

Thus, for the defenders of exclusively adaptive evolution, there remains the last refuge, correlative variability, and to this they have to resort every time that an attempt to construct the whole process of evolution and speciation exclusively on the basis of struggle for existence and natural selection is made. That correlative variability exists there can be no doubt, and, furthermore, modern genetics gives us an entirely new point of view for its understanding in the concept of "pleiotropic action" of genes (see the last section). Yet, to explain all the innumerable instances of neutral, non-adaptive specific differences of such a kind by this type of variability still means to explain nothing, but to be satisfied in every case with a simply unprovable hypothesis.

I tried to show above that the process of intraspecific differentiation need not necessarily be accompanied by an invariably adaptive change of the forms differentiating. That such adaptive change is possible in some cases is sufficiently easy to imagine, but we do not have any basis for attributing the whole process of intra-specific and inter-specific differentiation to adaptive evolution. Such occurrences will be merely special cases of the more general process of speciation and splitting.

The strict parallelism between the intensity of intra-specific (and, in the last analysis, interspecific) differentiation and the splitting up of the whole population of the species into separate colonies *isolated* from each other, which we saw at least in the example of island forms of land mollusks, definitely and clearly indicates that in the process of splitting up of the species the predominant role belongs not to selection, but to isolation, which finds its material in the enormous *heterogeneity* of natural populations, the existence of which I tried to prove above.

Naturally, it should not be thought that the factors named exhaust the full essence of the phenomenon of differentiation in the realization of which a series of still other processes not here considered also participates. But it is absolutely necessary to appreciate clearly the fact that never can an adaptive change, related to the action of selection in the struggle for existence but not related to isolation, initiate speciation. *Not selection, but isolation is the actual source, the real cause of the origin of species.*

Finally, even in those cases in which we are in a position to establish the presence of truly adaptive differences between species, genera, etc., it is necessary to be very cautious in accepting the idea that these differences are of a primary character, i.e., in recognizing that precisely they have given rise to a splitting up of the initial single form into two, thus leading to the process of *speciation.* It should not be forgotten that, as we just saw, every species in the course of its existence *obligately* undergoes as a whole Waagen's adaptive mutation, should a favorable mutation, in our sense, arise, and thereby acquires an adaptive trait absent in its kin. A *new* species-distinguishing *adaptive* trait is established, but it is not the *cause* of the splitting-up of close forms, but on the contrary, its species-characteristic nature is a *result* of still earlier inter-specific differentiation.

Actually, this process must occur fairly often, for it is difficult to imagine that not one favorable mutation should originate in the course of even a slight duration of a species. Such mutation would thus give rise to an adaptive specific trait; it is precisely in this way that probably an enormous majority of adaptive differences between closely related species are established.

The longer the course of evolution which has been run by the forms after the differentiation has already begun, the more adaptive differences involving many different organs must have been accumulated in them, and, actually, we see that in the establishment of differences between the higher systematic categories (families, orders, etc.), the differences in adaptive traits come more and more to the fore.

And thus, in the evolutionary development of the organic world, *two processes* proceed side by side, their paths sometimes crossing, but they are still strictly separate in their causes as well as in the consequences resulting from them: one is the process of differentiation, of splitting-up, leading in the end to *speciation*—isolation is its basis; the other leads to *adaptation*, to the progressive evolution of organic life, and its cause lies in the struggle for existence and the resulting *natural selection*.

IV. GENOTYPIC MILIEU

It is impossible to end an essay on the role of natural selection in the evolutionary process without touching upon one important question, which has seriously preoccupied biologists in recent times and which is not completely clear even to many specialists in genetics. What is the role of selection in the progressive process of evolution? Is it only a passive factor, eliminating, eradicating the less fit genes, and protecting, on the contrary, those which have an advantage in the struggle for existence? Or does it *create* its own material, actively entering into the evolutionary process, directing variability into definite channels?

Lately an animated and occasionally bitter dispute concerning these questions has been carried on, particularly among American geneticists. The first point of view, presented most clearly by Johannsen and a whole series of geneticists (including Morgan and his students) bases itself on the grounds of the constancy of the gene and its independence of all environmental conditions, including selection. The opposite opinion, in favor of an active participation of selection in the creation of the material for its application, was defended by a small group of geneticists, headed by Castle (1912, 1914). At the basis of their views was the concept of the variability of the gene, of the fluctuation of its hereditary powers, of the mutual effects of alleles one on the other.

At present the dispute seems to be over and decided in favor of the first view, that is, for the perfect constancy of the gene. The arguments and facts advanced by the second group of investigators in favor of the views held by them were found to be explainable by means of a completely different interpretation, in agreement with the concept of the purity and constancy of genes. In the enormous majority of cases the whole matter comes down to genetic heterogeneity, to impurity of the material with which the work was carried out. And the classical experiments of Johannsen (1913) showing failure of selection to be effective in pure lines, and a whole series of analogous experiments with animals and plants, demonstrating the impossibility of changing a character by selection from a genetically homogeneous population, remain in full force at the present time.

Nevertheless, in the concept so thoroughly established, a weakness is beginning to show up, which Morgan himself brought up in his extraordinarily important, and extremely rich with consequences, notion of the plurality of action or manifestation of genes, which subsequently received the designation "pleiotropy."

An extremely clear and convincing picture of pleiotropic action of genes is to be found published in the article of Timofeeff-Ressovsky (1925), appearing in the preceding issue of this journal.[10] It is therefore possible to touch upon this question here only lightly, and mainly in so far as it is related to selection.

The concept of pleiotropic action of genes consists of the idea that every gene may influence not only the specific character corresponding to it, but a whole series of others; generally speaking, the entire soma. In so far as we now accept the proven localization of genes in the chromosomes, and in so far as *all cells* of the body receive *the full set of chromosomes*, so in the ultimate differentiation of the cells determining some specific trait all genes can be influential, affecting by their action one or another form of manifestation of genes specifically corresponding to a trait.

In this way, the former notion of the mosaic structure of the organism consisting of various, independent characters, conditioned by various, independent genes, is discarded. The genes remain pure and qualitatively independent of each other, but their *manifestations*, that is, the traits they condition, are now a complex result of the manifold interaction of all the genes comprising the genotype of the organism. And each individual is in the literal sense an *"in-dividuum"*—not divisible. It is not divisible not only in its soma, not only in the physiological functioning of its various parts, but indivisible in the manifestation of its genotype, its hereditary structure. Each inherited trait, the hereditary structure of each cell of its body, is determined by not just some one gene, but by their whole aggregate, their complex. True, every gene has a specific mani-

[10] I.e. *Zhurnal Eksperimental'noi Biologii*. Ed.

festation, its "trait." But in its expression this trait depends on the action of the whole genotype.

Each gene does not act isolatedly from the whole genotype, is not independent of it, but acts, manifests itself, *within it*, in relation to it. *The very same gene will manifest itself differently, depending on the complex of the other genes in which it finds itself.* For it, this complex, this genotype, will be the *genotypic milieu,* within the surroundings of which it will be externally manifested. And as *phenotypically* every character depends for its expression on the surrounding external environment, and is the reaction of the organism to the given external influences, so *genotypically* each character depends for its expression on the structure of the whole genotype, and is a reaction to definite internal influences.

This is a point of view that is necessary to master clearly in order to appreciate fully the total significance of natural selection in the evolutionary process. As we saw in preceding sections, because of the continuous process of origin of new mutations and because of their accumulation due to the action of free crossing, we must visualize the genetical structure of the species as consisting of an enormous number of genotypes more or less distinct from each other. The very same gene, entering into various genotypical combinations, will, each time, enter into a different "genotypical milieu," and consequently each time its external manifestation will be *hereditarily* modified, its appearance will *vary hereditarily,* will *fluctuate hereditarily.*

In combination with one genotype a given trait conditioned by a definite gene will manifest itself more strongly; in combination with another, more weakly; and this variability will be hereditary. This interpretation of the significance of the genotypic milieu for the hereditary variability of traits opens completely new perspectives for the understanding of a whole series of phenomena in the field of genetics and evolution.

The concept of pleiotropic action of genes at once releases genetics from the extremely heavy ballast accumulated recently in the shape of all kinds of special genes, "enhancers," "weakeners," or "modifiers" of other genes, the number of which has now grown to absolutely threatening proportions. In the light of the notion of the genotypic milieu, the existence of such "supplementary" genes becomes completely understandable and even unavoidable. Gene A, specific for the corresponding trait A, at the same time acts on trait B, conditioned by its gene B, and this action may express itself either as an *enhancement* of the trait B, or its *weakening,* or, finally, as a change—*modification.*

Beyond this, the concept of the *genotypic* milieu allows us to understand still another complex and involved phenomenon, which perplexes thoughtful geneticists not a little. At the same time that we see that *qualitative characters* are commonly determined by one gene, relatively seldom depending on several equivalent polymeric genes, for *quantitative* (metric or meristic) characters, polymerism is the common rule, and manifests itself in that the quantitative characters always show a significant amplitude of oscillation in their expression, and these oscillations are heritable, genotypic.

Some unintelligible difference between quantitative and qualitative characters is thus created, leading some investigators to the assumption of the existence of difference in principle between the two kinds of variation. Under the point of view developed here, there is no such a difference, nor can there be, in principle between qualitative and quantitative variation. The heritable variability of quantitative traits (which, naturally, must be distinguished sharply from non-heritable, phenotypic variability) is determined by the action of this very genotypic milieu on the character studied, as has been said above. And here we, usually, deal with a whole series of "enhancers" and "weakeners," influencing the expression of the basic gene. Inasmuch as there is little probability for the expectation that in one genotype all the "enhancers," or, on the contrary, all the "weakeners" would be concentrated, the extreme deviations from the norm will actually seldom be found, whereas average values will be encountered more often, and the whole series of forms will then follow the law of random errors, that is, a binomial distribution, as is, indeed, commonly observed.

Any of these forms, in so far as it is conditioned by the *genotypic milieu,* that is, by the particular heritable complex, may be genetically isolated, and in this way the impression that a whole series of heritable forms determined by many polymeric genes, in agreement with the well-known hypothesis of Lang (1910) concerning rabbit ears, obtains. And here, the cross between the extreme plus and minus variants will produce a certain average combination of enhancers and weakeners, which will subsequently stay more or less on an average level (permanently intermediate heredity) distributed anew according to the law of random deviations.

On the background of the concept of *genotypic milieu*, a completely new field for the activity of *natural selection* is also opened up. According to the view now prevalent, selection in a population acts until all individuals become homozygous for the selected character (a pure or pure-bred line), after which the action of selection automatically ceases, and no selection of extreme plus or minus fluctuations is powerful enough to displace the character from its average value.

Yet, already in Johannsen's (1913) own experiments on the selection for percentage of aborted grains in the *D* pure line of barley, and subsequently in those of a whole series of other investigators, instances have been encountered where after a whole series of generations in which selection has remained completely powerless, it suddenly begins to be active again, sharply increasing the intensity of expression of the selected trait.

The usual interpretation of these facts is that a new mutational change took place in the experimental cultures exactly in the direction of selection, and that this heritable change was immediately utilized by selection and thereupon extended in subsequent generations.

However, such an explanation must appear unconvincing to any unbiased person. In view of the extraordinary rarity, in general, of new mutations, the probability that in a given culture a new mutation will appear precisely in the necessary direction borders on almost complete improbability. The explanation cited appears far-fetched, though the difficulty disappears as soon as the concept of the genotypical milieu is adopted.

Any newly arising mutation may appear in connection with the selected feature either as an "enhancer" or a "weakener." In the case of an "enhancer," selection will pick it up and spread this gene in subsequent generations through the whole population, enhancing the selected trait. In this way selection does not cease with the passage of the selected character into the homozygous condition, but is extended further for an indefinitely long time, acting on the whole genotype.

Exactly this process occurs also in nature under the influence of natural selection. It no longer merely selects a given mutation, nor only selects genes favored by it; its influence extends a great deal further over the total complex of genes, over the whole "genotypic milieu," on the background of which a given gene will manifest itself in various ways. In selecting one trait, one gene, selection indirectly also selects a definite genotypic milieu, a genotype, most favorable for the manifestation of the given character.

By removing thus unfavorable combinations of genes, selection aids the realization of more advantageous genotypes, of a more advantageous genotypic milieu. *Selection results in the enhancement of the trait,* and in this sense *it actively participates in the evolutionary process.*

Finally, the notion of pleiotropic action of genes in the genotypic milieu makes comprehensible an area dealing with correlative variability and the genotypic correlation between characters, which is still a mystery. If two characters are conditioned pleiotropically, by one gene, they will always be coexistent: such an extreme, complete bond between the two traits is *correlative variability.* But when selection, selecting a definite character, afterwards reinforces it, indirectly selecting a corresponding genotype, then among the different traits included in such simultaneous selection, definite, although less durable bonds, characteristic of the idea of *genotypic correlation,* may establish themselves.

The brief suggestions cited are sufficient to show what an enormous significance the elaboration of the proper concept of pleiotropic action of genes has for our theoretical thought, and to what important results the application of this principle in the genetic analysis of the evolutionary process may lead us.

RESULTS

Let us summarize:

1. In nature the process of mutation proceeds in precisely the same way as it does in the laboratory, or among domestic animals and cultivated plants. Only a series of special conditions hampers its observation in the natural state.

2. Among mutations arising, a very great number is less viable than the normal forms, but this in no case may be considered as the general rule, since there undoubtedly exist some whose viability is not at all reduced.

3. A species population, in the conditions of free crossing, is a stable aggregate, within which the apparatus which stabilizes the frequencies of the component allelic pairs (the laws of Hardy and Pearson) is an intrinsic feature of the very conditions of free crossing.

4. Each newly arising mutation is absorbed by the species in the heterozygous state and remains in it in this state (in the absence of selection) for an indefinite time, without changing its frequency.

5. From year to year, from generation to generation, more and more new mutations arise, either similar to the preceding ones, or completely new. They are constantly absorbed into the basic species, which continually preserves its external homogeneity. This heterozygosity saturates the species in all directions, recombining and spreading in accordance with the laws of chance (in so far as the various genes are not linked with one another), and gradually "contaminates" the majority of the individuals of the species.

6. With a sufficiently great number of newly arising mutations (and this is connected with the "age" of the species), almost all its individuals prove to be contaminated by various members of heterozygous and recessive mutations.

7. Because of the law of combining probabilities, even though the probability of appearance of a particular mutation in a population will ordinarily be extremely small, the probability of appearance of any one of them in the homozygous condition increases proportionally to the number of mutations absorbed by the species, and in this way with a sufficient accumulation of them, the species begins to manifest more and more often heritable deviations, begins to show instability in its characters, i.e., it begins "to age."

8. The most favorable conditions for the appearance of genotypic variability are provided when a numerous species breaks up into a series of small isolated colonies (island forms of land mollusks).

9. Isolation, together with continuously arising genotypic variability, is a basic factor in intraspecific (and hence, inter-specific) differentiation. Most commonly, this isolation is spatial but sometimes it may be temporal, and, perhaps, environmental (ecological isolation).

10. Natural selection is in essence an antagonist of free crossing. It is a dynamic principle.

11. Norton's table shows that every evolutionary process evoked by selection, regardless of whether the genes are dominant or recessive, always proceeds to the end, to the complete replacement of the less adapted form by the more adapted one. It also demonstrates that selection takes up and eventually fixes every, even the most insignificant, improvement in the organism.

12. Adaptive evolution, without isolation, always leads to the complete transformation of the species (Waagen's mutations) but never can lead to the subdivision of the species into two parts, speciation.

13. The cessation of the action of selection leads to the formation of polymorphic species.

14. Selection, like free crossing, promotes the accumulation of recessive (and less viable) genes in the heterozygous condition in the population.

15. The strong prevalence in number among some of the investigated forms of recessive mutations over dominant ones, demonstrates the continuous accumulation within the species under natural conditions of precisely *recessive* genes, the process being conditioned by the specific action of free crossing as well as selection.

16. The significance of selection and free crossing in relation to the newly originating mutations is markedly different from the above: free crossing, in accumulating mutations, leads to the differentiation of forms, while selection, in eliminating harmful mutations, purifies the species of inordinate variability and in general leads to monomorphism of the species.

17. We have no basis for denying the possibility of *nonadaptive* evolution. On the contrary, in many cases one must assume that the existing adaptive differences between closely related forms were not the cause of their divergence, but, on the contrary, that the particular nature of these adaptive characters was a consequence of a still earlier individualization of these forms. The more ancient such separation, the more adaptive characters will distinguish each form from others.

18. The concept of the multiple action of genes (pleiotropy) advanced by Morgan is extremely important for an understanding of the action of selection. This leads us to the concept of the *genotypic milieu,* as a complex of genes, internally and hereditarily influencing the manifestation of every gene in its character. Every individual is *indivisible* not only in so far as its soma is concerned, but also in the manifestation of every one of its genes.

19. The concept of pleiotropic action of genes clarifies a series of difficult and involved questions of genetics: enhancers, weakeners, modifiers, constant polymerism of quantitative characters.

20. Selection, in selecting not only the gene determining the selected character, but the whole genotype (the genotypic milieu) leads to the reinforcement of the selected character and in this

sense actively participates in the evolutionary process.

21. The concept of pleiotropic action of genes yields a new theoretical basis for the phenomenon of correlated variability and for the genotypical correlation of characters.

CONCLUSION

I shall end with this my analysis of certain aspects of the evolutionary process from the standpoint of modern genetics. This analysis, naturally, must provoke objections from some biologists. Have we the right to oversimplify the problem so much? Is it not a gross error to isolate separate evolutionary aspects, to make an analysis of separate components of the infinitely complex and unitary evolutionary process? After all, nature is not an urn containing marbles with which we carry out our experiments in probability theory, nor does life flow along a bed of mathematical formulas. And do we have a logical right to base a *systematic* process of evolution on the *chance* appearance of mutations?

Undoubtedly in this essay we deal with a very primitive, crudest attempt at approaching an understanding of some evolutionary aspects in accordance with our modern genetic views. But we cannot approach such a complex phenomenon as the evolutionary process otherwise than by a preliminary breakdown into its component elements, viewing different aspects separately, analyzing it into parts and carrying the analysis to the logical end possible. We do not yet have the perspective, we are not yet able to give a relative estimate of the potency of each of the separate factors having a share in this complex process. Therefore, it is still too early to speak of a synthetic formulation of the evolutionary process. Only after we have disentangled the basic principles and regularities underlying the evolution of organisms in the widest sense of the word, as well as the phenomena of speciation, only then will we finally be able to attempt a reconstruction of the definitive structure of evolution and a consideration of its separate parts and finer details.

The concept developed in the present article regarding the effects of free crossing and natural selection is an elementary analysis of the significance of these factors from the genetical point of view. This analysis is based on the premise of ideal regularity of both processes, on taking complete independence of genes for granted, on the assumption of a rigorous operation of Mendelian laws of segregration. Actually, we know that in nature these processes are far from being even and regular, as was represented above. But nevertheless *it is precisely these regularities that underlie all of the irregularities which we in fact encounter in nature.* Many of these irregularities have become comprehensible to us now, and we are thoroughly convinced that in the not far distant future the enormously greater part which remains as yet incomprehensible will reveal its secrets to us. But only after we have unravelled the basic principles, those of speciation as well as of the whole evolutionary process, shall we be able to take stock of all of these deviations and seeming irregularities. And I am convinced that many facts presented to us now as insoluble enigmas will themselves provide the answers.

And there is nothing that is, in principle, inadmissible in that we place the *random* appearances of mutations at the basis of the systematic process of evolution, for the theory of probability shows that chance is subject to the same kind of laws as everything on earth. And to construct a systematic process of evolution on the chance play of various separately arising mutations is in no way less legitimate and logical, than to construct a systematic theory of the expansion of gases on the play of random collisions of the molecules of the gas against the walls of the container. And it should not be forgotten that in our considerations we always have to deal with mass phenomena, with large numbers. But here the law of large numbers, first formulated already at the beginning of the seventeenth century by Jacob Bernouilli, is involved. It is the same law that provides the basis of the constancy of the annual number of suicides, which holds stable at a given historical moment one or another level of the number of annual births of twins and triplets, etc. And here again statistics will say: no, these results exist not only on paper, but they are as real and valid as many of our physical theories based on just such statistical regularities.

REFERENCES

CASTLE, W. E. 1912. The inconstancy of unit-characters. *Amer. Nat.* **46**.

——. 1914. Piebald rats and selection. *Carnegie Inst. Publ.* **195**.

COUTAGNE, G. 1895. Rechérches sur le Polymorphisme des Mollusques de France. Lyon.

CRAMPTON, H. E. 1916. Studies on the variation, distribution and evolution of the genus Partula. *Carnegie Inst. Publ.* **228**.

——. 1925. Contemporaneous organic differentiation in the species of Partula living in Moorea, Society Islands. *Amer. Nat.* **59.**

East, E. M., and D. F. Jones. 1919. Inbreeding and outbreeding. Philadelphia.

Fischer, E. 1901. Experimentelle Untersuchungen über die Vererbung erworbener Eigenschaften. *Allg. Zeit. f. Entomol.* **6.**

——. 1902. Weitere Untersuchungen über die Vererbung erworbener Eigenschaften. *Allg. Zeit. f. Entomol.* **7.**

Fish, H. 1914. On the progressive increase of homozygosis in brother-sister matings. *Amer. Nat.* **48.**

Garrett, A. 1884. The terrestrial Mollusca inhabiting the Society Islands. *Acad. Nat. Sci. Phila. Jour.* **9.**

Gulick, J. 1872. Diversity of evolution under one set of external conditions. *Jour. Linn. Soc. Lond.* (*Zool.*) **2.**

——. 1888. Divergent evolution through cumulative segregation. *Jour. Linn. Soc. Lond.* (*Zool.*) **20.**

Guyer, M. F., and E. A. Smith. 1920. Studies on Cytolysins. II. Transmission of induced eye defects. *Jour. Exp. Zool.* **31.**

Guyer, M. F., *and* E. A. Smith. 1924. Further studies on inheritance of eye defects induced in rabbits. *Jour. Exp. Zool.* **38.**

Hardy, G. H. 1908. Mendelian proportions in a mixed population. *Science* **28.**

Heincke, F. 1898. Naturgeschichte des Herings. Berlin.

Herbst, C. 1919, 1924. Beiträge zur Entwicklungsphysiologie der Färbung und Zeichnung der Tiere. *Heidelberg Akad. Wissensch., Abhandl.* **7,** and *Arch. f. mikr. Anat. u. Entwicklungsmech.* **102.**

Jennings, H. 1912. Production of pure homozygotic organisms from heterozygotes by self-fertilization. *Amer. Nat.* **46.**

——. 1914. Formulae for the results of inbreeding. *Amer. Nat.* **48.**

——. 1916. The numerical results of diverse systems of breeding. *Genetics* **1.**

——. 1917. The numerical results of diverse systems of breeding, with respect to two pairs of characters, linked or independent, with special relation to the effects of linkage. *Genetics* **2.**

Johannsen, W. 1913. Elemente der exakten Erblichkeitslehre. Jena.

Jordan, K. 1905. Der Gegensatz zwischen geographisches und nichtgeographisches Variation. *Zeit. f. wiss. Zool.* **83.**

Kammerer, P. 1907, 1909, 1913. Vererbung erzwungener Fortpflanzungsanpassen. *Arch f. Entwicklungsmech. d. Org.* **25, 28, 36.**

Koltzov, N. K. 1924. The newest attempts to demonstrate the inheritance of acquired characters. *Russ. Evg. Zhur.* **2.** In Russian.

Korzhinsky, S. 1892. Flora of Eastern European Russia. Tomsk. In Russian.

Kosminsky, P. 1924. Der Gynandromorphismus bei *Lymantria dispar* L. unter Einwirkung ausserer Einflusse. *Biol. Zentralbl.* **44.**

Lang, A. 1910. Die Erblichkeitsverhältnisse der Ohrenlänge der Kaninchen nach Castle und das Problem der intermediären Vererbung und Bildung konstanter Bastarderassen. *Zeit. f. ind. Abst. u. Vererb.* **4.**

Lotsy, J. 1916. Evolution by means of hybridization. The Hague.

Marchal, Él., and Ém. Marchal. 1906. Recherches expérimentales sur la sexualité des spores chez les mousses dioiques. *Acad. Roy. Belg.; Classe de Sci., Mém.* Ser. **2, 1.**

Mayer, A. G. 1902. Some species of Partula from Tahiti. *Harv. Mus. Comp. Zool., Mem.* **26.**

Millière, P. 1859–1868. Iconographie et description de Chenilles et Lépidoptères Inédits. Lyon.

Morgan, T. H. 1924. Human inheritance. *Amer. Nat.* **58.**

Morgan, T. H., C. B. Bridges, and A. H. Sturtevant. 1925. The genetics of Drosophila. 'SGravenhage.

Neumayr, M., and K. M. Paul. 1875. Die Congerienund Paludinenschichten Westslavoniens und deren Faunen. *Abhandl. Geolog. Riechsanst.* **7.**

Osborn, H. F. 1912. The continuous origin of certain unit characters observed by a paleontologist. *Amer. Nat.* **46.**

Pearl, R. 1913. A contribution towards an analysis of the problem of inbreeding. *Amer. Nat.* **47.**

——. 1914 *a.* On the results of inbreeding a Mendelian population: a correction and extension of previous conclusions. *Amer. Nat.* **48.**

——. 1914 *b.* On a general formula for the constitution of the n-th generation of a Mendelian population in which all matings are brother × sister. *Amer. Nat.* **48.**

Pearson, K. 1900. Grammar of science. 2nd ed. London.

——. 1904. On a generalized theory of alternative inheritance with special reference to Mendel's laws. *Phil. Trans. Roy. Soc.* **203.**

Petersen, W. 1903. Entstehung der Arten durch physiologische Isolierung. *Biol. Centralbl.* **23.**

Philipchenko, Yu. 1919. Expression of Mendel's law from the point of view of genotypic structure. *Russ. Acad. Sci., Izv.* **13.** In Russian.

——. 1924. Ueber Spaltungprozesse innerbalb einer Population bei Panmixie. *Zeit. f. ind. Abst. und Vererb.* **35.**

Plate, L. 1913. Vererbungslehre. Leipzig.

Punnett, R. C. 1915. Mimicry in butterflies. Cambridge.

Robbins, R. B. 1917–1918. Some applications of mathematics to breeding problems. *Genetics* **2** and **3.**

——. 1918. Random mating with the exception of sister by brother mating. *Genetics* **3.**

Romanes, G. J. 1886. Physiological selection. *Jour. Linn. Soc. Lond.* (*Zool.*) **19.**

Romashov, D. D. 1925. Mathematical expression of Mendel's law. *Zhurn. Eksp. Biol.* **1.** In Russian.

Semenov-Tyanshansky, A. P. 1910. Taxonomical boundaries of species and their subdivisions. *Imp. Akad. Nauk. Zapiski Fiz. Mat. Otd.*, ser. 8, **25.** In Russian.

Standfuss, M. 1906. Die Resultate 30-jahriger Experimente mit Bezug auf Artenbildung und Umgestaltung in der Tierwelt. *Schweiz. Naturf. Gesell., Verhandl.* **88.**

Sturtevant, A. H. 1925. The effects of unequal crossing-over at the bar locus in Drosophila. *Genetics* **10.**

TIETZE, H. 1923. Ueber das Schicksal gemischter Populationen nach den Mendel'schen Vererbungsgesetzen. *Zeitsch f. angew. Mathem. u. Mech.* **3.**

TIMOFEEFF-RESSOVSKY, N. V. 1925. Ueber den Einfluss des Genotypus auf das phänotipische Auftreten eines einzelnen Gens. *Jour. f. Psych. u. Neur.* **31.**

——. 1926. Investigation of the phenotypical manifestation of heritable factors. *Zhurn. Eksp. Biol.* **1.** In Russian.

TOWER, W. L. 1906. An investigation of evolution in chrysomelid beetles of the genus Leptinotarsa. *Carnegie Inst. Publ.* **48.**

DE VRIES, H. 1901, 1903. Die Mutationstheorie. Leipzig.

WAAGEN, W. 1869. Die Formenreihe des *Ammonites subradiatus*. *Geognost.-palänot. Beiträge* **2.**

WAGNER, M. 1868. Die Darwin'sche Theorie und das Migrationsgesetz der Organismen. Leipzig.

——. 1889. Die Einstehung der Arten durch räumliche Sonderung. Basel.

WEISMANN, A. 1872. Ueber den Einfluss der Isolierung auf die Artbildung. Leipzig.

——. 1904. Vorträge über Descendenztheorie. Jena.

WENTWORTH, E. N., and B. L. REMICK. 1916. Some breeding properties of the generalized Mendelian population. *Genetics* **1.**

WETTSTEIN, F. VON. 1924. Morphologie and Physiologie des Formwechsels der Moose auf genetischer Grundlage. *Zeit. ind. Abst. und Vererb.* **33.**

WINKLER, H. 1916. Ueber die experimentelle Erzeugung von Pflanzen mit abweichenden Chromosomenzahlen. *Zeit. f. Botanik.* **8.**

WRIGHT, S. 1921*a*. Correlation and causation. *Jour. Agric. Res.* **20.**

——. 1921*b*. Systems of mating. *Genetics* **6.**

19

ON THE GENETIC CONSTITUTION OF WILD POPULATIONS

S. S. Chetverikov

This article was translated expressly for this Benchmark volume by Roy A. Jameson, from "Uber die Genetisch Beschaffenheit Wilder Populationen," Proc. 5th Intern. Congr. Genetics 1927, *Springer-Verlag, 1927, pp. 1499-1500*

From the excellent studies undertaken by T. H. Morgan and his students we have learned that a process of genetic variation (mutation) operates in populations. By and large this process is independent from external conditions and must occur in nature just as it does in the laboratory. But what fate lies to the gene variations (mutations) which develop in wild populations? Hardy and Pearson's equations of equilibrium (Gleichgewichts satze) demonstrate that there exists a mechanism in free breeding which automatically maintains the genotypic make up of wild populations in equilibrium and which simultaneously retains all newly appearing gene variations (if only they are not lethal or

[*Editor's Note:* This translation by Roy A. Jameson places particular emphasis on the science and language of the time the paper was prepared and published.]

strong selectively disadvantageous) in the population in the heterozygous state. By this means the wild populations remain phenotypically uniform and yet are permeated genotypically by heterozygous genes. Selection, however, operates oppositely, providing for the phenotypic uniformity of the population by gradually eliminating those inferior gene variations which are manifested.

The attempt to allow the unobservable (heterozygous) traits to express themselves by inbreeding of the progeny of particular wild individuals succeeded completely and showed that among various *Drosophila* species the wild populations are completely heterozygous with regard to the most different traits. Particularly fruitful were the experiments with a *Drosophila melanogaster* population from nature which, briefly, gave the following results:

Among the offspring of 239 wild female *Drosophila* 32 different inheritable characteristics emerged. Of these several occured in a wholely unexpected frequency. Thus we encountered the gene "polychaeta" (increased number of thoraicic bristles) in the offspring of 50% of all investigated females, and among some of the latter (15) this gene had already expressed itself in wild populations (homozygous). Almost as commonly observed (40%) was the gene "ramuli", which was manifest in small accessory lateral veins in regions of the second (and occasionally also the first) cross veins. Very strange and monstrous genes also appeared occasionally: thus among the first 239 investigated females one gene was found which resulted in the development of legs instead of the usual antennae.

All of these factors lead to the conclusion that the usual "wild" populations are inordinately heterozygous in the most varied directions. They therefore present a rich material of inheritable variations which could be useful in a changing environment and which must thus play a decisive role in the evolutionary process. In practical regard our investigations show first that in the inbreeding of wild populations we possess an excellent method of obtaining dozens of new gene variations. Secondly, these investigations show how easily a mistake can be committed in experiments with the inheritable incursion of external factors. An inbreeding of wild individuals according to established practice produces gene variations whose origin do not lie in the experiment but rather in the wild heterozygotes themselves.

Part IV

HOW NATURE WORKS

Editor's Comments on Papers 20 Through 25

20 FISHER
Excerpt from *The Fundamental Theorem of Natural Selection*

21 HALDANE
A Mathematical Theory of Natural and Artificial Selection. Part VII. Selection Intensity as a Function of Mortality Rate

22 HALDANE
A Mathematical Theory of Natural Selection. Part VIII. Metastable Populations

23 HALDANE
A Mathematical Theory of Natural and Artificial Selection. Part IX. Rapid Selection

24 WRIGHT
Excerpts from *Evolution in Mendelian Populations*

25 WRIGHT
The Roles of Mutation, Inbreeding, Crossbreeding and Selection in Evolution

Only Darwin and Mendel have more secure places in the development of evolutionary genetics than those of Fisher, Haldane, and Wright. Additionally, each made fundamental contributions to other fields—Fisher to theoretical statistics, Haldane to biochemistry and physiology, and Wright to physiological genetics. Each continued his studies many years after evolutionary genetics had diversified into a number of specialized fields. What is striking about reading their papers is that they understood so very much about the plants and animals they were working with and writing about. Their models were not made up of whole cloth but were studies initiated because the biological data required theoretical bases and the theoretical models provided a means of asking additional questions about nature. Therefore, Fisher, Haldane, and Wright were always able to give empirical results which supported their separate syntheses. Their ideas developed during the early days of genetics and they stimulated each other by their extensive productivity and great facility to express their own positions.

In *The Genetical Theory of Natural Selection* Fisher treated natural selection as a general study with underlying principles which could be applied whether or not selection was the only, primary, or a marginal source of evolution in most natural populations. Fisher felt his fundamental theorem of natural selection offered great generality to the subject by considering the effects of the additive genetic variance under panmixia. The theorem states, "The rate of increase in fitness of any organism at any time is equal to its genetic variance in fitness at that time." By genetic variance, Fisher means the proportion of the variance that is attributable to the additive effects of genes. He analogized the theorem to the widely general second law of thermodynamics and regarded it as occupying a similarly important place in biology. By whatever degree the theorem is true it has predictive value for the direction and the rate of change of a complex biological phenomena. The fundamental theorem has stimulated theoretical and experimental studies and its ultimate value lies not with the exceptions and refinements but with its general usefulness. Fisher also discussed the evolution of dominance and of mimicry, drawing heavily on the influence of other modifying factors. He developed the role of mutation and selection, noting that beneficial mutations are initially subject to considerable chance but only a slight recurrence rate is necessary to assure their ultimate pervasion of the species, even with very low selection. Additionally, advantageous mutations resemble neutral mutations in gene ratio distribution. The decay of variance without selection and with random mating is very slow and that decay will be countered by very low mutation rates. Fisher suggests, as had Darwin, that abundant species should be more variable than rare species He pointed out that factors which are favored by selection of the heterozygote will tend to accumulate while other factors will be eliminated. He emphasized the greater likelihood of the selective advantage of minor rather than major genes.

Fisher developed his understanding of factor interaction by suggesting that genes were modified by a larger number of other genes and that selection would favor those modifiers which produced individuals which more closely resembled the adaptive forms. This approach was used to explain the evolution of dominance and to describe the evolution of mimicry. He noted that stable equilibria are established by the interactions of two or more factors and discussed linkage and quantitative meristic characters emphasizing the interaction of coordinated structures. A large part of the book is devoted to man and how the activities and structure of industrial society influence the evolution of man. We have selected the statement of Fisher on his fundamental theorem for this volume (Paper 20).

Haldane's first three papers developing a mathematical theory of natural and artificial selection were presented in Part III of this volume. He dealt with a simple situation encumbered with numerous assumptions and considering various mating situations. With random mating he noted that replacement of one allele by another took the same number of generations whether the allele was dominant or recessive, but the dominant increased most rapidly at first. In fact, rare recessive characteristics do not respond to low selection, but partial dominance vastly increases the effectiveness of selection. Multiple factors making approximately equal contributions to the characteristic will result in a very slow selective influence.

In Part IV of "A Mathematical Theory of Natural and Artificial Selection," Haldane examined the selection when the generations overlap and the age schedules of birth and death are stable and are different for the individuals possessing each allele. The rates of change do not differ significantly from that found in the discrete generation cases. This part has been reprinted in *Demographic Genetics* (Vol. 3, "Benchmark Papers in Genetics"). Part V of "A Mathematical Theory," reprinted in *Stochastic Models in Population Genetics* (Vol. 7, "Benchmark Papers in Genetics") approached the problem of mutation and selection. Haldane summarized as follows:

> To sum up, if selection acts against mutation, it is ineffective provided that the rate of mutation is greater than the coefficient of selection. Moreover, mutation is quite effective where selection is not, namely in causing an increase of recessives when these are rare. It is also more effective than selection in weeding out rare recessives provided that it is not balanced by back mutation of dominants. Mutation therefore determines the course of evolution as regards factors of negligible advantage or disadvantage to the species. It can only lead to results of importance when its frequency becomes large (p. 842).

In Part VI of "A Mathematical Theory," Haldane considered the role of isolation in evolution. This part is reprinted in the Benchmark Volume *Genetics of Speciation* and is only summarized here. Haldane considered the case of no amphimixis, dominants or recessives favored, the other immigrating, no dominance, sex-linked, and various multiple factor situations. In every case a critical value exists (different in each case) where the selection coefficient must be greater than the immigration coefficient or the selected type will be completely swamped by immigration.

Part VII of "A Mathematical Theory" (Paper 21) considered the influence, through changes in mortality, of various degrees of competition on selection intensity and showed that "the intensity of selection may diminish and become negative at high rates of elimination" (p. 131) while, as long as selection was slight, the gain from selection was extremely low.

Part VIII (Paper 22) expanded on Haldane's understanding of the evolutionary process. When two or more genes are considered, their individual disadvantageous character may be advantageous because of the interactions between the genes. With only two factors and no over-dominance the stable equilibria are at fixation or loss. With single gene heterozygotic vigor the stable equilibrium includes the heterozygotes. When there are *m* genes with heterozygotic advantages there are more stable points.

In Part IX (Paper 23) the selection index was allowed to differ being more in a population with low density, less and continuous in a population with continuous high densities. Haldane found that the results might be different depending on whether or not the character was the result of dominance or recessiveness. Part X appeared in *Genetics* and was concerned with artificial selection. Haldane considered various aspects of the problems of working with an F_1 obtained by mating two pure lines. He considered the case where selection is against multiple dominants or multiple recessives are eliminated with random mating, self-fertilization, and brother-sister matings, and provided mathematical expressions for the effects. The paper should prove useful to many experimentalists, but it is not included here because it is not directly applicable.

In 1932 Haldane's small book *The Causes of Evolution* reviewed the information presented in his papers and expanded on the positions of Fisher and Wright. He placed considerable emphasis on the role of polyploidy and hybridization and considered evidence from paleontology and other fields. Haldane believed that species are usually in genetic equilibrium and that evolution is usually extremely slow. Selection coefficients are low and balanced by mutation. Change is slow and under these conditions chance plays a significiant role. The interaction of multiple genetic factors slows selection rates. Most of variance is from rare genes maintained by mutation, but these are necessary for a source of variation for selection. Isolation is important for fixation in small populations and in the process of speciation. The appendix of this book provides an excellent summary of Haldane's position in the early 1930s.

Sewall Wright developed his theory of evolution as a result of his study of the genetics of mammals. He was particularly impressed with the nonadditive interaction effects of factors and the improbability of selecting for these except through the interaction of the role of inbreeding, isolation and diffusion. In his small mammal studies he demonstrated profound differentiation of the inbred strains in their interaction systems. He thus approached evolution from the standpoint of determining the structure of the natural population which would allow evolutionary advance.

Wright considered the variation and distribution of gene frequencies and classified the factors of evolution. He noted that the subdivision of the species into small local populations provides the most effective means for the species to find its way from one adaptive peak to another. The subpopulations are partially isolated permitting some random differentiation and some selection. There is a balance among the pressures allowing high frequency multiple allele systems at many or all loci. Additionally there is a systematic balance of selection and random processes at each loci. "The course of evolution of the species as a whole is then determined by interdemic selection" (Wright, 1960:370). In 1931 he concluded: "Finally in a large population, divided and subdivided into partially isolated local races of small size, there is a continually shifting differentiation among the latter (intensified by local differences in selection but occurring under uniform and static conditions) which inevitably brings about an indefinitely continuing, irreversible, adaptive, and much more rapid evolution of the species" (p. 158). Only the introduction and very thorough summary of the 1931 paper is presented here (Paper 24) because the entire paper has been included in *Stochastic Models in Population Genetics* (Vol. 7, "Benchmark Papers in Genetics").

In 1932 Wright presented a condensed version of his views to the Sixth International Genetics Congress, where he pointed out that inbreeding balanced by occasional crossbreeding resulted in a largely nonadaptive differentiation of local races and the course of evolution was controlled less by mutation and direct selection and more by trial and error and by "a determination of long time trend by intergroup selection." Thus the earliest statements of Wright's "Shifting Balance Theory" of evolution are clear. The adaptive nature of evolution takes advantage of randomly occurring interactive gene complexes.

There are some obvious major differences in the positions taken by Fisher, Haldane, and Wright, and there are a number of specific differences which have been discussed in various papers (Wright, 1930; Haldane, 1931). All three are mathematically consistent and based on Mendelian principles, but they emphasize different biological positions. Haldane emphasized the role of individual mutations and mass selection. Wright and Fisher emphasized the evolution of panmictic populations, while Wright considered the results of inbreeding which accompanies isolation. Fisher emphasized the role of evolution of additive genetic characteristics, while Wright emphasized the importance of interaction effects and of the selective diffusion from differentiated local populations.

Nature is very diverse and the positions taken by Haldane, Fisher, and Wright are each represented in nature. The relative importance is not nearly as significant as is the notation that each position is an understanding of nature which laid a fundamental groundwork for extensive later investigations. These three made benchmark contributions to the foundations of genetics, considered in a number of separate volumes of the Benchmark series. This *Evolutionary Genetics* volume provides the early historical basis and background for these several other volumes, each of which emphasizes more recent literature.

20

Reprinted through the permission of the publisher from *The Genetical Theory of Natural Selection,* Dover Publications, 1958, pp. 22, 30–41

THE FUNDAMENTAL THEOREM OF NATURAL SELECTION

R. A. Fisher

The life table and the table of reproduction. The Malthusian parameter of population increase. Reproductive value. The genetic element in variance. Natural Selection. The nature of adaptation. Deterioration of the environment. Changes in population. Summary.

One has, however, no business to feel so much surprise at one's ignorance, when one knows how impossible it is without statistics to conjecture the duration of life and percentage of deaths to births in mankind. DARWIN, 1845. (*Life and Letters*, ii, 33.)

In the first place it is said—and I take this point first, because the imputation is too frequently admitted by Physiologists themselves—that Biology differs from the Physico-chemical and Mathematical sciences in being 'inexact'. HUXLEY, 1854.

The genetic element in variance

Let us now consider the manner in which any quantitative individual measurement, such as human stature, may depend upon the individual genetic constitution. We may imagine, in respect of any pair of alternative genes, the population divided into two portions, each comprising one homozygous type together with half of the heterozygotes, which must be divided equally between the two portions. The difference in average stature between these two groups may then be termed the *average excess* (in stature) associated with the gene substitution in question. This difference need not be wholly due to the single gene, by which the groups are distinguished, but possibly also to other genes statistically associated with it, and having similar or opposite effects. This definition will appear the more appropriate if, as is necessary for precision, the population used to determine its value comprises, not merely the whole of a species in any one generation attaining maturity, but is conceived to contain all the genetic combinations possible, with frequencies appropriate to their actual probabilities of occurrence and survival, whatever these may be, and if the average is based upon the statures attained by all these genotypes in all possible environmental circumstances, with frequencies appropriate to the actual probabilities of encountering these circumstances. The statistical concept of the excess in stature of a given gene substitution will then be an exact one, not dependent upon chance as must be any practical estimate of it, but only upon the genetic nature and environmental circumstances of the species. The excess in a factor will usually be influenced by the actual frequency ratio $p : q$ of the alternative genes, and may also be influenced, by way of departures from random mating, by the varying reactions of the factor in question with other factors; it is for this reason that its value for the purpose of our argument is defined in the precise statistical manner chosen, rather than in terms of the average sizes of pure genotypes, as would be appropriate in specifying such a value in an experimental population, in which mating is under control, and in which the numbers of the different genotypes examined is at the choice of the experimenter.

For the same reasons it is also necessary to give a statistical definition of a second quantity, which may be easily confused with that just defined, and may often have a nearly equal value, yet which must be distinguished from it in an accurate argument; namely the *average effect* produced in the population as genetically constituted, by the substitution of the one type of gene for the other. By whatever rules mating, and consequently the frequency of different gene combinations, may be governed, the substitution of a small proportion of the genes of one kind by the genes of another will produce a definite proportional effect upon the average stature. The amount of the difference produced, on the average, in the total stature of the population, for each such gene substitution, may be termed the average effect of such substitution, in contra-distinction to the average excess as defined above. In human stature, for example, the correlation found between married persons is sufficient to ensure that each gene tending to increase the stature must be associated with other genes having a like effect, to an extent sufficient to make the average excess associated with each gene substitution exceed its average effect by about a quarter.

If a is the magnitude of the *average excess* of any factor, and α the magnitude of the *average effect* on the chosen measurement, we shall now show that the contribution of that factor to the genetic variance is represented by the expression $2pqa\alpha$.

The variable measurement will be represented by x, and the relation of the quantities α to it may be made more clear by supposing that for any specific gene constitution we build up an 'expected' value, X, by adding together appropriate increments, positive or negative, according to the natures of the genes present. This expected value will not necessarily represent the real stature, though it may be a good approximation to it, but its statistical properties will be more intimately involved in the inheritance of real stature than the properties of that variate itself. Since we are only concerned with variation we may take as a primary ingredient of the value of X, the mean value of x in the population, and adjust our positive and negative increments for each factor so that these balance each other when the whole population is considered. Since the increment for any one gene will appear p times to that for its alternative gene q times in the whole population, the two increments must be of opposite sign and in the ratio $q : (-p)$. Moreover, since their difference must be α, the actual values cannot but be $q\alpha$ and $(-p\alpha)$ respectively.

The value of the *average excess* a of any gene substitution was obtained by comparing the average values of the measurement x in two moieties into which the population can be divided. It is evident that the values of α will only be properly determined if the same average difference is maintained in these moieties between the values of X, or in other words if in each such moiety the sum of the deviations, $x - X$, is zero. This supplies a criterion mathematically sufficient to determine the values of α, which represent in the population concerned the average effects of the gene substitutions. It follows that the sum for the whole population of the product $X(x - X)$ derived from each individual must be zero, for each entry $q\alpha$ or $(-p\alpha)$ in the first term will in the total be multiplied by a zero, and this will be true of the items contributed by every factor severally. It follows from this that if X and x are now each measured from the mean of the population, the variance of X, which is the mean value of X^2, is equal to the mean value of Xx. Now the mean value of Xx will involve α for each Mendelian factor; for X will contain the item $q\alpha$ in the p individuals of one moiety and $(-p\alpha)$ in the q individuals of the other, and since the average values of x in these two moieties differ by a and each individual contains two genes at each locus, the mean value of Xx must be the sum for all factors of the quantities $2pqa\alpha$. Thus the variance of X is shown to be

$$W = \Sigma(2pqa\alpha)$$

the summation being taken over all factors, and this quantity we may distinguish as the *genetic* variance in the chosen measurement x. That it is essentially positive, unless the effect of every gene severally is zero, is shown by its equality with the variance of X. An extension of this analysis, involving no difference of principle, leads to a similar expression for cases in which one or more factors have more than two different genes or allelomorphs present.

The appropriateness of the term genetic variance lies in the fact that the quantity X is determined solely by the genes present in the individual, and is built up of the average effects of these genes. It therefore represents the genetic potentiality of the individual concerned, in the aggregate of the mating possibilities actually open to him, in the sense that the progeny averages (of x, as well as of X) of two males mated with an identical series of representative females will differ by exactly half as much as the genetic potentialities of their sires differ. Relative genetic values may therefore be determined experimentally by the diallel method, in which each animal tested is mated to the same series of animals of the opposite sex, provided that a large number of offspring can be obtained from each such mating, and that the mates are representative of the actual population. Without obtaining individual values, the genetic variance of the population may be derived from the correlations between relatives, provided these correlations are accurately obtained. For this purpose the square of the parental correlation divided by the grandparental correlation supplies a good estimate of the fraction, of the total observable variance of the measurement, which may be regarded as genetic variance.

It is clear that the actual measurements, x, obtained in individuals may differ from their genetic expectations by reason of fluctuations due to purely environmental circumstances. It should be noted that this is not the only cause of difference, for even if environmental fluctuations were entirely absent, and the actual measurements therefore determined exactly by the genetic composition, these measurements, which may be distinguished as *genotypic*, might still differ from the genetic values, X. A good example of this is afforded by dominance, for if dominance is complete the genotypic value of

the heterozygote will be exactly the same as that of the corresponding dominant homozygote, and yet these genotypes differ by a gene substitution which may materially affect the genetic potentiality represented by X, and be reflected in the average measurement of the offspring. A similar cause of discrepancy occurs when gene substitutions in different factors are not exactly additive in their average effects. The genetic variance as here defined is only a portion of the variance determined genotypically, and this will differ from, and usually be somewhat less than, the total variance to be observed.

It is consequently not a superfluous refinement to define the purely genetic element in the variance as it exists objectively, as a statistical character of the population, different from the variance derived from the direct measurement of individuals.

Often more than two genes may alternatively occupy the same locus. These are termed multiple allelomorphs. In extending the notion of genetic excess to such cases, it is convenient to define the genetic excess associated with a single gene. Thus if we suppose that the genotypic value X has been ascertained for an entire natural population, the genetic composition of each individual of which is known, we may let $\bar{X}$ stand for the general mean, and x for the deviation of any genotypic value, so that

$$x = X - \bar{X}.$$

Choosing any particular factor, we may pick out all the individuals carrying any one gene, counting the homozygotes twice, and find the average value of x for this selected group of individuals.

Thus if out of a population of N individuals there are n_{11} homozygotes, and n_{lk} heterozygotes formed by combination with any other chosen allelomorph, the total of the values of x from the homozygotes may be represented by $S(x_{11})$, and that from any class of heterozygotes containing the chosen gene by $S(x_{lk})$. Then

$$\frac{2S(n_{11}) + \sum_{k=2}^{s}{}' S(n_{lk})}{2n_{11} + \sum_{k=2}^{s}{}' n_{lk}} = a_1$$

where a_1 may be spoken of as the average genotypic excess of the particular gene chosen. Σ is used for summation over allelomorphs of the same factor. If p_1 is the proportion of this kind of gene among all homologous kinds which might occupy the same locus, it is evident that

$$\sum_{k=1}^{s}(p_k a_k) = 0.$$

We may now introduce a second quantity which has sometimes been confused with that defined above, and which may indeed have a nearly equal value, yet which must be distinguished from it in an accurate argument; namely the average effect produced in the population genetically constituted by the substitution of one gene for another. With multiple allelomorphism it is convenient to define this quantity also by the effect of substituting any chosen gene for a random selection of the genes homologous with it. By whatever rules mating, and consequently the frequency of different combinations, may be governed, the substitution of a small proportion of genes of one kind for the others will produce a definite proportional effect upon the average measurement. The amount of the difference produced, on the average, in the total measurement of the population, for each such gene substitution, may be termed the *average effect* of the gene substituted in contradistinction to the average excess defined above. In human stature, for example, the correlation found between married persons is sufficient to ensure that each gene must be associated with other genes having a like effect to an extent sufficient to make the average excess associated with each gene exceed its average effect by about a quarter.

If α is the magnitude of the average effect of a given gene, its measurement by direct substitution implies the requirement that the values of α are the best additive system for predicting the genotypic value from the actual genes present in any individual. Subject to the condition

$$\sum_{k=1}^{s}{}'(p\alpha) = 0$$

let ξ represent the value of the genotype as best predicted from the genes present, so that

$$\xi = \Sigma(\alpha)$$

where Σ stands for summation over all the genetic factors affecting stature. Then if we make

$$S(x - \xi)^2$$

a minimum for variation of all quantities α, we find varying any one of them, α_1,

$$2S(x_{11} - \xi_{11}) + \sum_{k=2}^{s}{}' S(x_{ik} - \xi_{ik}) = \lambda p_1$$

where λ is some constant undetermined. These equations are sufficient to determine all values of α.

Consider now the quantity

$$S\{\xi(x-\xi)\} = S(\xi x) - S(\xi^2);$$

if we substitute for the first ξ in the expression on the left its expression in terms of α, the coefficient of any particular value such as α_1 is

$$2S(x_{11} - \xi_{11}) + \sum_{k=2}^{s} S(x_{ik} - \xi_{1k})$$

which we have shown to be equal to λp_1. Hence

$$S(\xi x) - S(\xi^2) = \sum_{k=1}^{s}{}'(p_k \alpha_k) = 0.$$

It follows that the variance of the genetic value ξ is equal to the co-variance of the genetic and genetypic values ξ and x.

If now we substitute for ξ in the expression $S(\xi x)$ the coefficient of α_1 is

$$2S(x_{11}) + \sum_{k=2}^{s}{}' S(x_{lk}) = 2Np_1 a_1$$

by definition of a. The total of the contributions from any set of homologous genes is therefore

$$2N \sum_{k=1}^{s} (p_k a_k \alpha_k)$$

and for all factors affecting stature it is

$$2N\Sigma\Sigma'(pa\alpha).$$

Dividing by N, the total number of individuals involved, it is now seen that the genetic variance of diploid individuals is given by

$$\Sigma\Sigma'(2pa\alpha).$$

If only two allelomorphs are present, we have

$$p_1 + p_2 = 1$$
$$p_1 a_1 + p_2 a_2 = 0$$

whence, if $a_1 - a_2 = d$, then

$$a_1 = p_2 d \quad \text{and} \quad a_2 = -p_1 d.$$

Similarly, where $\alpha_1 = p_2 \delta$ and $\alpha_2 = -p_2\delta$, then

$$\begin{aligned} \Sigma'(2pa\alpha) &= 2p_1 a_1 \alpha_1 + 2p_2 a_2 \alpha_2 \\ &= 2p_1 p_2^2 \, d\delta + 2p_1^2 p_2 \, d\delta \\ &= 2p_1 p_2 \, d\delta \end{aligned}$$

in accordance with the expression obtained in the first edition of this book for the case of only two allelomorphs. In the formula there given the factor 2 for diploids is omitted, through treating the population as one of the $2N$ loci instead of N individuals, as seems to be in every way preferable.

Natural Selection

Any group of individuals selected as bearers of a particular gene, and consequently the genes themselves, will have rates of increase which may differ from the average. The excess over the average of any such selected group will be represented by a, and similarly the average effect upon m of introducing the gene in question will be represented by α. Since m measures fitness by the objective fact of representation in future generations, the quantity

$$\Sigma'(2pa\alpha)$$

will represent the contribution of each factor to the genetic variance in fitness. The total genetic variance in fitness being the sum of these contributions, which is necessarily positive, or, in the limiting case, zero. Moreover, any increase dp in the frequency of the chosen gene will be accompanied by an increase $2\alpha\, dp$ in the average fitness of the species, where α may, of course, be negative. But the definition of a requires that

$$\frac{d}{dt} \log p = a$$

or

$$dp = (pa)dt$$

hence

$$(2\alpha)dp = (2pa\alpha)dt$$

which must represent the rate of increase of the average fitness due to the change in progress in frequency of this one gene. Summing for all allelomorphic genes, we have

$$dt\Sigma'(2pa\alpha)$$

and taking all factors into consideration, the total increase in fitness is

$$\Sigma\alpha\, dp = dt\Sigma\Sigma'(2pa\alpha) = W\, dt\, .$$

If therefore the time element dt is positive, the total change of fitness $W dt$ is also positive, and indeed the rate of increase in fitness due to all changes in gene ratio is exactly equal to the genetic variance of fitness W which the population exhibits. We may consequently state the fundamental theorem of Natural Selection in the form:

The rate of increase in fitness of any organism at any time is equal to its genetic variance in fitness at that time.

The rigour of the demonstration requires that the terms employed should be used strictly as defined; the ease of its interpretation may be increased by appropriate conventions of measurement. For example, the frequencies p should strictly be evaluated at any instant by the enumeration, not necessarily of the census population, but of all individuals having reproductive value, weighted according to the reproductive value of each.

Since the theorem is exact only for idealized populations, in which fortuitous fluctuations in genetic composition have been excluded, it is important to obtain an estimate of the magnitude of the effect of these fluctuations, or in other words to obtain a standard error appropriate to the calculated, or expected, rate of increase in fitness. It will be sufficient for this purpose to consider the special case of a population mating and reproducing at random. It is easy to see that if such chance fluctuations cause a difference δp between the actual value of p obtained in any generation and that expected, the variance of δp will be

$$\frac{pq}{2n},$$

where n represents the number breeding in each generation, and $2n$ therefore is the number of genes in the n individuals which live to replace them. The variance of the increase in fitness, $\Sigma 2\alpha dp$, due to this cause, will therefore be

$$\frac{1}{2n}(2pq\alpha^2).$$

Now, with random mating, the chance fluctuation in the different gene ratios will be independent, and the values of a and α are no longer distinct, it follows that, on this condition, the rate of increase of fitness, when measured over one generation, will have a standard error due to random survival equal to

$$\frac{1}{T}\sqrt{\frac{W}{2n}}$$

where T is the time of a generation. It will usually be convenient for each organism to measure time in generations, and if this is done it will be apparent from the large factor $2n$ in the denominator, that the random fluctuations in W, even measured over only a single generation, may be expected to be very small compared to the average rate of progress. The regularity of the latter is in fact guaranteed by the same circumstance which makes a statistical assemblage of particles, such as a bubble of gas obey, without appreciable deviation, the laws of gases. A visible bubble will indeed contain several billions of molecules, and this would be a comparatively large number for an organic population, but the principle ensuring regularity is the same. Interpreted exactly, the formula shows that it is only when the rate of progress, W, when time is measured in generations, is itself so small as to be comparable to $1/n$, that the rate of progress achieved in successive generations is made to be irregular. Even if an equipoise of this order of exactitude, between the rates of death and reproduction of different genotypes, were established, it would be only the rate of progress for spans of a single generation that would be shown to be irregular, and the deviations from regularity over a span of 10,000 generations would be just a hundredfold less.

It will be noticed that the fundamental theorem proved above bears some remarkable resemblances to the second law of thermodynamics. Both are properties of populations, or aggregates, true irrespective of the nature of the units which compose them; both are statistical laws; each requires the constant increase of a measurable quantity, in the one case the entropy of a physical system and in the other the fitness, measured by m, of a biological population. As in the physical world we can conceive of theoretical systems in which dissipative forces are wholly absent, and in which the entropy consequently remains constant, so we can conceive, though we need not expect to find, biological populations in which the genetic variance is absolutely zero, and in which fitness does not increase. Professor Eddington has recently remarked that 'The law that entropy always increases—the second law of thermodynamics—holds, I think, the supreme position among the laws of nature'. It is not a little instructive that so similar a law should hold the supreme position among the biological sciences. While it is possible that both may ultimately be absorbed by some more general principle, for the present we should note that the laws as they stand present profound differences—(1) The systems considered in thermodynamics are permanent; species on the contrary are liable to extinction, although biological improvement must be expected to occur up to the end of their existence. (2) Fitness, although measured by a uniform method, is qualitatively different for every different organism, whereas entropy, like temperature, is taken to have the same meaning for all physical systems. (3) Fitness may be increased or decreased by changes in the environment, without reacting quantitatively upon that environment. (4) Entropy changes are exceptional in the physical world in being irreversible, while irreversible evolutionary changes form no exception among biological phenomena. Finally, (5) entropy changes lead to a progressive disorganization of the physical world, at least from the human standpoint of the utilization of energy, while evolutionary changes are generally recognized as producing progressively higher organization in the organic world.

The statement of the principle of Natural Selection in the form of a theorem determining the rate of progress of a species in fitness to survive (this term being used for a well-defined statistical attribute of the population), together with the relation between this rate of progress and its standard error, puts us in a position to judge of the validity of the objection which has been made, that the principle of Natural Selection depends on a succession of favourable chances. The objection is more in the nature of an innuendo than of a criticism, for it depends for its force upon the ambiguity of the word chance, in its popular uses. The income derived from a Casino by its proprietor may, in one sense, be said to depend upon a succession of favourable chances, although the phrase contains a suggestion of improbability more appropriate to the hopes of the patrons of his establishment. It is easy without any very profound logical analysis to perceive the difference between a succession of favourable deviations from the laws of chance, and on the other hand, the continuous and cumulative action of these laws. It is on the latter that the principle of Natural Selection relies.

In addition to the genetic variance of any measurable character there exists, as has been seen, a second element comprised in the total genotypic variance, due to the heterozygote being in general not equal to the mean of the two corresponding homozygotes. This component, ascribable to dominance, is also in a sense capable of exerting evolutionary effects, not through any direct effect on the gene ratios, but through its possible influence on the breeding system. For if, in general, heterozygotes were favoured as compared with homozygotes, it is evident that the offspring of outcrosses would be at an advantage compared with those of matings between relatives, or of self-fertilization, and any heritable tendencies favouring such matings might come to be eliminated, with consequent increase in the proportion of heterozygotes.

This indirect and conditional factor in selection seems to have been able to produce effects of considerable importance, such as the separation of the sexes, self-sterility in many plants, and flowers made attractive by colour, scent and nectar. A first step to the understanding of these effects of dominance has been made in Chapter III, but the author would emphasize that in his opinion no satisfactory selective model has been set up competent even to derive a distylic species like the primrose from a monostylic species of the same genus. Possibly, therefore, the course of evolutionary change has been complex and circuitous.

Such effects ascribable to the dominance component of the genotypic variation are not in reality additional to the evolutionary changes accounted for by the fundamental theorem; for in that theorem they are credited to the gene-substitutions needed, for example, to develop bigger or brighter flowers; although the selective advantage conferred by these may be wholly due to dominance deviations in fitness recognizable in numerous other factors.

The nature of adaptation

In order to consider in outline the consequences to the organic world of the progressive increase of fitness of each species of organism, it is necessary to consider the abstract nature of the relationship which we term 'adaptation'. This is the more necessary since any *simple* example of adaptation, such as the lengthened neck and legs of the giraffe as an adaptation to browsing on high levels of foliage, or the conformity in average tint of an animal to its natural background, lose, by the very simplicity of statement, a great part of the meaning which the word really conveys. For the more complex the adaptation, the more numerous the different features of conformity, the more essentially adaptive the situation is recognized to be. An organism is regarded as adapted to a particular situation, or to the totality of situations which constitute its environment, only in so far as we can imagine an assemblage of slightly different situations, or environments, to which the animal would on the whole be less well adapted; and equally only in so far as we can imagine an assemblage of slightly different organic forms, which would be less well adapted to that environment. This I take to be the meaning which the word is intended to convey, apart altogether from the question whether organisms really are adapted to their environments, or whether the structures and instincts to which the term has been applied are rightly so described.

21

Reprinted from *Cambridge Phil. Soc. Proc.* **27**:131–136 (1930)

A MATHEMATICAL THEORY OF NATURAL AND ARTIFICIAL SELECTION. PART VII. SELECTION INTENSITY AS A FUNCTION OF MORTALITY RATE.

J. B. S. Haldane

The assumption is often made that when competition is extremely intense at any stage in a life cycle, natural selection is bound to be intense also. This assumption will be examined quantitatively and it will be shown that the intensity of selection may diminish and become negative at high rates of elimination, while at its best its increase is extremely slow.

The intensity of competition is measured by the ratio, z, of organisms eliminated, to survivors. This may be small, e.g. $z = 0{\cdot}1$ or less for the period between birth and maturity in civilised human societies. It may exceed 10^6 in marine organisms producing many million eggs per year, or spermatozoa of which 10^9 are ejaculated at a time. But in few cases can it exceed 10^{12}.

Confining ourselves for the moment to a population consisting of two types A and B, the intensity of selection is measured by the coefficient of selection k, where the ratio of A to B is increased $1 + k$ times as the result of selection. k is taken to be small throughout the argument.

Consider a character whose measure x is normally distributed, according to Gauss' law, in the A and B groups, the standard deviation being the same in each, and the differences between the means and the standard deviations being small in comparison with the standard deviations. For example, Johansen (1926) found the mean breadths of 8·091 and 8·152 mm. and standard deviations of ·400 and ·405 mm. in two lines, BB and GG, of beans, the difference being clearly significant in the first case, doubtfully so in the second. If all individuals in which the variate x falls below a certain value are eliminated by selection, we can readily calculate the proportion of the whole population eliminated, and the proportion of A to B among the survivors.

Conditions are not grossly dissimilar under natural selection. We may imagine a variate, to be called viability, which is normally distributed and such that only those individuals possessing more than a certain viability survive. The large size of its standard deviation compared to the difference of the mean values would signify the relatively large part played by chance in natural selection. The best studied case is that of pollen-tube growth, described by Buchholz and Blakeslee (1929). Here those tubes which arrive first at the ovules are selected. The distribution of growth rates is definitely skew, but the skewness is not likely to affect the general character of the result if the two types compared are sufficiently similar. Where viability depends on a greater variety of accidental causes, as is generally the case, the distribution is likely to be more normal.

Without loss of generality we can put the initial numbers of A and B equal, and take the mean value of x as zero and its standard deviation as unity. We suppose the mean value of x for the A type to be λ and its standard deviation to be $1+\mu$, the corresponding values for B being $-\lambda$ and $1-\mu$. The ratio of the frequency of any value of x in the population to that in a strictly normal population is $1-(1+x^2)(\lambda+\mu x)^2+$ higher powers of λ and μ. Hence the mixed population is normal to the second order of small quantities provided that λx and μx^2 are small. x will rarely exceed 7 even in a population of 10^{12}.

Then provided that the population is numerous compared with both the numbers surviving and eliminated, the survivors will be those members for which $x > X$, X being given by

$$\frac{1}{\sqrt{2\pi}}\int_X^\infty e^{-\frac{1}{2}x^2}\,dx = \frac{1}{z+1}.$$

The proportion of the A type exceeding this value is

$$\frac{1}{\sqrt{2\pi}\,(1+\mu)}\int_X^\infty e^{\frac{-(x-\lambda)^2}{2(1+\mu)^2}}\,dx = \frac{1}{\sqrt{2\pi}}\int_{(X-\lambda)/(1+\mu)}^\infty e^{-\frac{1}{2}t^2}\,dt$$

$$= \frac{1}{\sqrt{2\pi}}\int_X^\infty e^{-\frac{1}{2}t^2}\,dt + \frac{1}{\sqrt{2\pi}}\int_{(X-\lambda)/(1+\mu)}^X e^{-\frac{1}{2}t^2}\,dt$$

$$= \frac{1}{z+1} + \frac{(\lambda+\mu X)}{\sqrt{2\pi}}e^{-\frac{1}{2}X^2}, \text{ approximately,}$$

so that

$$k = \frac{2(\lambda+\mu X)(z+1)}{\sqrt{2\pi}}e^{-\frac{1}{2}X^2},$$

the value of X being found as above.

First consider the case when $\mu = 0$, i.e. the standard deviations are equal. The value of $q = k/2\lambda$ is plotted against $\log_{10} z$ in the figure (calculated from Pearson's (1924) tables). When $z = 1$, $q = (2/\pi)^{\frac{1}{2}} = \cdot 798$. When z is large, we may put

$$\frac{1}{1+z} = \frac{e^{-\frac{1}{2}X^2}}{\sqrt{2\pi}\,X} \text{ approximately,}$$

whence

$$q = \sqrt{\log_e \frac{z^2}{2\pi}} \text{ approximately.}$$

So the intensity of selection only increases extremely slowly with z. Thus q is only doubled when z increases from 1 to about 6·4, or from 10 to 1800, and only increased 9 times over the whole range from 1 to 10^{12}. On the other hand when z is small, q approximates to $z\sqrt{-\log_e 2\pi z^2}$, a small quantity of the order of z, and is roughly proportional to z over small ranges. For example, when $z = 10^{-4}$, $q = \cdot 0004$, and when $z = 10^{-2}$, $q = \cdot 03$. To sum up, the efficiency of selection increases very rapidly with z until about 80 °/ₒ of the population is eliminated, and thereafter very slowly.

The only experimental data known to me are those of Correns (1918) who measured the sex-ratio of *Melandrium* when pollinated with mixtures of male-producing and female-producing pollen, and used numbers of pollen-grains either less than that of ovules, so that $z = 0$, or greater, in various proportions. When z approximated to 6, k was 0·195; when z was about 142, k rose to ·710. The value of k thus increased only 3·6 times while z increased 24 times. According to Fig. 1 the increase of q should be only 1·8 times. But the values of both z and k are very uncertain, thus the value ·195 of k has a standard error of ·07. Figures well within the limit of experimental error would give complete agreement with the theory. Moreover μ is probably not zero nor is the distribution of growth rates normal. Certainly, however, k does not increase anything like proportionally to z, even when λ has the somewhat large value of 0·1, which as we shall see later will tend to exaggerate the rate of increase of k with z.

When μ is not zero the case is rather more complicated. If λ and μ have the same sign, i.e. the type with the largest mean has also the largest standard deviation, selection favours them unless X is negative and less than $-\lambda/\mu$. In this case the group of lower average viability will be favoured when competition is very slight, but their selective advantage will be extremely small at best. For example, Johansen's bean line *GG* had a mean breadth of 8·152 mm. with standard deviation ·415, while the corresponding figures for the line *MM* were 7·976 and ·348. Hence $\lambda = \cdot 101$ and $\mu = \cdot 076$. Selection for greater breadth would favour line *MM* slightly when z was less than 0·1, while for higher values *GG* would be considerably favoured, k being ·159 when $z = 1$.

In many cases the coefficient of variation of the two groups is approximately equal, i.e. $\mu = \lambda$. In four of Johansen's pure lines the coefficient of variation for length only varied between 5·0 °/ₒ and 4·4 °/ₒ. In a family described on p. 136 two slightly impure genotypes had coefficients of 7·0 °/ₒ and 6·8 °/ₒ, the heterozygote a coefficient of 6·8 °/ₒ. It thus seems likely that in a large number of cases λ and μ will be very nearly equal. When this is the case q is small and negative for low mortalities, attaining a minimum value of $-0{\cdot}070$ when $z = \cdot 066$, i.e. with a mortality of 6·2 °/ₒ, vanishes when $z = \cdot 1886$, i.e. with a mortality of 15·9 °/ₒ, and then increases, being ·798 when $z = 1$ (50 °/ₒ mortality), 23·4 when $z = 10^6$, and 57·6 when $z = 10^{12}$. For large values of z, k varies as $\log z$.

Even though λ and μ are not quite equal, in a very large number, perhaps the majority of cases, λ/μ will lie between ·5 and 2, and the direction of selection will be reversed at a mortality of between 31 °/₀ and 2·3 °/₀. It is of interest to note that during the last fifty

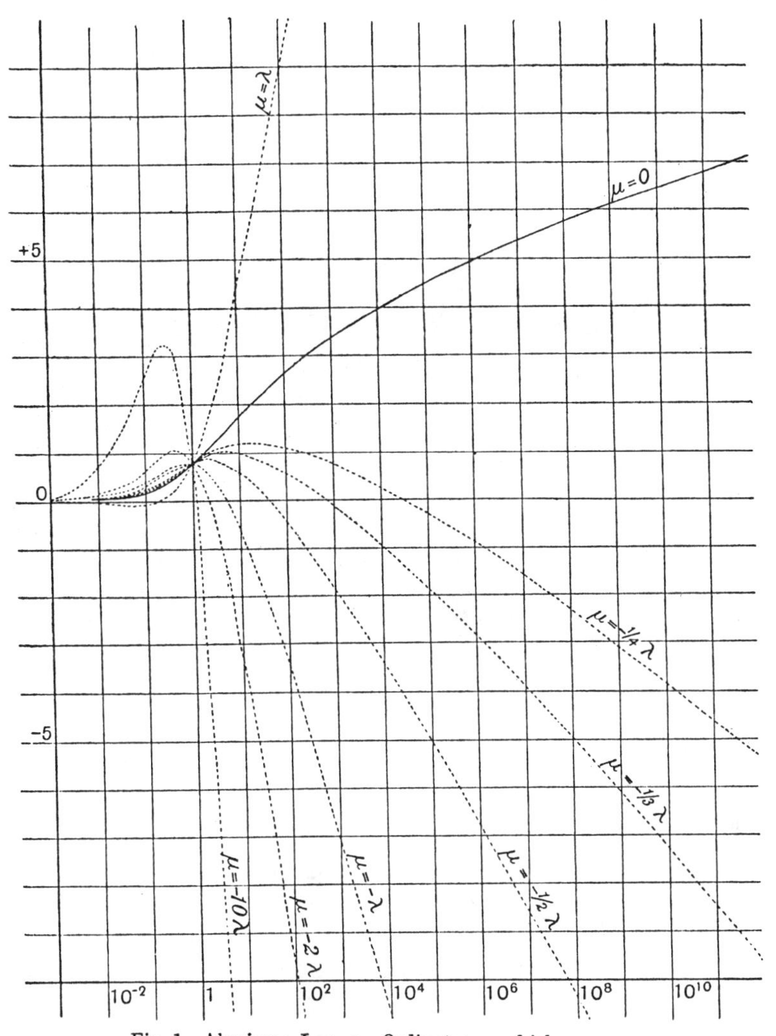

Fig. 1. Abscissa: $\text{Log}_{10} z$. Ordinate: q, which measures the intensity of selection.

years infantile mortality in most civilised countries has fallen from well above the critical value of 15·9 °/₀ to well below it. It seems probable, therefore, that the direction of selection for certain genes has been reversed.

If λ and μ have opposite signs, the group of highest average viability will be favoured until $\bar{X}$ exceeds $-\lambda/\mu$, i.e. until

$$z > \frac{\sqrt{2\pi}}{\int_{-\lambda/\mu}^{\infty} e^{-\frac{1}{2}x^2}\,dx} - 1.$$

If $-\lambda/\mu$ is fairly large, say greater than 3, this becomes approximately

$$z > \sqrt{2\pi}\,\frac{\lambda}{\mu}\,e^{\lambda^2/\mu^2}.$$

In practice however since z rarely exceeds 10^{12}, the direction of selection is not likely to be reversed if $-\mu < \lambda/7$.

In the figure q is plotted against z when

$$\mu = 0,\ \lambda,\ -10\lambda,\ -2\lambda,\ -\tfrac{1}{2}\lambda,\ -\tfrac{1}{3}\lambda,\ -\tfrac{1}{4}\lambda.$$

The maximum intensity of selection is reached when z is greater or less than unity according as $-\mu/\lambda$ is less or greater than unity.

But wherever it is not zero the results of slight and intense competition are in opposite directions, although the required competition may sometimes be too intense or the selection too slight to be of practical importance. This is in full accordance with the views of Bidder, who points out that, where "cataclasms" occasionally destroy the vast majority of a species, characters which are useless or worse under normal conditions may be selected. He specially mentions the case of a violent or erratic response of an animal by migration or otherwise to unfavourable environments, which would be likely to lower the average viability, but increase its dispersion.

It is easy to extend the above arguments to a population consisting of many genotypes. To take one example, suppose that $\mu = 0$, but λ is normally distributed with a standard deviation σ. Then the new frequency of any value of λ will be given by

$$df = \frac{1 + q\lambda}{\sqrt{2\pi}\,\sigma}\,e^{-\frac{1}{2}\lambda^2/\sigma^2}\,d\lambda.$$

The new mean value of λ will therefore be

$$\frac{\int_{-\infty}^{\infty} \lambda(1 + q\lambda)\,e^{-\frac{1}{2}\lambda^2/\sigma^2}\,d\lambda}{\int_{-\infty}^{\infty} (1 + q\lambda)\,e^{-\frac{1}{2}\lambda^2/\sigma^2}\,d\lambda}$$

$$= \frac{q}{\sqrt{2\pi}\,\sigma}\int_{-\infty}^{\infty} \lambda^2 e^{-\frac{1}{2}\lambda^2/\sigma^2}\,d\lambda, \text{ approximately,}$$

$$= q\sigma^2.$$

All the results here given apply only to the results of a single act of selection. The way in which the population will change depends on the way in which the mean viability and its dispersion are inherited, and on the system of mating. The effects of these have been considered in former papers of this series.

The theory can readily be extended to cover cases where λ and μ are no longer small, but the results are no longer elegant or simple. In particular the proportion of types in the original population must be taken into account. When the difference of the means is large compared with the standard deviations, A being more viable than B, it is convenient to take κ as the intensity of selection, where u is the ratio of A to B before, and ue^{κ} after selection. κ and k are of course equal when both are small. It is clear that

$$\kappa = \log_e \left(\frac{1+z}{1+uz}\right),$$

approximately, so that the intensity of selection is proportional to z when this is small, but becomes very large when $z = 1/u$. Such intense selection occasionally occurs in nature, for example between normal types and semi-lethal mutants, but its results as between competing types of organism would be very rapid, and it is not of much interest in a study of evolution. In general when λ is not small, the value of q for any value of z will be increased.

REFERENCES.

G. P. Bidder (1930). "The importance of cataclasms in evolution." *Nature*, **125**, p. 783.

J. T. Buchholz and A. F. Blakeslee (1929). "Pollen-tube growth in crosses between balanced chromosomal types of *Datura Stramonium*." *Genetics*, **14**, p. 538.

C. Correns. "Fortsetzung der Versuche zur experimentellen Verschiebung des Geschlechtsverhältnisses." *Sitzber. K. Preuss. Akad. Wiss.* 1918, p. 1175.

W. Johansen (1926). *Elemente der exakten Erblichkeitslehre*, p. 171.

K. Pearson (1924). *Tables for Statisticians and Biometricians.*

22

Reprinted from *Cambridge Phil. Soc. Proc.* **27**:137–142 (1930)

A MATHEMATICAL THEORY OF NATURAL SELECTION. PART VIII. METASTABLE POPULATIONS

J. B. S. Haldane

Almost every species is, to a first approximation, in genetic equilibrium; that is to say no very drastic changes are occurring rapidly in its composition. It is a necessary condition for equilibrium that all new genes which arise at all frequently by mutation should be disadvantageous, otherwise they will spread through the population. Now each of two or more genes may be disadvantageous, but all together may be advantageous. An example of such balance has been given by Gonsalez(1). He found that, in purple-eyed *Drosophila melanogaster*, arc wing or axillary speck (each due to a recessive gene) shortened life, but the two together lengthened it.

Consider the case of two dominant genes A, B, where the relative chances of producing offspring by the four phenotypes are as follows: AB 1, aaB $1-k_1$, Abb $1-k_2$, $aabb$ $1+K$. k_1 and k_2 are small and positive. K is small, and if negative its absolute value is less than k_1 or k_2.

Consider a random mating population where in the nth generation the genic ratios are $u_n A:1a$; $v_n B:1b$.

Then

$$u_{n+1}=\frac{(u_n^2+u_n)\{1-k_2(1+v_n)^{-2}\}}{u_n\{1-k_2(1+v_n)^{-2}\}+1+\{K-k_1(v_n^2+2v_n)\}(1+v_n)^{-2}},$$

whence $$\Delta u_n=\frac{u_n\{k_1(1+v_n)^2-K-k_1-k_2\}}{(1+u_n)(1+v_n)^2}, \text{ approximately.}$$

So, taking a generation as the unit of time,

$$\frac{du}{dt}=\frac{u\{k_1(1+v)^2-K-k_1-k_2\}}{(1+u)(1+v)^2}, \text{ approximately.}$$

Let $x=1/(1+u)$ (the proportion of recessive genes) and $y=1/(1+v)$, so that $1>x>0$, $1>y>0$.

Then $$\frac{dx}{dt}=x^2(1-x)[(K+k_1+k_2)y^2-k_1].$$

Similarly $$\frac{dy}{dt}=y^2(1-y)[(K+k_1+k_2)x^2-k_2].$$

Clearly $x=0$, $y=0$; and $x=1$, $y=1$ are the only stable equilibria, though Fisher (2) appears to regard a mixed population as stable in such a case. Putting

$$\frac{k_1}{K+k_1+k_2}=a^2, \quad \frac{k_2}{K+k_1+k_2}=b^2,$$

we have
$$\frac{dy}{dx}=\frac{y^2(1-y)(x^2-a^2)}{x^2(1-x)(y^2-b^2)}.$$

So
$$\int_{y_0}^{y}\frac{(s^2-b^2)\,ds}{s^2(1-s)}=\int_{x_0}^{x}\frac{(s^2-a^2)\,ds}{s^2(1-s)},$$

whence
$$f(y,b)-f(x,a)=c=f(y_0,b)-f(x_0,a),$$

where x_0, y_0 represent the initial conditions, and

$$f(x,a)=a^2/x-a^2\log x+(a^2-1)\log(1-x).$$

Each value of $f(y_0,b)-f(x_0,a)$ determines a trajectory passing to $(0, 0)$ or $(1, 1)$, which represent populations composed entirely of double dominants or double recessives respectively. The minimum value of $f(x, a)$ occurs when $x=a$ and is

$$a-a^2\log a+(a^2-1)\log(1-a),$$

and $f(x, a)$ is always real and positive, becoming infinite when $x=0$ or 1. If $c>f(b,b)-f(a,a)$, there are two values of x corresponding to each value of y, but some values of x are excluded. Hence the trajectories fall into four families divided by the two branches of the curve whose equation is

$$f(y,b)-f(x,a)=f(b,b)-f(a,a).$$

This consists of two trajectories running from $(0, 1)$ and $(1, 0)$ to (a, b) and two from (a, b) to $(0, 0)$ and $(1, 1)$. These are represented by the dotted lines in Fig. 1, where $a=\frac{1}{2}$, $b=\frac{1}{4}$. The former divides the whole area into two portions. Populations in the one tend to the values $x=y=0$, in the other to the values $x=y=1$. Some examples of trajectories are given. It is clear that a population consisting mainly of *AABB* or *aabb* tends, as the result of selection, to return to those compositions. If the signs of K, k_1, and k_2 be changed, the same trajectories will be described in the reverse direction.

If the original population is *AABB*, the factors *A* and *B* will generally have a small tendency to mutate to *a* and *b* respectively. Let p_1 and p_2 be the probabilities that *A* will mutate to *a* and *B* to *b* in the course of a generation. These appear to be generally small numbers of the order of 10^{-6} or less. The population is in equilibrium when $x=p_1/k_1$, $y=p_2/k_2$ (Haldane (3)). In general x will be much smaller than a, and y than b, but from time to time chance fluctuations may isolate a population where this is no longer the case. Its representative point will lie in the area whose stable type is *aabb*, and the whole population will be transformed into this type, apart from rare exceptions due to back mutation. In such a population modifying factors will be selected in such a way as to increase the viability of the *aabb* type, i.e. the value of K. But even so it may be expected to be swamped by hybridisation on coming into contact with the original *AABB* population, unless one of two things has happened.

aabb may possess or develop characters which render mating with *AABB* rare. For example, it may have a different flowering time if a plant, or a different psychology if an animal. In this case the species will divide into two. Or chromosome changes may occur to cause close linkage of A and B when the populations are crossed. Thus if the loci of A and B are in the same chromosome an inversion of the portion containing them will lead to their behaving as a single factor on crossing. In this case if K is positive the whole species will be transformed into the type *aabb*. A species which is liable to transformations of this kind may be called metastable. Possibly metastability is quite a general phenomenon, but it is only rarely that the circumstances arise which favour a change of the type considered.

In a population which is mainly self-fertilised, conditions are probably more favourable. Were self-fertilisation universal, the proportion of *aaBB* zygotes, when mutation and selection were in equilibrium, would be $\frac{1}{2}p_1/k_1$. So that of *aabb* would be $p_1p_2/(4k_1k_2)$ or less. This is presumably a small number, probably of the order 10^{-9}, and when such individuals occur, they will generally be wiped out by chance. But their probability of spreading through the population, though small, will be finite, and roughly equal to $2K$ (Haldane (3)). Hence, within a geologically short period we may expect evolution to occur in such cases.

The theory may be extended in two different ways. We may consider m genes. In this case any population can be represented by a point in m-dimensional space, all populations being represented by the points of a regular orthotope, or hypercube. Each of the 2^m apices of this figure represents a homozygous population. Clearly the condition for stability of any such population is that no change in a *single* factor should yield a more viable type. In other words, no adjacent apices can both represent stable populations. The maximum number of stable populations is thus 2^{m-1}, represented by the vertices of the polytope arising from the omission of alternate vertices of the regular orthotope. This is not regular but only semi-regular if $m > 3$. In general the numbers of stable genotypes will be much smaller than this, and may not exceed 1.

If there is more than one stable population the orthotope is divided into two or more regions analogous to the two areas of Fig. 1. A population in any given region tends to the same point of stable equilibrium. The regions are separated by a variety (surface or hyper-surface) of $m-1$ dimensions. If we take as our variables x_1, x_2, x_3, etc. not the proportions of recessive genes, but their squares, i.e. the proportion of recessive zygotes, we have

$$\frac{dx_1}{dt} = x_1^{\frac{3}{2}}(1 - x_1^{\frac{1}{2}}) f_1(x_2, x_3, x_4, \ldots),$$

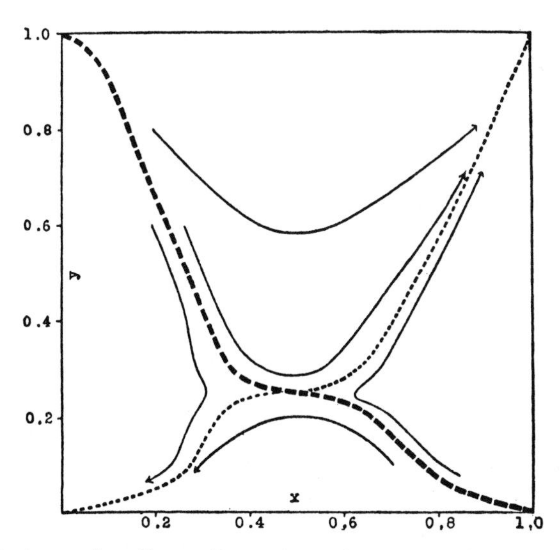

Fig. 1. Abscissa and ordinate. Proportions of genes *a* and *b* in a population. Trajectories of points representing populations are represented by continuous lines, and boundaries between families of trajectories by dotted lines.

where $f_1(x_2, x_3, x_4, \ldots)$ is linear in each of x_2, x_3, etc. and has 2^{m-1} constant coefficients; and $m-1$ similar equations. The $(m-1)$-dimensional space defined by x_2, x_3, etc. is thus divided into two regions, in one of which x_1 increases with time, whilst it diminishes in the other. These are not necessarily connected, as is obvious in the case where there are only three variables, and $f_1(x_2, x_3)$ may define a hyperbola which divides the unit square into three regions, in two of which dx_1/dt has the same sign. Hence in the course of a trajectory dx_1/dt may change sign several times. I have been unable to obtain the general equation for the trajectories or for the boundaries of the regions in which they lie.

So far we have only considered cases of complete dominance. If the heterozygotes are exactly intermediate in viability between the corresponding homozygous types, we have, in the terminology of the case first considered,

$$\frac{dx}{dt} = \tfrac{1}{2}x(1-x)\{(K+k_1+k_2)y-(K+k_1)\},$$

$$\frac{dy}{dt} = \tfrac{1}{2}y(1-y)\{(K+k_1+k_2)y-(K+k_2)\}.$$

Thus

$$\frac{dy}{dx} = \frac{y(1-y)(x-a)}{x(1-x)(y-b)}, \quad \text{where } a = \frac{K+k_2}{K+k_1+k_2}, \quad b = \frac{K+k_1}{K+k_1+k_2}.$$

Hence
$$\frac{x^a(1-x)^{1-a}}{x_0{}^a(1-x_0)^{1-a}} = \frac{y^b(1-y)^{1-b}}{y_0{}^b(1-y_0)^{1-b}}.$$

By an argument similar to that used above we can show that the trajectories fall into two families, separated by one branch of the curve whose equation is

$$\left(\frac{x}{a}\right)^a\left(\frac{1-x}{1-a}\right)^{1-a} = \left(\frac{y}{b}\right)^b\left(\frac{1-y}{1-b}\right)^{1-b}.$$

The general case where heterozygotes are of any arbitrary viability is rather complicated. But where a heterozygote has a greater viability than any genotype differing from it in respect of a single gene only, there will be a stable population including some of these heterozygotes. Thus if *aabb* has a viability $1+K$, *AABb* of $1+K_2$, all other genotypes having unit viability,

$$\frac{dx}{dt} = xy(1-x)\{K_1xy - K_2(1-x)(1-y)\},$$

$$\frac{dy}{dt} = y\{K_1x^2y - K_2(1-x)^2(2y-1)\}.$$

The stable equilibria are at $x=1$, $y=1$ and $x=0$, $y=\frac{1}{2}$. But I have not been able to integrate these equations, since the variables are not readily separable. Nevertheless it is clear that the trajectories fall into two groups bounded by a curve passing through $(0, 1)$ and $\left(\frac{K_2}{K_1+2K_2}, \frac{1}{2}\right)$.

In the case of m genes, if heterozygotes have an advantage as such there may be points of stable equilibrium anywhere in the m-dimensional space, but it seems fairly clear that their number cannot exceed 2^{m-1}.

It is suggested that in many cases related species represent stable types such as I have described, and that the process of species formation may be a rupture of the metastable equilibrium. Clearly such a rupture will be specially likely where small communities are isolated. I have to thank Mr C. H. Waddington for calculating and drawing the figure.

REFERENCES.

(1) B. M. GONSALEZ, *Am. Nat.*, 57, p. 289 (1923).
(2) R. A. FISHER, *The Genetical Theory of Natural Selection*, p. 102 (1930).
(3) J. B. S. HALDANE, *Proc. Camb. Phil. Soc.*, 23, p. 838 (1927).

23

Reprinted from *Cambridge Phil. Soc. Proc.* **28**:244–248 (1932)

A MATHEMATICAL THEORY OF NATURAL AND ARTIFICIAL SELECTION. PART IX. RAPID SELECTION

J. B. S. Haldane

In Part I of this series (1) it was proved that in a random mating population in which the ratio between the numbers of two genes in the nth generation was u_n, and the proportion of recessives therefore $(u_n + 1)^{-2}$, the coefficient of selection being k, then

$$u_{n+1} = \frac{u_n (u_n + 1)}{u_n + 1 - k}. \tag{1}$$

Hence, if $x_n = 1 + 1/u_n$, we have

$$x_{n+1} = x_n - \frac{k (x_n - 1)^2}{x_n}.$$

It has since been shown (2) that, if y is a given function of x, $y_r = \left(\frac{d}{dx}\right)^r y$, and $x_{n+1} = x_n + ky$, then

$$n = \int_{x_0}^{x_n} \Big[k^{-1} y^{-1} + \tfrac{1}{2} y^{-1} y_1 - \frac{k}{12} (y^{-1} y_1^2 + y_2) + \frac{k^2}{24} (y^{-1} y_1^3 + 2 y_1 y_2) - \frac{k^3}{720} (19 y^{-1} y_1^4 + 59 y_1^2 y_2 + y y_2^2 - 2 y y_1 y_3 - y^2 y_4) + \dots \Big] dx,$$

provided that y is regular and does not vanish in the interval considered, and that the series converges uniformly. It was also pointed out that the coefficient of $y^{-1} y_1^r$ is the $(r+1)$th term in the expansion of $\{\log (1 + k)\}^{-1}$.

In our case $y = -(x - 1)^2/x$, hence

$$n = \int_{x_0}^{x_n} \Big[\frac{-x}{k (x - 1)^2} + \frac{x + 1}{2x (x - 1)} + \frac{k (x^2 + 2x + 3)}{12 x^3} + \frac{k^2 (x^2 - 1)(x^2 + 2x + 5)}{24 x^5} + \frac{k^3 (x - 1)^2 (19 x^4 + 76 x^3 + 220 x^2 + 360 x + 105)}{720 x^7} + \dots \Big] dx$$

$$= C + \frac{1}{k} \left[\frac{1}{x_n - 1} - \log (x_n - 1) \right] + \tfrac{1}{2} \Big[2 \log (x_n - 1) - \log x_n \Big] + \frac{k}{12} (\log x_n - 2 x_n^{-1} - \tfrac{3}{2} x_n^{-2}) + \frac{k^2}{24} (\log x_n - 2 x_n^{-1} - 2 x_n^{-2} + \tfrac{2}{3} x_n^{-3} + \tfrac{5}{4} x_n^{-4}) + \frac{k^3}{720} (19 \log x_n - 38 x_n^{-1} - \tfrac{87}{2} x_n^{-2} + \tfrac{4}{3} x_n^{-3} + \tfrac{395}{4} x_n^{-4} - 30 x_n^{-5} - \tfrac{35}{2} x_n^{-6}) + \dots, \tag{2}$$

where C does not vary with n. Now since $x_n > 1$, the error due to neglecting all terms except the logarithmic terms and $\frac{1}{k(x_n-1)}$ is very small when x_n and x_0 are large, and is always less than

$$\frac{7k}{24}+\frac{25k^2}{288}-\frac{347k^3}{8640}+\ldots.$$

Although I have been unable to prove that this series converges, it will be seen that its sum can be neglected without serious error, even when $|k|>1$. The remaining terms, when $1>k\geqslant -1$, give

$$n=C+\frac{1}{k(x_n-1)}+\frac{\log x_n}{\log(1-k)}+\frac{1-k}{k}\log\frac{x_n}{x_n-1}, \text{ approximately.}$$

This summation is only justified in theory if $1>k\geqslant -1$, but it will be seen later that it holds even when $k<-1$. Since $x_n=1+1/u_n$, we can write the above equation as

$$n=\frac{u_n-u_0}{k}+\log\left(\frac{1+1/u_n}{1+1/u_0}\right)\Big/\{\log(1-k)\}+\frac{1-k}{k}\log\left(\frac{1+u_n}{1+u_0}\right). \quad (3)$$

Though series (2) is doubtless more accurate, equation (3) is quite satisfactory for ordinary purposes. k cannot exceed 1, and the equation is exact when $k=1$, or is infinitesimal, and approximate for intermediate values. When k is negative it is also approximate. It can also be shown to attain any desired degree of accuracy if both u_n and u_0 are sufficiently large or small. Its accuracy can readily be tested by substituting values of u_0 and u_1 calculated from equation (1) for any values of u_0 and k, and then calculating n, which should equal unity were equation (3) exact. If $k=\frac{1}{2}$,

$u_0=1$, $u_1=\frac{4}{3}$, $n=1{\cdot}0135$,

$u_0=100$, $u_1=100{\cdot}4975$, $n=1{\cdot}0011$,

$u_0={\cdot}01$, $u_1={\cdot}019804$, $n=1{\cdot}0032$.

Hence in this case the error probably does not exceed 2 °/₀. If $k=0{\cdot}9$,

$u_0=1$, $u_1=1{\cdot}\dot{8}\dot{1}$, $n=0{\cdot}9670$,

$u_0=100$, $u_1=100{\cdot}8991$, $n=1{\cdot}0068$,

$u_0={\cdot}01$, $u_1={\cdot}09\dot{1}\dot{8}$, $n=1{\cdot}0277$.

The error is thus under 4 °/₀. Finally, if $k=-4$,

$u_0=1$, $u_1={\cdot}\dot{3}$, $n=1{\cdot}1042$,

$u_0=100$, $u_1=96{\cdot}1905$, $n=0{\cdot}9606$,

$u_0={\cdot}01$, $u_1={\cdot}002016$, $n=1{\cdot}0361$.

The error here exceeds 10 °/₀ over a certain range of values of u. But it is to be noted that these values of k refer to extremely intense selection. Thus, when $k = -4$, five recessives survive for every dominant, and when $k = 0{\cdot}9$, ten dominants for every recessive. Such intense selection can hardly ever, in the course of evolution, have been the direct cause of large changes in a population.

Elton (3) has expressed the opinion that occasional intense selection, for example during periodic famines and plagues, may be more efficient than less intense selection acting in every generation. This view can be examined quantitatively. Consider two populations, in the first of which selection of intensity k occurs in every generation, whilst in the second selection occurs with intensity mk in every mth generation, where $|mk| < 1$. Then from equation (2) it follows that in the first population the number of generations needed to change u_0 to u_n is approximately

$$n = \frac{1}{k}(u_n - u_0 + \log u_n - \log u_0) - \tfrac{1}{2}[\log(u_n^2 + u_n) - \log(u_0^2 + u_0)],$$

and in the second population, approximately,

$$n' = \frac{1}{k}(u_n - u_0 + \log u_n - \log u_0) - \frac{m}{2}[\log(u_n^2 + u_n) - \log(u_0^2 + u_0)],$$

assuming that squares of km can be neglected. Since $m > 1$, the second time is shorter than the first if $u_n(u_n + 1) > u_0(u_0 + 1)$, regardless of the sign of k, i.e. if $u_n > u_0$. Thus when selection is favouring dominants, it is more efficient if concentrated in a series of cataclysms, but when it is favouring recessives the opposite is the case. But unless $|mk|$ is fairly large the difference is unimportant.

We can also compare the time taken, with selection of the same intensity, to change u from a to b, when dominants are favoured (k positive), with the time taken for the change from b to a when recessives are favoured (k negative). From equation (3), putting $u_0 = a$, $u_n = b$, the time needed for the first change is

$$n = \frac{b-a}{k} + \log\left(\frac{1+1/b}{1+1/a}\right)\Big/\{\log(1-k)\} + \frac{1-k}{k}\log\left(\frac{1+b}{1+a}\right).$$

Changing the sign of k, and putting $u_0 = b$, $u_n = a$, we find that the time needed for the second change is

$$n' = \frac{b-a}{k} - \log\left(\frac{1+1/b}{1+1/a}\right)\Big/\{\log(1+k)\} + \frac{1+k}{k}\log\left(\frac{1+b}{1+a}\right).$$

Hence

$$n' - n = \left[\frac{-\log(1-k) - \log(1+k)}{\log(1-k)\log(1+k)}\right]\log\left(\frac{1+1/b}{1+1/a}\right) + 2\log\left(\frac{1+b}{1+a}\right).$$

This is positive, since k is positive and $b > a$. Hence on the above convention selection appears to be more effective when dominants are favoured. But it is illegitimate to regard a selection measured by $-k$ as the inverse of one measured by k unless both are very small. Thus when $k = \frac{1}{2}$, two dominants survive for every recessive, but when $k = -\frac{1}{2}$, only one-and-a-half recessives survive for every dominant. To obtain a fairer comparison we put $1 - k = e^{-\kappa}$, and change the sign of κ when selection is reversed. Equation (3) now becomes

$$n = \frac{u_n - u_0}{1 - e^{-\kappa}} - \frac{1}{\kappa} \log\left(\frac{1 + 1/u_n}{1 + 1/u_0}\right) + \frac{1}{e^{\kappa} - 1} \log\left(\frac{1 + u_n}{1 + u_0}\right). \quad (4)$$

Hence the time taken to change u from a to b when κ is positive is

$$n = \frac{b - a}{1 - e^{-\kappa}} - \frac{1}{\kappa} \log\left(\frac{1 + 1/b}{1 + 1/a}\right) + \frac{1}{e^{\kappa} - 1} \log\left(\frac{1 + b}{1 + a}\right),$$

and the time for the converse change is

$$n' = \frac{b - a}{e^{\kappa} - 1} - \frac{1}{\kappa} \log\left(\frac{1 + 1/b}{1 + 1/a}\right) - \frac{1}{1 - e^{-\kappa}} \log\left(\frac{1 + b}{1 + a}\right),$$

so that

$$n - n' = b - a + \log\left(\frac{1 + b}{1 + a}\right),$$

which is positive. Thus selection is more rapid if recessives are favoured. The difference is however only significant if b is large, that is to say recessives very rare in one of the populations considered. This result was to be expected, since if all dominants are killed off, i.e. $k = \kappa = -\infty$, selection is complete in one generation, whilst if all recessives are killed off, i.e. $k = 1$, $\kappa = \infty$, selection is a relatively slow process. The number κ occurring in equation (4) is the difference of the Malthusian parameters, as defined by Fisher (4), of the dominants and recessives. It approximates to k when both are small.

The problem which is here solved is the simplest, though perhaps the most important, of a large number. When selection acts at different rates in the two sexes, or when it acts on a sex-linked character, or one determined by several genes, or by one gene in a polyploid, we have to solve two or more simultaneous non-linear finite difference equations. When generations are not separate, we have, in general, to solve a set of at least four simultaneous non-linear integral equations. These equations have been stated in other parts of this series, and have been approximately solved when selection is not intense. But their complete solution is desirable for a discussion of problems raised by eugenics and artificial selection.

Summary.

Equations (3) and (4) describe the changes undergone by a Mendelian population mating at random, and under intense selection.

REFERENCES.

HALDANE (1). *Trans. Camb. Phil. Soc.* 23 (1924).

—— (2). *Proc. Camb. Phil. Soc.*, 28 (1932).

ELTON (3). *Ecology and Evolution.*

FISHER (4). *The Genetical Theory of Natural Selection.*

24

Reprinted from *Genetics* **16**:97-100, 155-159 (1931)

EVOLUTION IN MENDELIAN POPULATIONS

SEWALL WRIGHT
University of Chicago, Chicago, Illinois

Received January 20, 1930

TABLE OF CONTENTS

THEORIES OF EVOLUTION

One of the major incentives in the pioneer studies of heredity and variation which led to modern genetics was the hope of obtaining a deeper insight into the evolutionary process. Following the rediscovery of the Mendelian mechanism, there came a feeling that the solution of problems of evolution and of the control of the process, in animal and plant breeding and in the human species, was at last well within reach. There has been no halt in the expansion of knowledge of heredity but the advances in the field of evolution have, perhaps, seemed disappointingly small. One finds the subject still frequently presented in essentially the same form as before 1900, with merely what seems a rather irrelevant addendum on Mendelian heredity.

The difficulty seems to be the tendency to overlook the fact that the evolutionary process is concerned, not with individuals, but with the species, an intricate network of living matter, physically continuous in space-time, and with modes of response to external conditions which it appears can be related to the genetics of individuals only as statistical consequences of the latter. From a still broader viewpoint (compare LOTKA 1925) the species itself is merely an element in a much more extensive evolving pattern but this is a phase of the matter which need not concern us here.

The earlier evolutionists, especially LAMARCK, assumed that the somatic effects of physiological responses of individuals to their environments were transmissible to later generations, and thus brought about a directed evolution of the species as a whole. The theory remains an attractive one to certain schools of biologists but the experimental evidence from genetics is so overwhelmingly against it as a general phenomenon as to render it unavailable in present thought on the subject.

DARWIN was the first to present effectively the view of evolution as primarily a statistical process in which random hereditary variation merely furnishes the raw material. He emphasized differential survival and fecundity as the major statistical factors of evolution. A few years later, the importance of another aspect of group biology, the effect of isolation, was brought to the fore by WAGNER. Systematic biologists have continued to insist that isolation is the major species forming factor. As with natural selection, a connection with the genetics of individuals can be based on statistical considerations.

There were many attempts in the latter part of the nineteenth century to develop theories of direct evolution in opposition to the statistical viewpoint. Most of the theories of orthogenesis (for example, those of EIMER and of COPE) implied the inheritance of "acquired characters." NÄGELI postulated a slow but self contained developmental process within protoplasm; practically a denial of the possibility of a scientific treatment of the problem. Differing from these in its appeal to experimental evidence and from the statistical theories in its directness, was DE VRIES' theory of the abrupt origin of species by "mutations." A statistical process, selection or isolation, was indeed necessary to bring the new species into predominance, but the center of interest, as with Lamarckism, was in the physiology of the mutation process.

The rediscovery of Mendelian heredity in 1900 came as a direct consequence of DE VRIES' investigations. Major Mendelian differences were naturally the first to attract attention. It is not therefore surprising that the phenomena of Mendelian heredity were looked upon as confirming DE VRIES' theory. They supplemented the latter by revealing the possibilities of hybridization as a factor bringing about an extensive recombination of mutant changes and thus a multiplication of incipient species, a phase emphasized especially by LOTSY. JOHANNSEN'S study of pure lines was interpreted as meaning that DARWIN'S selection of small random variations was not a true evolutionary factor.

A reaction from this viewpoint was led by CASTLE, who demonstrated the effectiveness of selection of small variations in carrying the average of a stock beyond the original limits of variation. This effectiveness turned out to depend not so much on variability of the principal genes concerned as on residual heredity. As genetic studies continued, ever smaller differences were found to mendelize, and any character, sufficiently investigated, turned out to be affected by many factors. The work of NILSSON-EHLE, EAST, SHULL, and others established on a firm basis the multiple factor hypothesis in cases of apparent blending inheritance of quantitative variation.

The work of MORGAN and his school securely identified Mendelian heredity with chromosomal behavior and made possible researches which further strengthened the view that the Mendelian mechanism is the general mechanism of heredity in sexually reproducing organisms. The only exceptions so far discovered have been a few plastid characters of plants. That differences between species, as well as within them, are Mendelian, in the broad sense of chromosomal, has been indicated by the close parallelism between the frequently irregular chromosome behavior and the genetic phenomena of species crosses (FEDERLEY, GOODSPEED and CLAUSEN, etc.). Most of DE VRIES' mutations have turned out to be chromosome aberrations, of occasional evolutionary significance, no doubt, in increasing the number of genes and in leading to sterility of hybrids and thus isolation, but of secondary importance to gene mutation as regards character changes. As to gene mutation, observation of those which have occurred naturally as well as of those which MULLER, STADLER, and others have recently been able to produce wholesale by X-rays, reveals characteristics which seem as far as possible from those required for a directly adaptive evolutionary process. The conclusion nevertheless seems warranted by the present status of genetics that any theory of evolution must be based on the properties of Mendelian factors, and beyond this, must be concerned largely with the statistical situation in the species.

[*Editor's Note:* Material has been omitted at this point.]

SUMMARY

The frequency of a given gene in a population may be modified by a number of conditions including recurrent mutation to and from it, migration, selection of various sorts and, far from least in importance, mere chance variation. Using q for gene frequency, v and u for mutation rates to and from the gene respectively, m for the exchange of population with neighboring groups with gene frequency q_m, s for the selective advantage of the gene over its combined allelomorphs and N for the effective number in the breeding stock (much smaller as a rule than the actual number of adult individuals) the most probable change in gene frequency per generation may be written:

$$\Delta q = v(1 - q) - uq - m(q - q_m) + sq(1 - q)$$

and the array of probabilities for the next generation as $[(1-q-\Delta q)a + (q+\Delta q)A]^{2N}$. The contribution of zygotic selection (reproductive rates of aa, Aa and AA as $1-s^1:1-hs^1:1$) is $\Delta q = s^1q(1-q)[1-q+h(2q-1)]$. In interpreting results it is necessary to recognize that the above coefficients are continually changing in value and especially that the selection coefficient of a particular gene is really a function not only of the relative frequencies and momentary selection coefficients of its different allelomorphs but also of the entire system of frequencies and selection coefficients of non-allelomorphs. Selection relates to the organism as a whole and its environment and not to genes as such. The mutation rate to a gene (v) can usually be treated as of negligible magnitude assuming the prevalence of multiple allelomorphs.

In a population so large that chance variation is negligible, gene frequency reaches equilibrium when $\Delta q = 0$. Among special cases is that of opposing mutation rates $\left(q = \frac{v}{u+v}\right)$, of selection against both homozygotes $\left(q = \frac{1-h}{1-2h}\right)$, of mutation against genic selection $\left(q = 1 - \frac{u}{s}\right)$, of mutation against zygotic selection ($q = 1 - \frac{u}{hs^1}$ unless h approaches 0, when $q = 1 - \sqrt{\frac{u}{s}}$), of selection and migration $\left(q = 1 - \frac{m}{s}(1 - q_m)\right)$ or

$-\frac{mq_m}{s}$ if s is much greater than m, $q = q_m\left(1+\frac{s}{m}(1-q_m)\right)$ if s is much smaller than m, while the values $q=\sqrt{q_m}$ or $1-\sqrt{1-q_m}$ when $s=\pm m$ illustrate the intermediate case).

Gene frequency fluctuates about the equilibrium point in a distribution curve, the form of which depends on the relations between population number and the various pressures. The general formula in the case of a freely interbreeding group, assuming genic selection, is

$$y = Ce^{4Nsq}q^{4N[mq_m+v]-1}(1 - q)^{4N[m(1-q_m)+u]-1}.$$

The correlation between relatives is affected by the form of the distribution of gene frequencies through FISHER's "dominance ratio." It appears that this is less than 0.20 in small populations under low selection but may even approach 1 in large populations under severe selection against recessives.

In a large population in which gene frequencies are always close to their equilibrium points, any change in conditions other than population number is followed by an approach toward the new equilibria at rates given by the Δq's. Great reduction in population number is followed by fixation and loss of genes, each at the rate $1/4N$ per generation, where N refers to the new population number. This applies either in a group of monoecious individuals with random fertilization or, approximately, in one equally divided between males and females (9.6 percent instead of 12.5 percent, however, under brother-sister mating, $N=2$). More generally with an effective breeding stock of N_m males and N_f females, the rates of fixation and of loss are each approximately $(1/16N_m+1/16N_f)$ until mutation pressure at length brings equilibrium in a distribution approaching first the form $y=C(1-q)^{-1}$ with decay at rate u and ultimately $Cq^{-1}(1-q)^{-1}$. The converse process, great increase in the size of a long inbred population, is followed by a slow approach toward the new equilibrium at a rate dependent in the early stages on mutation pressure.

With respect to genes which are indifferent to selection, the mean frequency is always $q=v/(u+v)$. The variance of characters, dependent on such genes, is proportional (at equilibrium) to population number up to about $N=1/4u$. Beyond this, there is approach of variance to a limiting value.

In the presence of selection (s considerably greater than 2u) the mean frequency at equilibrium varies between approximate fixation of the favored genes ($q=1-u/s$) in large populations and approximate, if not complete, fixation of mutant allelomorphs ($q=v/(u+v)$) in small populations, the rate of change from one state to the other being the mutation rate (u). A consequence is a slow but increasing tendency to decline in vigor in inbred stocks, to be distinguished from the relatively rapid but soon completed fixation process, described above as occurring at rate 1/2N. The variance of characters in this as in the preceding case, is approximately proportional to population number up to a certain point (N less than 1/4s) and above this rapidly approaches a limiting value. Variance is inversely proportional to the severity of selection in large populations unless the selection is very slight but in small populations is little affected by selection unless the latter is very severe (s greater than 1/4N).

Evolution as a process of cumulative change depends on a proper balance of the conditions, which, at each level of organization—gene, chromosome, cell, individual, local race—make for genetic homogeneity or genetic heterogeneity of the species. While the basic factor of change—the infrequent, fortuitous, usually more or less injurious gene mutations, in themselves, appear to furnish an inadequate basis for evolution, the mechanism of cell division, with its occasional aberrations, and of nuclear fusion (at fertilization) followed at some time by reduction make it possible for a relat vely small number of not too injurious mutations to provide an extensive field of actual variations. The type and rate of evolution in such a system depend on the balance among the evolutionary pressures considered here. In too small a population (1/4N much greater than u and s) there is nearly complete fixation, little variation, little effect of selection and thus a static condition modified occasionally by chance fixation of new mutations leading inevitably to degeneration and extinction. In too large a freely interbreeding population (1/4N much less than u and s) there is great variability but such a close approach to complete equilibrium of all gene frequencies that there is no evolution under static conditions. Change in conditions such as more severe selection, merely shifts all gene frequencies and for the most part reversibly, to new equilibrium points in which the population remains static as long as the new conditions persist. Such evolutionary change as occurs is an extremely slow adaptive process. In a population of intermediate size (1/4N of the order of u) there is continual random shifting of gene frequencies and a consequent shifting of selection coefficients which leads to a relatively rapid, continuing, irreversible, and largely fortuitous, but not degenerative series of changes, even under static conditions. The rate is rapid only in comparison with the preceding cases, however, being limited by mutation pressure and thus requiring periods of the order of 100,000 generations for important changes.

Finally in a large population, divided and subdivided into partially isolated local races of small size, there is a continually shifting differentiation among the latter (intensified by local differences in selection but occurring under uniform and static conditions) which inevitably brings about an indefinitely continuing, irreversible, adaptive, and much more rapid evolution of the species. Complete isolation in this case, and more slowly in the preceding, originates new species differing for the most part in nonadaptive respects but is capable of initiating an adaptive radiation as well as of parallel orthogenetic lines, in accordance with the conditions. It is suggested, in conclusion, that the differing statistical situations to be expected among natural species are adequate to account for the different sorts of evolutionary processes which have been described, and that, in particular, conditions in nature are often such as to bring about the state of poise among opposing tendencies on which an indefinitely continuing evolutionary process depends.

LITERATURE CITED

BERNSTEIN, F., 1925 Zusammenfassende Betrachtungen über die erblichen Blutstrukturen des Menschen. Z. indukt. Abstamm.-u. VererbLehre **37**: 237–269.

CALDER, A., 1927 The role of inbreeding in the development of the Clydesdale breed of horse. Proc. Roy. Soc. Edinb. **47**: 118–140.

EAST, E. M., 1918 The role of reproduction in evolution. Amer. Nat. **52**: 273–289.

FISHER, R. A., 1918 The correlation between relatives on the supposition of Mendelian inheritance. Trans. Roy. Soc., Edinb. 52 part **2**: 399–433.

1922 On the dominance ratio. Proc. Roy. Soc., Edinb. **42**: 321–341.

1928 The possible modification of the response of the wild type to recurrent mutations. Amer. Nat., **62**: 115–126.

1929 The evolution of dominance; reply to Professor Sewall Wright: Amer. Nat. **63**: 553–556.

1930 The genetical theory of natural selection. 272 pp. Oxford: Clarendon Press.

HAGEDOORN, A. L., and HAGEDOORN A. C., 1921 The relative value of the processes causing evolution. 294 pp. The Hague: Martinus Nijhoff.

HALDANE, J. B. S., 1924–1927 A mathematical theory of natural and artificial selection. Part I. Trans. Camb. Phil. Soc. **23**: 19–41. Part II. Proc. Camb. Phil. Soc. (Biol. Sci) **1**: 158–163. Part III. Proc. Camb. Phil. Soc. **23**: 363–372. Part IV. Proc. Camb. Phil. Soc. **23**: 607–615. Part V. Proc. Camb. Phil. Soc. **23**: 838–844.

HARDY, G. H., 1908 Mendelian proportions in a mixed population. Science **28**: 49–50.

JENNINGS, H. S., 1916 The numerical results of diverse systems of breeding. Genetics **1**: 53–89.

JONES, D. F., 1917 Dominance of linked factors as a means of accounting for heterosis. Genetics **2**: 466–479.

KEMP, W. B., 1929 Genetic equilibrium and selection. Genetics **14**: 85–127.

LOTKA, A. J., 1925 Elements of Physical Biology. Baltimore: Williams and Wilkins.

MCPHEE, H. C., and WRIGHT S., 1925, 1926 Mendelian analysis of the pure breeds of livestock. III. The Shorthorns. J. Hered., **16**: 205–215. IV. The British dairy Shorthorns. J. Hered. **17**: 397–401.

MORGAN, T. H., BRIDGES, C. B., STURTEVANT, A. H., 1925 The genetics of Drosophila. 262 pp. The Hague: Martinus Nijhoff.

MULLER, H. J., 1922 Variation due to change in the individual gene. Amer. Nat. **56**: 32–50.

1928 The measurement of gene mutation rate in Drosophila, its high variability, and its dependence upon temperature. Genetics **13**: 279–357.

1929 The gene as the basis of life. Proc. Int. Cong. Plant Sciences **1**: 897–921.

ROBBINS, R. B., 1918 Some applications of mathematics to breeding problems. III. Genetics **3**: 375–389.

SMITH, A. D. B., 1926 Inbreeding in cattle and horses. Eugen. Rev. **14**: 189–204.

WAHLUND, STEN., 1928 Zusammensetzung von Populationen und Korrelationserscheinungen vom Standpunkt der Vererbungslehre aus betrachtet. Hereditas **11**: 65–106.

WEINBERG, W., 1909 Über Vererbungsgesetze beim Menschen. Z. indukt. Abstamm.-u. Vererb-Lehre **1**: 277–330.

1910 Weiteres Beiträge zur Theorie der Vererbung. Arch. Rass.-u. Ges. Biol. **7**: 35–49, 169–173.

WRIGHT, S., 1921 Systems of mating. Genetics **6**: 111–178.

1922 Coefficients of inbreeding and relationship. Amer. Nat. **61**: 330–338.

1922a The effects of inbreeding and crossbreeding on guinea-pigs. III. Crosses between highly inbred families. U. S. Dept. Agr. Bull. No. 1121.

1923 Mendelian analysis of the pure breeds of live stock. Part I. J. Hered. **14**: 339–348. Part II, J. Hered. **14**: 405–422.

1929 FISHER's theory of dominance. Amer. Nat. **63**: 274–279.

1929a The evolution of dominance. Comment on Doctor FISHER's reply. Amer. Nat. **63**: 556–561.

1929b Evolution in a Mendelian population. Anat. Rec. **44**: 287.

WRIGHT, S., and MCPHEE, H. C., 1925 An approximate method of calculating coefficients inbreeding and relationship from livestock pedigrees. J. Agric. Res. **31**: 377–383.

25

Reprinted from *Proc. 6th Intern. Congr. Genetics* 1:356–366 (1932)

THE ROLES OF MUTATION, INBREEDING, CROSSBREEDING AND SELECTION IN EVOLUTION

Sewall Wright, University of Chicago, Chicago, Illinois

The enormous importance of biparental reproduction as a factor in evolution was brought out a good many years ago by EAST. The observed properties of gene mutation—fortuitous in origin, infrequent in occurrence and deleterious when not negligible in effect—seem about as unfavorable as possible for an evolutionary process. Under biparental reproduction, however, a limited number of mutations which are not too injurious to be carried by the species furnish an almost infinite field of possible variations through which the species may work its way under natural selection.

Estimates of the total number of genes in the cells of higher organisms range from 1000 up. Some 400 loci have been reported as having mutated in Drosophila during a laboratory experience which is certainly very limited compared with the history of the species in nature. Presumably, allelomorphs of all type genes are present at all times in any reasonably numerous species. Judging from the frequency of multiple allelomorphs in those organisms which have been studied most, it is reasonably certain that many different allelomorphs of each gene are in existence at all times. With 10 allelomorphs in each of 1000 loci, the number of possible combinations is 10^{1000} which is a very large number. It has been estimated that the total number of electrons and protons in the whole visible universe is much less than 10^{100}.

However, not all of this field is easily available in an interbreeding population. Suppose that each type gene is manifested in 99 percent of the individuals, and that most of the remaining 1 percent have the most favorable of the other allelomorphs, which in general means one with only a slight differential effect. The average individual will show the effects of 1 percent of the 1000, or 10 deviations from the type, and since this average has a standard deviation of $\sqrt{10}$ only a small proportion will exhibit more than 20 deviations from type where 1000 are possible. The population is thus confined to an infinitesimal portion of the field of possible gene combinations, yet this portion includes some 10^{40} homozygous combinations, on the above extremely conservative basis, enough so that there is no reasonable chance that any two individuals have exactly the same genetic constitution in a species of millions of millions of individuals persisting over millions of generations. There is no difficulty in accounting for the probable genetic uniqueness of each individual human being or other organism which is the product of biparental reproduction.

If the entire field of possible gene combinations be graded with respect to adaptive value under a particular set of conditions, what would be its nature? Figure 1 shows the combinations in the cases of 2 to 5 paired allelomorphs. In the last case, each of the 32 homozygous combinations is at one remove from 5 others, at two removes from 10, etc. It would require 5 dimensions to represent these relations symmetrically; a sixth dimension is needed to represent level of adaptive value. The 32 combina-

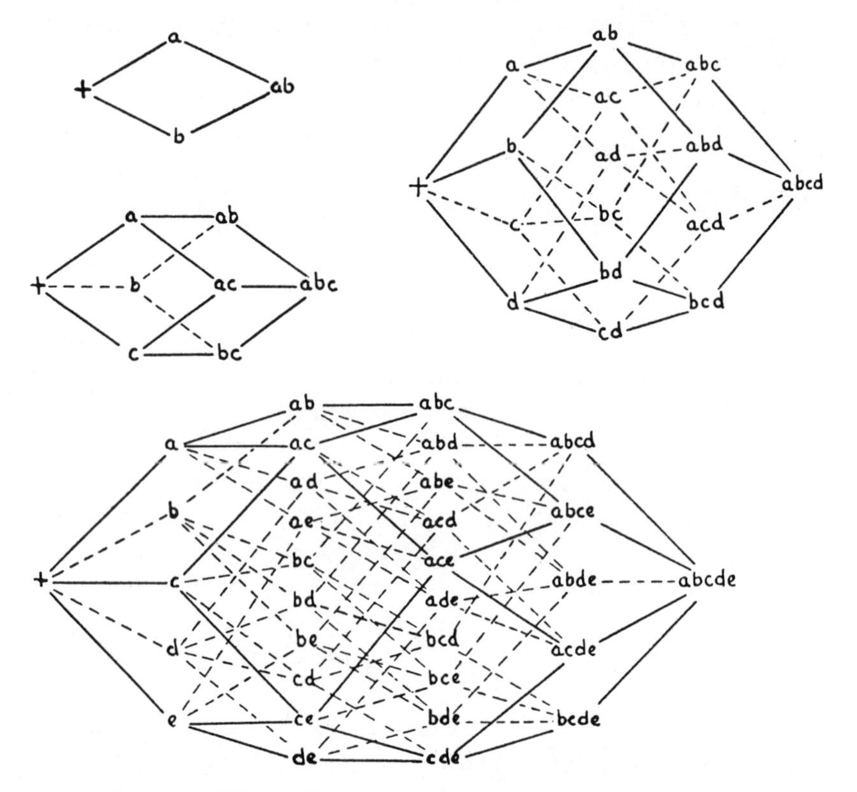

FIGURE 1.—The combinations of from 2 to 5 paired allelomorphs.

tions here compare with 10^{1000} in a species with 1000 loci each represented by 10 allelomorphs, and the 5 dimensions required for adequate representation compare with 9000. The two dimensions of figure 2 are a very inadequate representation of such a field. The contour lines are intended to represent the scale of adaptive value.

One possibility is that a particular combination gives maximum adaptation and that the adaptiveness of the other combinations falls off more or less regularly according to the number of removes. A species whose individuals are clustered about some combination other than the highest would move up the steepest gradient toward the peak, having reached which it would remain unchanged except for the rare occurrence of new favorable mutations.

But even in the two factor case (figure 1) it is possible that there may be two peaks, and the chance that this may be the case greatly increases with each additional locus. With something like 10^{1000} possibilities (figure 2) it may be taken as certain that there will be an enormous number of

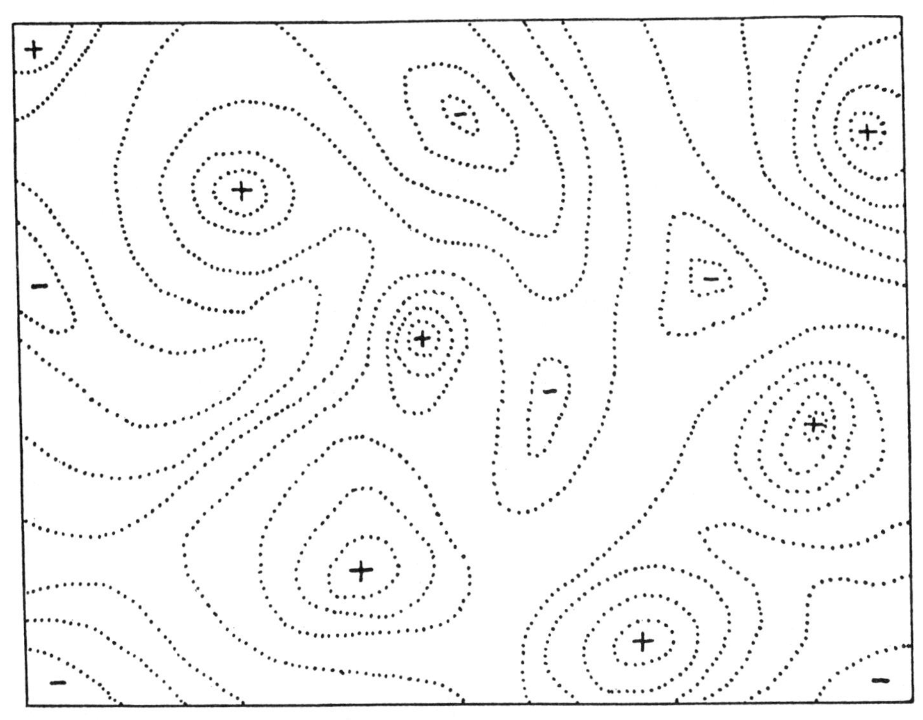

FIGURE 2.—Diagrammatic representation of the field of gene combinations in two dimensions instead of many thousands. Dotted lines represent contours with respect to adaptiveness.

widely separated harmonious combinations. The chance that a random combination is as adaptive as those characteristic of the species may be as low as 10^{-100} and still leave room for 10^{800} separate peaks, each surrounded by 10^{100} more or less similar combinations. In a rugged field of this character, selection will easily carry the species to the nearest peak, but there may be innumerable other peaks which are higher but which are separated by "valleys." The problem of evolution as I see it is that of a mechanism by which the species may continually find its way from lower to higher peaks in such a field. In order that this may occur, there must be some trial and error mechanism on a grand scale by which the species may explore the region surrounding the small portion of the field which it occupies. To evolve, the species must not be under strict control of natural selection. Is there such a trial and error mechanism?

At this point let us consider briefly the situation with respect to a single locus. In each graph in figure 3 the abscissas represent a scale of gene frequency, 0 percent of the type genes to the left, 100 percent to the right. The elementary evolutionary process is, of course, change of gene frequency, a

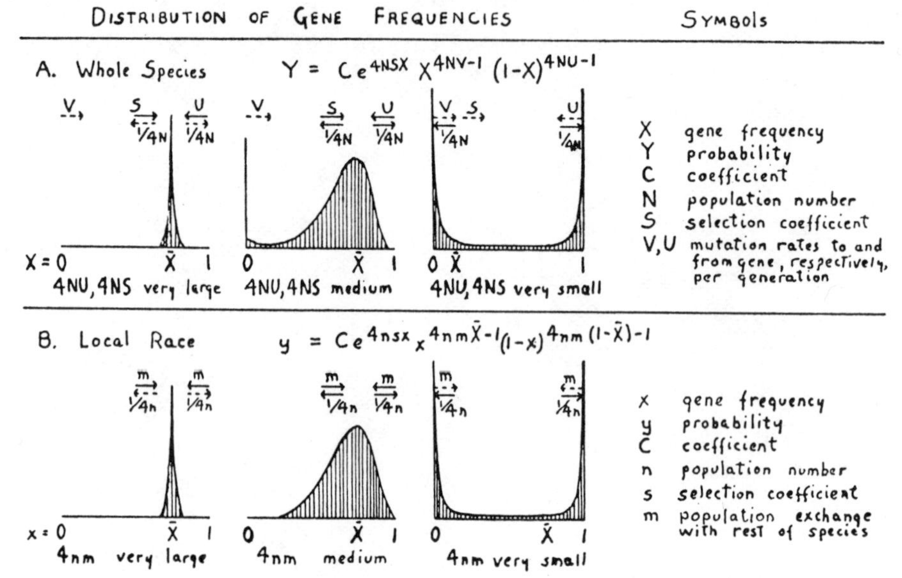

FIGURE 3.—Random variability of a gene frequency under various specified conditions.

practically continuous process. Owing to the symmetry of the Mendelian mechanism, any gene frequency tends to remain constant in the absence of disturbing factors. If the type gene mutates at a certain rate, its frequency tends to move to the left, but at a continually decreasing rate. The type gene would ultimately be lost from the population if there were no opposing factor. But the type gene is in general favored by selection. Under selection, its frequency tends to move to the right. The rate is greatest at some point near the middle of the range. At a certain gene frequency the opposing pressures are equal and opposite, and at this point there is consequently equilibrium. There are other mechanisms of equilibrium among evolutionary factors which need not be discussed here. Note that we have here a theory of the stability of species in spite of continuing mutation pressure, a continuing field of variability so extensive that no two individuals are ever genetically the same, and continuing selection.

If the population is not indefinitely large, another factor must be taken into account: the effects of accidents of sampling among those that survive and become parents in each generation and among the germ cells of these, in other words, the effects of inbreeding. Gene frequency in a given generation is in general a little different one way or the other from that in the preceding, merely by chance. In time, gene frequency may wander a long way from the position of equilibrium, although the farther it wanders the greater the pressure toward return. The result is a frequency distribution within which gene frequency moves at random. There is considerable spread even with very slight inbreeding and the form of distribution becomes U-shaped with close inbreeding. The rate of movement of gene frequency is very slow in the former case but is rapid in the latter (among unfixed genes). In this case, however, the tendency toward complete fixation of genes, practically irrespective of selection, leads in the end to extinction.

In a local race, subject to a small amount of crossbreeding with the rest of the species (figure 3, lower half), the tendency toward random fixation is balanced by immigration pressure instead of by mutation and selection. In a small sufficiently isolated group all gene frequencies can drift irregulary back and forth about their mean values at a rapid rate, in terms of geologic time, without reaching fixation and giving the effects of close inbreeding. The resultant differentiation of races is of course increased by any local differences in the conditions of selection.

Let us return to the field of gene combinations (figure 4). In an indefinitely large but freely interbreeding species living under constant conditions, each gene will reach ultimately a certain equilibrium. The species will occupy a certain field of variation about a peak in our diagram (heavy broken contour in upper left of each figure). The field occupied remains constant although no two individuals are ever identical. Under the above conditions further evolution can occur only by the appearance of wholly new (instead of recurrent) mutations, and ones which happen to be favorable from the first. Such mutations would change the character of the field itself, increasing the elevation of the peak occupied by the species. Evolutionary progress through this mechanism is excessively slow since the chance of occurrence of such mutations is very small and, after occurrence, the time required for attainment of sufficient frequency to be subject to selection to an appreciable extent is enormous.

The general rate of mutation may conceivably increase for some reason. For example, certain authors have suggested an increased incidence of cosmic rays in this connection. The effect (figure 4A) will be as a rule a spreading of the field occupied by the species until a new equilibrium is reached. There will be an average lowering of the adaptive level of the species. On the other hand, there will be a speeding up of the process discussed above, elevation of the peak itself through appearance of novel favorable mutations. Another possibility of evolutionary advance is that the spreading of the field occupied may go so far as to include another and

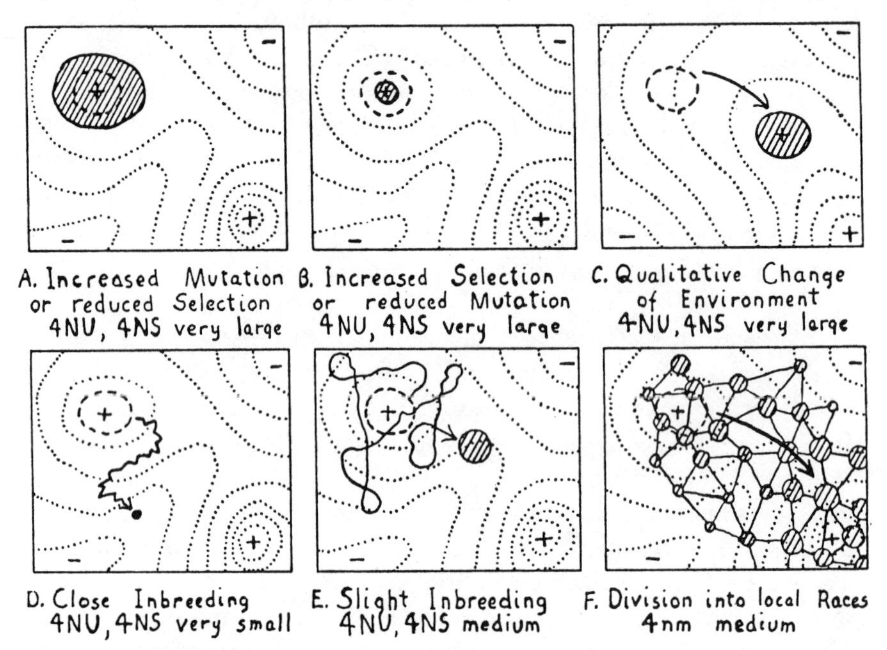

FIGURE 4.—Field of gene combinations occupied by a population within the general field of possible combinations. Type of history under specified conditions indicated by relation to initial field (heavy broken contour) and arrow.

higher peak, in which case the species will move over and occupy the region about this. These mechanisms do not appear adequate to explain evolution to an important extent.

The effects of reduced mutation rate (figure 4B) are of course the opposite: a rise in average level, but reduced variability, less chance of novel favorable mutation, and less chance of capture of a neighboring peak.

The effect of increased severity of selection (also 4B) is, of course, to increase the average level of adaptation until a new equilibrium is reached. But again this is at the expense of the field of variation of the species and reduces the chance of capture of another adaptive peak. The only basis for continuing advance is the appearance of novel favorable mutations which are relatively rapidly utilized in this case. But at best the rate is extremely slow even in terms of geologic time, judging from the observed rates of mutation.

Relaxation of selection has of course the opposite effects and thus effects somewhat like those of increased mutation rate (figure 4A).

The environment, living and non-living, of any species is actually in continual change. In terms of our diagram this means that certain of the high places are gradually being depressed and certain of the low places are becoming higher (figure 4C). A species occupying a small field under influence of severe selection is likely to be left in a pit and become extinct, the victim of extreme specialization to conditions which have ceased, but if under sufficiently moderate selection to occupy a wide field, it will merely be kept continually on the move. Here we undoubtedly have an important evolutionary process and one which has been generally recognized. It consists largely of change without advance in adaptation. The mechanism is, however, one which shuffles the species about in the general field. Since the species will be shuffled out of low peaks more easily than high ones, it should gradually find its way to the higher general regions of the field as a whole.

Figure 4D illustrates the effect of reduction in size of population below a certain relation to the rate of mutation and severity of selection. There is fixation of one or another allelomorph in nearly every locus, largely irrespective of the direction favored by selection. The species moves down from its peak in an erratic fashion and comes to occupy a much smaller field. In other words there is the deterioration and homogeneity of a closely inbred population. After equilibrium has been reached in variability, movement becomes excessively slow, and, such as there is, is nonadaptive. The end can only be extinction. Extreme inbreeding is not a factor which is likely to give evolutionary advance.

With an intermediate relation between size of population and mutation rate, gene frequencies drift at random without reaching the complete fixation of close inbreeding (figure 4E). The species moves down from the extreme peak but continually wanders in the vicinity. There is some chance that it may encounter a gradient leading to another peak and shift its allegiance to this. Since it will escape relatively easily from low peaks as compared with high ones, there is here a trial and error mechanism by which in time the species may work its way to the highest peaks in the general field. The rate of progress, however, is extremely slow since change of gene frequency is of the order of the reciprocal of the effective population size and this reciprocal must be of the order of the mutation rate in order to meet the conditions for this case.

Finally (figure 4F), let us consider the case of a large species which is subdivided into many small local races, each breeding largely within itself but occasionally crossbreeding. The field of gene combinations occupied by each of these local races shifts continually in a nonadaptive fashion (except in so far as there are local differences in the conditions of selection). The rate of movement may be enormously greater than in the preceding case since the condition for such movement is that the reciprocal of the population number be of the order of the proportion of crossbreeding instead of the mutation rate. With many local races, each spreading over a considerable field and moving relatively rapidly in the more general field about the controlling peak, the chances are good that one at least will come under the influence of another peak. If a higher peak, this race will expand in numbers and by crossbreeding with the others will pull the whole species toward the new position. The average adaptiveness of the species thus advances under intergroup selection, an enormously more effective process than intragroup selection. The conclusion is that subdivision of a species into local races provides the most effective mechanism for trial and error in the field of gene combinations.

It need scarcely be pointed out that with such a mechanism complete isolation of a portion of a species should result relatively rapidly in specific differentiation, and one that is not necessarily adaptive. The effective intergroup competition leading to adaptive advance may be between species rather than races. Such isolation is doubtless usually geographic in character at the outset but may be clinched by the development of hybrid sterility. The usual difference of the chromosome complements of related species puts the importance of chromosome aberration as an evolutionary process beyond question, but, as I see it, this importance is not in the character differences which they bring (slight in balanced types), but rather in leading to the sterility of hybrids and thus making permanent the isolation of two groups.

How far do the observations of actual species and their subdivisions conform to this picture? This is naturally too large a subject for more than a few suggestions.

That evolution involves nonadaptive differentiation to a large extent at the subspecies and even the species level is indicated by the kinds of differences by which such groups are actually distinguished by systematists. It is only at the subfamily and family levels that clear-cut adaptive differences become the rule (Robson, Jacot). The principal evolutionary mechanism in the origin of species must thus be an essentially nonadaptive one.

That natural species often are subdivided into numerous local races is indicated by many studies. The case of the human species is most familiar. Aside from the familiar racial differences recent studies indicate a distribution of frequencies relative to an apparently nonadaptive series of allelomorphs, that determining blood groups, of just the sort discussed above. I scarcely need to labor the point that changes in the average of mankind in the historic period have come about more by expansion of some types and decrease and absorption of others than by uniform evolutionary advance. During the recent period, no doubt, the phases of intergroup competition and crossbreeding have tended to overbalance the process of local differentiation, but it is probable that in the hundreds of thousands of years of prehistory, human evolution was determined by a balance between these factors.

Subdivision into numerous local races whose differences are largely nonadaptive has been recorded in other organisms wherever a sufficiently detailed study has been made. Among the land snails of the Hawaiian Islands, Gulick (sixty years ago) found that each mountain valley, often each grove of trees, had its own characteristic type, differing from others in "nonutilitarian" respects. Gulick attributed this differentiation to inbreeding. More recently Crampton has found a similar situation in the land snails of Tahiti and has followed over a period of years evolutionary changes which seem to be of the type here discussed. I may also refer to the studies of fishes by David Starr Jordan, garter snakes by Ruthven, bird lice by Kellogg, deer mice by Osgood, and gall wasps by Kinsey as others which indicate the role of local isolation as a differentiating factor. Many other cases are discussed by Osborn and especially by Rensch in recent summaries. Many of these authors insist on the nonadaptive character of most of the differences among local races. Others attribute all differences to the environment, but this seems to be more an expression of faith than a view based on tangible evidence.

An even more minute local differentiation has been revealed when the methods of statistical analysis have been applied. Schmidt demonstrated the existence of persistent mean differences at each collecting station in certain species of marine fish of the fjords of Denmark, and these differences were not related in any close way to the environment. That the differences were in part genetic was demonstrated in the laboratory. David Thompson has found a correlation between water distance and degree of differentiation within certain fresh water species of fish of the streams of Illinois. SUMNER's extensive studies of subspecies of Peromyscus (deer mice) reveal genetic differentiations, often apparently nonadaptive, among local populations and demonstrate the genetic heterogeneity of each such group.

The modern breeds of livestock have come from selection among the products of local inbreeding and of crossbreeding between these, followed by renewed inbreeding, rather than from mass selection of species. The recent studies of the geographical distribution of particular genes in livestock and cultivated plants by SEREBROVSKY, PHILIPTSCHENKO and others are especially instructive with respect to the composition of such species.

The paleontologists present a picture which has been interpreted by some as irreconcilable with the Mendelian mechanism, but this seems to be due more to a failure to appreciate statistical consequences of this mechanism than to anything in the data. The horse has been the standard example of an orthogenetic evolutionary sequence preserved for us with an abundance of material. Yet MATHEW's interpretation as one in which evolution has proceeded by extensive differentiation of local races, intergroup selection, and crossbreeding is as close as possible to that required under the Mendelian theory.

Summing up: I have attempted to form a judgment as to the conditions for evolution based on the statistical consequences of Mendelian heredity. The most general conclusion is that evolution depends on a certain balance among its factors. There must be gene mutation, but an excessive rate gives an array of freaks, not evolution; there must be selection, but too severe a process destroys the field of variability, and thus the basis for further advance; prevalence of local inbreeding within a species has extremely important evolutionary consequences, but too close inbreeding leads merely to extinction. A certain amount of crossbreeding is favorable but not too much. In this dependence on balance the species is like a living organism. At all levels of organization life depends on the maintenance of a certain balance among its factors.

More specifically, under biparental reproduction a very low rate of mutation balanced by moderate selection is enough to maintain a practically infinite field of possible gene combinations within the species. The field actually occupied is relatively small though sufficiently extensive that no two individuals have the same genetic constitution. The course of evolution through the general field is not controlled by direction of mutation and not directly by selection, except as conditions change, but by a trial and error mechanism consisting of a largely nonadaptive differentiation of local races (due to inbreeding balanced by occasional crossbreeding) and a determination of long time trend by intergroup selection. The splitting of species depends on the effects of more complete isolation, often made permanent by the accumulation of chromosome aberrations, usually of the balanced type. Studies of natural species indicate that the conditions for such an evolutionary process are often present.

LITERATURE CITED

CRAMPTON, H. E., 1925 Contemporaneous organic differentiation in the species of Partula living in Moorea, Society Islands. Amer. Nat. **59**:5-35.

EAST, E. M., 1918 The role of reproduction in evolution. Amer. Nat. **52**:273-289.

GULICK, J. T., 1905 Evolution, racial and habitudinal. Pub. Carnegie Instn. **25**:1-269.

JACOT, A. P., 1932 The status of the species and the genus. Amer. Nat. **66**:346-364.

JORDAN, D. S., 1908 The law of geminate species. Amer. Nat. **42**:73-80.

KELLOGG, V. L., 1908 Darwinism, today. 403 pp. New York: Henry Holt and Co.

KINSEY, A. C., 1930 The gall wasp genus Cynips. Indiana Univ. Studies. **84-86**:1-577.

MATHEW, W. D., 1926 The evolution of the horse. A record and its interpretation. Quart. Rev. Biol. **1**:139-185.

OSBORN, H. F., 1927 The origin of species. V. Speciation and mutation. Amer. Nat. **49**:193-239.

OSGOOD, W. H., 1909 Revision of the mice of the genus Peromyscus. North American Fauna **28**:1-285.

PHILIPTSCHENKO, J., 1927 Variabilität and Variation. 101 pp. Berlin.

RENSCH, B., 1929 Das Prinzip geographischer Rassenkreise und das Problem der Artbildung. 206 pp. Berlin: Gebrüder Borntraeger.

ROBSON, G. C., 1928 The species problem. 283 pp. Edinburgh and London: Oliver and Boyd.

RUTHVEN, A. G., 1908 Variation and genetic relationships of the garter snakes. U. S. Nat. Mus. Bull. **61**:1-301.

SCHMIDT, J., 1917 Statistical investigations with *Zoarces viviparus* L. J. Genet. **7**:105-118.

SEREBROVSKY, A. S., 1929 Beitrag zur geographischen Genetic des Haushahns in Sowjet-Russland. Arch. f. Geflügelkunde, Jahrgang **3**:161-169.

SUMNER, F. B., 1932 Genetic, distributional, and evolutionary studies of the subspecies of deer mice (Peromyscus). Bibl. genet. **9**:1-106.

THOMPSON, D. H., 1931 Variation in fishes as a function of distance. Trans. Ill. State Acad. of Sci. **23**:276-281.

WRIGHT, S., 1931 Evolution in Mendelian populations. Genetics **16**:97-159.

Editor's Comments on Linkages

This statement is an attempt to summarize some of the linkages between the papers in this volume and the events that followed with the papers that appear in other volumes in the series "Benchmark Papers in Genetics." This volume is not meant to cover the history of the early days of genetics. Omitted are large areas of the genetics investigation, such as cytological, biochemical, mutational, physiological, and microbiological studies. This volume describes a part of genetics that can be clearly identified as evolutionary genetics because the main questions concern the evolutionary process. After the 1930s the various parts of genetics became more specialized and many diverse aspects were considered for their own value, for their applications to human welfare, or for their interest to other disciplines and only peripherally did they consider evolutionary topics. We need to discuss, in this section, the linkages to these specialized fields.

One of the threads that appears to run throughout the development of evolutionary genetics is the considerable stimulus provided to research activity by controversy and by the competition among individuals for priority and prestige. Unfortunately there has also been occasional bitterness between individuals as a result of these divergent opinions. The controversy which accompanied the release of the *Origin of the Species* and the debates which occurred in the early 1860s have been noted. There was also an adversary proceeding to accompany the rediscovery of Mendel's work. In 1904, at a meeting of the British Association, Bateson and colleagues squared off against Weldon and Pearson in a meeting that must have been reminiscent of the Huxley-Wilberforce battle. When Pearson called for a truce the chairman closed the meeting with the statement ". . . let them fight it out." This entire sordid story is reviewed for the curious in some detail in Provine (1971).

Two other cases come to mind where evolutionary genetics controversies were "settled" by adversary proceedings. In the late 1920s, Clarence Darrow and William Jennings Bryant decided the legal fate of Mr. Scopes of Tennessee and the educational fate of many school children throughout the country in proceedings lost by both sides. In 1939 a meeting of the Lenin All-Union Academy of Agricultural Sciences pitted N. I. Vavilov and T. D. Lysenko in adversary proceedings on the issue of Mendelian versus the inheritance of acquired characters. This event was followed by death for thousands and hunger for the people of the USSR and a political struggle not yet resolved (Medvedev, 1969). Each of the above proceedings concerned the nature of inheritance and of evolution, and none of them can claim to expose man or scientist in his highest glory.

While the adversary process may be an unfortunate accompaniment of the controversial issue, it need not be so. Issues develop which reflect the divergent opinions of the scholars and the attention which the controversy draws provides for the dissemination of the information, and for the clarification of positions, and allows other individuals to participate in the development of the field. Controversy tends to be exhibited at significant points in the development of thought and thus identifies the focus for both individuals and agencies to direct resources. The direct descendant of evolutionary genetics is population genetics, where issues have played a continuing role. Crow and Denniston are preparing a Benchmark volume, *Issues in Population Genetics,* which considers the development of this field. Another direct outgrowth of early evolutionary studies was the development of the understanding of the species and the processes by which species arise. This literature is sampled in the Benchmark volume *Genetics of Speciation* (Vol. 9).

The belief in environmental influences on inheritance traced at the beginning of this volume from early times to Darwin persists in various forms and with divergent results. Kammerer (1924) designed a series of experiments to demonstrate the existence of the inheritance of acquired characters only to have the results obscured by inadequate procedures by his assistants. He committed suicide. Lysenko proposed that vernalization of seeds would produce inherited changes which would improve crops. Of course chemical, X-ray, and other radiation from the environment does produce genetic change, but these changes are not responsive to the inducing agent. That is, radiation does not produce mutations which are resistant to radiation but rather produces a variety of mutations of many kinds, some of which, by chance, reach proportions high enough to be retained or even fixed in the population.

The study of mutations, first described and named by de Vries, received considerable stimulus by the discoveries of H. J. Muller that X-rays would increase the rates of mutations. Later identification of the role of chemical mutagens by Auerbach provided a basis for an expanded analysis of the nature of mutations once the genetic code was identified in the 1950s. This information has been reviewed by Drake and Koch (1975) in the Benchmark volume *Mutagenesis* (Vol. 4). Additional volumes related to the role of DNA are in preparation including *Regulation Genetics*—Maas; *Genetic Recombination*—Hotchkiss; and *The Replication of Nucleic Acids*—Hanawalt.

The determination of how genes act relies heavily on early studies on pigments including studies of Cuenot on mice, Wright on guinea pigs, Beetle and Ephrussi on the eye colors of *Drosophila,* and various studies on the color of plants. From this beginning Wagner (1975) developed the Benchmark volume on *Genes and Proteins* (Vol. 2), which relates the end products of gene action to the action of the gene.

Genes occur on chromosomes: so obvious and clear today that it is difficult to think of particulate inheritance in any other terms. Phillips and Burnham (1977) have provided us with a Benchmark volume *Cytogenetics* (Vol. 6) that traces the relation between the genetic process and cytological events.

Also a direct outgrowth of evolutionary genetics are the studies of stochastic process. This very active field has been reviewed by Li (1977) in the Benchmark volume *Stochastic Models in Population Genetics* (Vol. 7). In preparation is a volume linking the development of ecology and genetics (Anderson and Schaal, *Ecological Genetics*).

Haldane, Fisher, and Wright considered their theories of evolution as they were supported by or provided for the understanding of the biology of man. Haldane and Fisher were driving forces in the application of genetics to man through their participation in the eugenics movement in England. Volume 1 in "Benchmark Papers in Genetics" was that of Ballonoff (1974) on *Genetics and Social Structure*, and recently Bajema (1976) has provided us with *Eugenics: Then and Now* (Vol. 5). The study of human population genetics leans heavily on demographic processes and the integration of these fields and has been considered in the Benchmark volume *Demographic Genetics* (Vol. 3), by Weiss and Ballonoff (1976). An additional volume in preparation is that of Schull on *Medical Genetics.*

Fisher and Wright are particularly concerned with the applications of genetic principles to agriculture, the first working at the Rothamsted Experimental Station in England and the second working for the Department of Agriculture in the United States. Haldane's interest in man in all his aspects was legend. Thus it is not surprising that the applications of genetics to agricultural practices leans so heavily on the research productivity of these three men. These specialized fields of genetics are reviewed in Benchmark volumes now in preparation (*Quantitative Genetics*—Dr. Comstock, *Plant Breeding*—Dr. Matzinger, and to be contracted *Animal Breeding*).

REFERENCES

Anderson, E. (1924) Studies on self-sterility. VI. The genetic basis of cross-sterility in *Nicotiana. Genetics* 9:13–40.

Baldwin, J. M. (1896) A new factor in evolution. *Amer. Naturalist* 30:441–451; 536–553.

Bateson, W. (1894) *Materials for the Study of Variation, Treated with Especial Regard to Discontinuity in the Origin of Species.* London: Macmillan and Co.

—— (1902) *Mendel's Principles of Heredity: A Defense.* Cambridge: Cambridge University Press.

——, and Saunders, E. R. (1902) *Reports to the Evolution Committee of the Royal Society.* Report 1. London: Harrison and Sons.

Blakeslee, A. F. (1936) Twenty-five years of genetics, 1910–1935. *Brooklyn Botanic Garden Memoirs* 4:29–40.

Chetverikov, S. S. (1927) On the genetic nature of wild populations. *Proc. 5th Intern. Congr. Genetics.* Berlin. pp. 1499–1500.

Clausen, J. (1927) Chromosome number and the relationship of species in the genus *Viola. Ann. Bot.* 41:677–714.

Conklin, E. G. (1944) Jean Baptiste-Pierre Antoine de Monet, Chevalier de Lamarck. *Genetics* 29:i–iv.

Correns, C. (1900) G. Mendel's Regel über das Verhalten der Nachkommenschaft der Rassenbastarde. *Ber. Deutsch Bot. Ges.* 18:158–168. English translation in *Genetics* 35 (1950), Suppl. 33–41.

Darwin, C. (1868) *The Variations of Animals and Plants under Domestication.* 2 vols. London: J. Murray.

—— (1871) *The Descent of Man, and Selection in Relation to Sex.* New York: D. Appleton and Company.

Darwin, E. (1796) *Zoonomia: Or the Laws of Organic Life.* London: T. & J. Swords.

Darwin, F. (1887) *The Life and Letters of Charles Darwin.* 2 Vols. New York: D. Appleton and Company. 1888.

Dobzhansky, Th. (1967) Sergei Sergeevich Tshetverikov. *Genetics* 55:1–3, frontispiece.

Dunn, L. C. (1921) Unit character variation in rodents. *J. Mammology* 2:125-140.
—— (1965) *A Short History of Genetics.* New York: McGraw-Hill Book Company.
East, E. M. and Hayes, H. K. (1912) Heterozygosis in evolution and in plant breeding. *U.S. Dept. Agr. Bur. Plant. Ind. Bull.* 243:1-58.
Fisher, R. A. (1918) The correlation between relatives on the supposition of mendelian inheritance. *Trans. Roy. Soc. Edin.* 54:399-433.
—— (1936) Has Mendel's work been rediscovered? *Annals of Science* 1:115-137.
Galton, F. (1876) A theory of heredity. *J. Anthropological Inst.* 5:329-348.
—— (1877a) Typical laws of heredity. *Nature* 15:512.
—— (1877b) Typical laws of heredity. *J. Roy. Institution of Great Britain* 8:282-301.
—— (1889) *Natural Inheritance.* London: Macmillan and Co.
Gaertner, C. F. von (1849) *Versuche und Beobachtungen über die Bastardezeugung im Pflanzenreich mit Hinweisung auf die ahnlichen Erscheinungen im Thierreiche.* Stuttgart. (Referred to by Olby, 1966.)
Glass, B. (1947) Maupertuis and the beginnings of genetics. *Quart. Rev. of Biol.* 22: 196-210.
Goldschmidt, R. (1934) Lymantria. *Bibliogr. Genetica* 11:1-186.
Haldane, J. B. S. (1926) A mathematical theory of natural and artificial selection. Part IV. *Proc. Cambridge Phil. Soc.* 23:607-615.
—— (1927) A mathematical theory of natural and artificial selection. Part V. *Proc. Cambridge Phil. Soc.* 23:838-44.
—— (1929) The species problem in the light of genetics. *Nature* 124:514-516.
—— (1930) A mathematical theory of natural and artificial selection. Part VI. *Proc. Cambridge Phil. Soc.* 26:220-230.
—— (1931) Mathematical Darwinism. A discussion of the Genetical Theory of Natural Selection. *The Eugenics Rev.* 23:115-117.
—— (1932) *The Causes of Evolution.* Longmans, Green & Co., Limited. Cornell Paperbacks, 1966.
—— (1934) A mathematical theory of natural and artificial selection. Part X. *Genetics* 19:412-29.
Huxley, T. H. (1887) On the reception of the Origin of Species. In: Darwin, F. (ed.), *The Life and Letters of Charles Darwin.* New York: D. Appleton and Company, pp. 533-558.
Iltis, H. (1932) *Life of Mendel.* English translation by E. and C. Paul, 1966. New York: Hafner Publishing Co.
Johannsen, W. (1909) Elemente der Exakten Erblichkeitslehre. Jena: G. Fischer.
Jones, D. F. (1939) Edward M. East. *Genetics* 14:2pp.
Kammerer, P. (1924) *The Inheritance of Acquired Characteristics.* New York: Boni and Liveright.
Koelreuter, J. G. (1761-1766) Vorlaufige Nachricht von einigen das Geschlecht der Pflanzen betreffenden Versuchen Beobachtungen, nebst Fortsetzungen 1, 2 und 3. Lipsiae. Reprinted in: Ostwald's Klassiker der exaketen Wissenschaften. No. 41. Leipzing. 1893. p. 43 (Reffered to in Olby, 1966).
Koestler, A. (1954) *The Case of the Midwife toad.* New York: Random House.
Kempthorne, O. (1974) Collected papers of R. A. Fisher (A review). *Soc. Biol.* 21: 98-101.
Lamarck, J. B. P. A. de Monet de. (1802) *Recherche sur l'organisation des corps vivants.* Paris: Maillard.
Lysenko, T. D. (1954) *Agrobiology; Essays on Problems of Genetics, Plant Breeding and Seed Growing.* Moscow: Foreign Languages Pub. House.
Mather, K. (1969) Ronald A. Fisher. *Genetics* 61:1-7, frontispiece.

Medvedev, Z. A. (1969) *The Rise and Fall of T. D. Lysenko.* New York: Columbia University Press.
Mendel, G. (1866) Versuche über Pflanzen Hybriden. *Verh. Naturforsch. ver. in Brunn* 4:3–47. English translation in *J. Roy. Horticultural Soc.* 26(1901):1–32.
Moran, P. A. P., and Smith, C. A. B. (1966) The correlation between relatives on the supposition of mendelian inheritance. *Eugenics Lab. Memoirs* XLI:1-62.
Morgan, T. H. (1932) The rise of genetics. *Science* 76:261-267; 285–288.
Naudin, C. (1865) Nouvelles recherches sur l'hybridite dans le vegetaux. *Arch. Mus. Hist. Nat., Paris* 1:26.
Newmann, H. H. (1915) Development and heredity in heterogenic teleost hybrids. *J. Exp. Zool.* 18:511-576.
Nilsson-Ehle, H. (1909) Kreuzungsuntersuchungen an Hafer und Weizen. *Lunds Universitets Arsskrift,* n.s. 2, Vol. 5, no. 2.
Norton, B., and Pearson, E. S. (1976) A note on the background to, and refereeing of, R. A. Fisher's 1918 paper 'On the correlation between relatives on the supposition of mendelian inheritance'. *Notes and Records of the Royal Society of London* 31:151-162.
Norton, H. T. J. (1928) Natural selection and mendelian variation. *Proc. London Math. Soc.* 28:1-45.
Olby, R. C. (1966) *Origins of Mendelism.* New York: Schocken Books.
Pearson, K. (1904) On a generalized theory of alternative inheritance with special reference to Mendel's laws. *Phil. Trans. Roy. Soc.* A203:53-86.
—— (1924) *The Life, Letters and Labours of Francis Galton.* Vol. I, II. Cambridge University Press.
—— (1930) *The Life, Letters and Labours of Francis Galton.* Vol. III. Cambridge University Press.
Pirie, N. W. (1966) John Burdon Sanderson Haldane, 1892-1964. *Biographical Memoirs Fellows. The Royal Society London* 12:218-249.
Provine, W. B. (1971) *The Origins of Theoretical Population Genetics.* Chicago: University of Chicago Press.
Punnett, R. C. (1905) *Mendelism.* Cambridge, England: Macmillan and Bowes.
—— (1950) Early days of genetics. *Heredity* 4:1-10.
Ramsbottom, J. (1938) Presidential address. Linnaeus and the species concept. *Proc. Linn. Soc. 150th Session,* pp. 192-219.
Roberts, H. F. (1929) *Plant Hybridization before Mendel.* Princeton: Princeton University Press.
Schmidt, J. (1917) Statistical investigations with *Zoarces viviparus. J. Genetics* 7: 105–118.
—— (1918) Racial studies in fishes. II. Experimental investigations with *Lebistes reticulatus* (Peters). *Regan. J. Genetics* 7:147-153.
Shull, G. H. (1908) The composition of a field of maize. *Rept. Amer. Breed. Assoc.* 4:296-301.
—— (1952) Erich von Tschermak-Seysenegg. *Genetics* 37:1–7, frontispiece.
Stern, C. (1962) Wilhelm Weinberg, 1862–1937. *Genetics* 47:1-5.
Sturtevant, A. H. (1965) *A History of Genetics.* New York: Harper & Row, Publishers.
Sumner, F. B. (1923) Results of experiments in hybridizing subspecies of *Peromyscus. J. Exper. Zool.* 38:245-292.
—— (1929) The analysis of a concrete case of intergradation between two subspecies. *PNAS* 15:110-120; 481-493.
—— (1930) Genetic and distributional studies of three subspecies of *Peromyscus. J. Genetics.* 23:275-376.

Tschermak, E. von (1900) Ulber küntsliche Kreuzung bei *Pisum sativum. Ber. Deutsche Bot. Ges.* 18:232-239. English translation in *Genetics* 35, Suppl. to No. 5, Part 2:42-47, 1950.
Vries, H. de. (1889) *Intracellulare Pangenesis.* Jena. Trans. by C. S. Gager. Chicago: Open Court Publ. Co. 1910.
—— (1900a) Sur la loi de disjonction des hybrides. *C. R. Acad. Sci. Paris* 130: 845-847. English translation in *Genetics* 35, Suppl. to No. 5, Part 2: 32, 1950.
—— (1900b) Das Spaltungsgesetz der bastarde. Vorlaufige Mitteilung. *Ber. Dtsch. Bot. Ges.* 18:90.
Wallace, A. R. (1889) *Darwinism: An Exposition of the Theory of Natural Selection.* London: Macmillan.
Weinberg, W. (1909) Uber vererbungsgesetze beim Menschen. *Zeitsch. für Induk. Abstamm. Vererbungslehre.* (Molecular and general genetics). 1:377-392; 440-460. 2:276-328.
—— (1910) Weitere Beitrage zur Theorie der Vererbung. *Arch. für Rassenund Gesellschaftsbiol.* 7:35-49.
Weismann, A. (1893) *The Germ-plasm, a Theory of Heredity.* London: Walter Scott, Ltd. Translated by W. Parker and H. Rönnfeldt.
Weldon, W. F. R. (1894) The study of animal variation. *Nature* 50:25-26.
—— (1895) Attempt to measure the death-rate due to the selective destruction of *Carcinus moenas* with respect to a particular dimension. *Proc. Royal Soc.* 47: 445-453.
—— (1902) Mendel's laws of alternative inheritance in peas. *Biometrika* 1:228-54.
—— (1903) Mr. Bateson's revisions of Mendel's theory of heredity. *Biometrika* 2:286-98.
White, M. J. D. (1965) J. B. S. Haldane. *Genetics* 52:1-7, frontispiece.
Wright, S. (1921) Systems of mating. *Genetics* 6:111-178.
—— (1930) The genetical theory of natural selection. A review. *J. Heredity* 21: 349-56.
—— (1937) The distribution of gene frequencies in populations. *PNAS* 23:307-320.
—— (1960) Genetics and twentieth century Darwinism. A review and discussion. *Am. J. Human Genetics* 12:365-372.
—— (1963) William Ernest Castle. *Genetics* 48:1-5, frontispiece.
—— (1965) Factor interaction and linkage in evolution. *Proc. Royal Soc.,* B162: 80-104.
—— (1966) Mendel's ratios. In: Stern, C., and Sherwood B. (eds.), *The Origin of Genetics. A Mendel Source Book.* San Francisco: Freeman.
—— (1969) Haldane's contribution to population and evolutionary genetics. *Proc. Twelfth Intern. Congr. Genetics* 3:445-451.
—— (1969) *Evolutions and the Genetics of Populations.* Vol. II. *The Theory of Gene Frequencies.* Chicago: University of Chicago Press.
—— (1977) Comparison of the impact of mendelism on evolutionary thought in England and America. *Proc. Ninth Intern. Biometric Conf.* (in press)
Yule, G. U. (1902) Mendel's laws and their probable relations to interracial heredity. *The New Phytologist* 1:193-207; 222-238.
Zirkle, C. (1935) *The Beginnings of Plant Hybridisation.* Morris Arboretum Monograph No. 1. Philadelphia: University of Philadelphia Press.
—— (1946) The early history of the idea of the inheritance of acquired characters and of pangenesis. *Trans. Amer. Phil. Soc.,* New Series 35:91-151.
—— (1959) Species before Darwin. *Proc. of the Amer. Phil. Soc.* 103:636-644.
—— (1964) Some oddities in the delayed discovery of mendelism. *J. Heredity* 55: 65-72.

AUTHOR CITATION INDEX

SUBJECT INDEX

About the Editor

DAVID L. JAMESON is Professor of Biology at the University of Houston, Houston, Texas. He was previously Director of the University of Houston Coastal Center, Director of the Clear Lake Graduate Center and Dean of the Graduate School. He has also taught at Pacific University, the University of Oregon, and San Diego State University. He reveived the B.S. from Southern Methodist University and the M.A. and Ph.D. from the University of Texas. He has done postdoctoral study at Oak Ridge, Brown University, American Museum of Natural History, University of Barcelona, and the University of Wisconsin. He has published papers on the Population Biology of Mammals and Amphibians and several books including a general biology text. He has served as Editor of *Copeia,* Secretary of the Society for the Study of Evolution, and as a member of the Council of the American Association for the Advancement of Science and of the Board of Governors of the American Institute of Biological Sciences. He was awarded a Fulbright Fellowship and in 1977 he served as a National Academy of Science Exchange Scholar at the Institute of Zoology, Bulgarian Academy of Science, Sofia, Bulgaria.